INORGANIC SYNTHESES

Volume 36

Editors-in-Chief

GREGORY S. GIROLAMI

University of Illinois at Urbana-Champaign

ALFRED P. SATTELBERGER

Argonne National Laboratory

●●●

INORGANIC SYNTHESES

Volume 36

Library of Congress Catalog Number: 39:23015

ISBN 978-1-118-74487-1

Printed in the United States of America

10 9 8 7 6 5 4 3 2 1

PREFACE

This volume of *Inorganic Syntheses* spans the preparations of a wide range of important inorganic, organometallic, and solid-state compounds. Continuing a long-standing tradition, we have emphasized useful compounds and methods. Reflecting our own personal research interests, transition metal halides, complexes with cyclopentadienyl and substituted cyclopentadienyl ligands, and compounds with metal–metal bonds are featured. We have also included a chapter on pedagogically important compounds that we hope will find their way into undergraduate inorganic chemistry teaching laboratories.

The volume is divided into six chapters. Chapter 1 contains the syntheses of some key early transition metal halide clusters and the very useful mononuclear molybdenum(III) synthon, $MoCl_3(THF)_3$. This set of procedures was submitted by Lou Messerle and Rinaldo Poli. Chapter 2 covers the synthesis of a number of cyclopentadienyl compounds, including a novel route to sodium and potassium cyclopentadienide, MC_5H_5. Special thanks are due to John Bercaw, Endy Min, and Ged Parkin for the syntheses of the bis(pentamethylcyclopentadienyl) compounds of groups 3, 4, 5, and 6. Chapter 3 details synthetic procedures for a range of metal–metal bonded compounds, including several with metal–metal multiple bonds. Special thanks here are due to Al Cotton, Carlos Murillo, and Dick Walton. Chapter 4 contains procedures for a range of early and late transition metal compounds, each a useful synthon for further synthetic elaboration. Chapter 5 deals with the synthesis of a number of main group compounds and ligands, while Chapter 6 covers teaching laboratory experiments. The editors are grateful to Marcetta Darensbourg for suggesting the teaching chapter.

We would like to thank everyone who submitted syntheses for Volume 36 and the checkers who dedicated considerable time and effort in checking the procedures. We acknowledge the long delay in getting this volume published and thank the contributors and checkers for their patience. To those contributors who will not see their syntheses in this volume, we apologize for not being able to find an individual willing or able to check their syntheses. We wish to extend special thanks to Vera Mainz for her expert help in the preparation of the cumulative indices that appear at the end of this volume. We undertook this large project in the hope that readers will find this material to be useful aids to locating recipes that have appeared since the last cumulative index, which summarized content up through volume 30 of *Inorganic Syntheses*.

Finally, we would like to thank our friend and colleague Tom Rauchfuss for his tireless encouragement and advice, and our mentors Dick Andersen and Geoff Wilkinson (for GSG) and Ward Schaap and John Fackler (for APS) who taught us both the joys and challenges of synthetic inorganic chemistry.

ALFRED P. SATTELBERGER
Argonne National Laboratory

GREGORY S. GIROLAMI
University of Illinois at Urbana-Champaign

DEDICATION

This volume is dedicated to the memory of eight eminent chemists who made outstanding contributions to inorganic chemistry in general and to *Inorganic Syntheses* in particular. We also note the recent passing of two other inorganic chemists, Bill Lipscomb and Gordon Stone, who were not former volume editors but whose contributions to inorganic chemistry were significant. Each was a talented synthetic chemist in his own right, and all helped shape the discipline we know and love.

GEORGE THERALD MOELLER (EDITOR-IN-CHIEF, VOLUME V, 1957)

Therald Moeller was born in North Bend, Oregon, on April 3, 1913, and died in Broken Arrow, Oklahoma, on November 24, 1997, at the age of 84. In 1934, he graduated from Oregon State College (now Oregon State University) in Corvallis as the top student of his senior class, having majored in chemical engineering. In 1938, he received his Ph.D. degree in inorganic and physical chemistry from the University of Wisconsin, Madison, for a thesis titled "A Study of the Preparation and Certain Properties of Hydrous Lanthanum Oxide Sols," carried out under the direction of Francis C. Krauskopf. Therald was Instructor in Chemistry at Michigan State College (now Michigan State University) at East Lansing (1938–1940), but in 1940 he moved to the University of Illinois, Urbana-Champaign. In 1969, Therald became Chair of the Department of Chemistry at Arizona State University in Tempe, serving in this capacity until 1975. He retired as Professor Emeritus in 1983.

Therald became an internationally recognized authority on the chemistry of the rare earth elements (lanthanides) and published 94 research papers and books in this area alone. During his 45 years of teaching and research, he guided the laboratory research of 43 Ph.D. students, 20 postdoctoral fellows, and 11 M.S. and 25 B.S. students for a total of 99 research students in inorganic chemistry. Of these, at least 39 became professors themselves at universities in the United States, Taiwan, Spain, India, Japan, Brazil, England, and Finland. Several became Department Chairs and one a College President.

Therald's 281 publications include 22 books and laboratory manuals authored or edited by him (32 books if Spanish, Russian, Japanese, Italian, and Polish editions are counted). One of these texts, *Inorganic Chemistry, An Advanced Text* (Wiley, 1952), was the "bible" of inorganic chemistry for decades, enjoying widespread adoption (I used it in my first inorganic chemistry class in 1956). Until its appearance, few U.S. universities taught inorganic courses more advanced than the freshman level because only foreign texts were available, and none of these were satisfactory. As soon as Therald's text appeared, universities began to teach advanced inorganic chemistry, which immensely influenced its development in the United States and around the world. Along with John C. Bailar Jr. (Editor-in-Chief, Volume IV, 1953), Therald cofounded the ACS Division of Inorganic Chemistry (1956), serving as its Chair in 1961–1962. He also served for many years as a member of the Board of Directors of *Inorganic Syntheses*, Inc.

EUGENE GEORGE ROCHOW (EDITOR-IN-CHIEF, VOLUME VI, 1960)

Gene Rochow was born in Newark, New Jersey, on October 4, 1909, and died in Fort Myers, Florida, on March 21, 2002, at the age of 92. He spent his childhood in Maplewood, New Jersey, where he displayed an interest in electricity and the early use of silicon as a crystal detector in radio sets. Gene followed his brother Theodore as a chemistry assistant both in high school and at Cornell University, Ithaca, New York. He was a lecture and laboratory assistant to Louis M. Dennis, Chairman of the Chemistry Department at Cornell, under whom he received his B.S. (1931) and Ph.D. degrees (1935), the latter for a thesis titled "Contributions to the Chemistry of Fluorine." Gene also worked as a special assistant to Alfred Stock, who spent several months in 1932 at Cornell as Baker Lecturer. From Stock he first learned about the chemistry of silicon hydrides, and in fact was responsible for drawing the diagrams for Stock's famous book, *The Hydrides of Boron and Silicon*, which was written during that time.

After a summer job as Research Chemist at the Halowax Corporation, New York City (1931–1932), and as a Lecture Assistant at Cornell (1932–1935), Gene found summer employment with the Hotpoint Company, a General Electric Company subsidiary. He later became a Research Associate at the General Electric Research Laboratory, Schenectady, New York (1935–1948). His most notable discovery there was a process to produce methylchlorosilanes, the precursors to silicones, from methyl chloride and a silicon/copper alloy, a process still used on a large scale today. He continued his research on silicone production until his transfer to Richland, Washington, where he conducted research on nuclear fission as a source of domestic energy. When the U.S. Government requested that GE work on nuclear propulsion for naval vessels, Gene, a Quaker, left in 1948 to teach chemistry at Harvard University, where he remained until retiring in 1970. His 1949 Baekeland address called for conservation, recycling, and the use of less wasteful alternatives long before these became fashionable. In the early 1950s, he became the first to apply broad-line NMR to the study of dynamic motion in silicone polymers. His interest in the differences in the chemistry of the group

14 elements led to work that culminated in the Allred–Rochow electronegativity scale, which can be found in many current textbooks.

Gene was the author or coauthor of several influential books, including *Chemistry of the Silicones* (1946, 1951), *General Chemistry—A Topical Introduction* (1954), *The Chemistry of Organometallic Compounds* (1956), *Unnatural Products* (1960), *Organometallic Chemistry* (1964), *Metalloids* (1966), *Chemistry—Molecules That Matter* (1974), and *Modern Descriptive Chemistry* (1977). The holder of numerous U.S. and foreign patents on chemical processes and organometallic substances, Gene received many awards, including the Baekeland Medal, American Chemical Society (1949); the Meyer Award, American Ceramic Society (1951); the Perkin Medal, Society of Chemical Industry (London, 1962); election to the American Academy of Arts and Sciences (1962); the Honor Scroll, American Institute of Chemists (1964); the Frederick Stanley Kipping Award, American Chemical Society (1965); the Chemical Pioneers Award, American Institute of Chemistry (1968); the Award for Excellence in Teaching, Manufacturing Chemists Association (1970); the James Flack Norris Award for the Teaching of Chemistry, American Chemical Society (1973); and the Alfred Stock Medal, German Chemical Society (1983).

JACOB KLEINBERG (EDITOR-IN-CHIEF, VOLUME VII, 1963)

Jake Kleinberg was born on February 14, 1914, in Passaic, New Jersey, and died on January 12, 2004, in Lawrence, Kansas, at the age of 89. He lost his father at the age of 3 and put himself through college by working part-time. Although initially enrolled at the City College of New York, he transferred to Randolph-Macon College, Ashland, Virginia, from which he received his B.S. degree in 1934. He earned his M.S. (1937) and Ph.D. degrees (1939) from the University of Illinois, Urbana-Champaign, under Ludwig F. Audrieth (Editor-in-Chief, Volume III, 1950), one of the founders and most prolific contributors to *Inorganic Syntheses* and a member of its Board of Editors (1934–1967). His thesis involved the ammonolysis of esters and the use of sulfamic acid in the separation of the rare earths. Jake was Assistant Professor at James Millikin University, Decatur, Illinois (1940–1943), and the College of Pharmacy at the University of Illinois, Chicago (1943–1946). In 1946, he joined the chemistry faculty of the University of Kansas, becoming Professor in 1951 and serving as Department Chairman from 1963 to 1970 before retiring in 1984. He was a Resident Lecturer for the National Science Foundation's summer institutes for high school chemistry teachers, led two committees that selected two Chancellors, and was President of the local chapter of Phi Beta Kappa.

Jake was the author of 95 scientific articles. He was the first to synthesize the ReH_9^{2-} anion (although its composition would remain mysterious for many years), and carried out many studies of the electrochemical reduction of both inorganic and organic substances. The titles of his books reflect his primary interests: *Unfamiliar Oxidation States and Their Stabilization* (1950); *Non-Aqueous Solvents: Applications as Media for Chemical Reactions* (with Ludwig F. Audrieth, 1953); *Inorganic Chemistry* (with William James Argersinger Jr. and Ernest Griswold, 1960); *University Chemistry* (with John C. Bailar Jr. and Therald Moeller, 1965); *Chemistry with Inorganic Qualitative Analysis* (with Therald Moeller and John C. Bailar Jr., 1965); *Introductory Analytical Chemistry* (with Alexander I. Popov and Ronald T. Pflaum, 1966); and *Radiochemistry of Iodine* (with Milton Kahn, 1977).

The winner of the ACS Midwest Award and the Amoco Foundation Award for Distinguished Teaching, Jake was a consultant for the Los Alamos National Laboratory and a member of the Editorial Board of *Chemical Reviews* (1951–1953), the *Journal of Inorganic & Nuclear Chemistry* (founded in 1955), and *Inorganic Chemistry* (1961–1964).

HENRY FULLER HOLTZCLAW JR. (EDITOR-IN-CHIEF, VOLUME VIII, 1966)

Henry Holtzclaw was born on July 30, 1921, in Stillwater, Oklahoma, and died in Lincoln, Nebraska, on May 24, 2001, after a long illness at the age of 79. He earned his A.B. degree in chemistry from the University of Kansas, Lawrence, in 1942, where his father was a Professor of Economics. While still a student, he was employed at the Eastman Kodak Company in Rochester, New York (summers of 1941 and 1942). He obtained his M.S. (1946) and Ph.D. degrees (1947) from the University of Illinois, Urbana-Champaign, under John C. Bailar Jr. (Editor-in-Chief, Volume IV, 1953). He participated in the Manhattan Project with the Tennessee Eastman Corporation in Oak Ridge, Tennessee (1944–1945), and his doctoral thesis was entitled "Polarographic Reduction of Cobaltic Coordination Complexes." In 1947, Henry joined the Chemistry Department of the University of Nebraska, Lincoln, and rose through the ranks to become Professor. He was appointed Foundation Professor of Chemistry in 1967 and Dean of Graduate Studies from 1976 to 1985. He spent a sabbatical leave as Guest Professor to teach and carry out research at the Universität Konstanz in Germany. He retired from the University of Nebraska in 1988.

Henry's research interests encompassed mass spectroscopy, proton magnetic resonance, and polarography of coordination compounds especially metal complexes of chelates such as β-diketonates. He was the coauthor of three popular freshman chemistry textbooks, some of which went through as many as 10 editions: *College Chemistry: With Qualitative Analysis* (1963), *General Chemistry* (1972), and *Basic Laboratory Studies in College Chemistry with Semi-Micro Quantitative Analysis* (1986).

Henry served as Chair of the Test of English as a Foreign Language Research Committee of the Educational Testing Services in the 1980s. In 1995, he received the James A. Lake Academic Freedom Award in recognition of his role as Chairman of a committee investigating a faculty member who helped lead a student anti-war demonstration in 1971.

WILLIAM LEE JOLLY (EDITOR-IN-CHIEF, VOLUME XI, 1968)

Bill Jolly was born in Chicago, Illinois, on December 27, 1927, and passed away at the age of 86 in Berkeley, California, on January 10, 2014. Bill received his B.S. (1948) and M.S. (1949) degrees from the University of Illinois at Urbana-Champaign, where he studied phosphate and hydrazine chemistry under the inorganic chemist Ludwig Audrieth (Editor-in-Chief, Volume III, 1950). He then moved to the University of California at Berkeley, where he obtained his Ph.D. degree under Wendell Latimer for work on the physical properties of germanium compounds. He was appointed for 1 year as an instructor at Berkeley in 1952 and then served as the Head of the Physical Chemistry and Inorganic Chemistry Division at the Radiation Laboratory in Livermore from 1952 until 1955. In the latter year, he returned to Berkeley as an Assistant Professor in the Chemistry Department, and was promoted to Associate Professor in 1957 and to Professor in 1962. He became Professor Emeritus in 1991.

Bill's research interests included thermodynamic and spectroscopic studies of liquid ammonia solutions, the synthesis of main group hydrides, and the chemistry of sulfur–nitrogen compounds, especially S_4N_4. In addition to work on the mechanism of hydrolysis of the borohydride ion, Bill developed improved routes to germane, stannane, arsine, and stibine. In the late 1960s and for the next 15 years, he carried out extensive and widely cited X-ray photoelectron spectroscopy (ESCA) studies of the chemical structure and bonding of inorganic compounds (especially those containing nitrogen and phosphorus) and organometallic compounds (especially metal carbonyls).

Among his other achievements, Bill wrote a highly entertaining history of the Chemistry Department at Berkeley called *From Retorts to Lasers* (1987). He was an expert in the chemistry of photography and invented a developer for the "solarization" of film, a technique that creates partly negative, partly positive photographic images. Among his awards was a Guggenheim Fellowship in 1959–1960, which he spent at the Chemical Institute of the University of Heidelberg, Germany. He was elected a fellow of the American Association for the Advancement of Science in 1984.

He was a prodigious author, especially in the area of preparative inorganic chemistry, where his textbooks were widely used and influential. Among his books are *Synthetic Inorganic Chemistry* (1960), *The Inorganic Chemistry of Nitrogen* (1964), *Preparative Inorganic Reactions* (editor, 7 volumes, 1964–1971), *The*

Chemistry of the Non-Metals (1966), *The Synthesis and Characterization of Inorganic Compounds* (1970), *Metal-Ammonia Solutions* (compiler, 1972), *Encounters in Experimental Chemistry* (1972, 1985), *Principles of Inorganic Chemistry* (1976), *Modern Inorganic Chemistry* (1985, 1991, 1998), and *Solarization Demystified* (unpublished, 1997).

JOHN KEEN RUFF (CO-EDITOR-IN-CHIEF, VOLUME XIV, 1973)

John was born on February 19, 1932, in New York City and died on January 6, 2004, in Athens, Georgia, of cancer at the age of 71. He received his B.S. degree in 1954 from Haverford College in Haverford, Pennsylvania, where he worked on hormones in his honors work. He obtained his Ph.D. degree from the University of North Carolina, Chapel Hill, in 1959 for a dissertation, "Light-Scattering of Aqueous Aluminum Nitrate and Gallium Perchlorate Solutions," that was supervised by S. Young Tyree (Editor-in-Chief, Volume IX, 1967). He then worked for 10 years at Redstone Arsenal, a research unit at Huntsville, Alabama, operated by the Rohm & Haas Company. In 1969, he moved to the University of Georgia, Athens, where he was a faculty member for 27 years.

John specialized in fluorine, boron, sulfur, phosphorus, and metal carbonyl chemistry. With M. Frederick Hawthorne, he discovered a series of amine complexes of aluminum trihydride and showed that some of them give aluminum metal when heated; this process later became useful for the formation of aluminum thin films by chemical vapor deposition. One of his notable achievements was the discovery that the PPN cation, bis(triphenylphosphoranylidene) ammonium, forms air-stable salts with many air-sensitive anions such as $[Co(CO)_4]^-$. He also discovered that cesium fluoride can serve as a catalyst for the synthesis of organic fluoroxy compounds (R_FOF) by the fluorination of acyl halides.

He wrote three editions (1995, 1998, 2001) of the laboratory manual *Experiments in General, Organic and Biological Chemistry* (coauthor Bobby Stanton). John was awarded a Sloan Research fellowship in 1969 and was a longtime member of the Atlanta Yacht Club.

DUWARD FELIX SHRIVER (EDITOR-IN-CHIEF, VOLUME XIX, 1979)

Duward ("Du") F. Shriver was born on November 20, 1934, in Glendale, California, and died on March 6, 2013, in Evanston, Illinois. He was raised on Oahu in the Hawaiian Islands, received his undergraduate degree in 1958 from the University of California, Berkeley, working with William L. Jolly (Editor-in-Chief, Volume XI, 1968), and his Ph.D. degree in 1961 from the University of Michigan, working with Robert W. Parry (Editor-in-Chief, Volume XIII, 1972). Du spent his entire academic career at Northwestern, beginning in 1961. He was named Morrison Professor of Chemistry in 1987 and served as Chemistry Department Chair from 1992 to 1995.

Du published more than 400 scientific articles spanning inorganic and organometallic synthesis, bioinorganic, solid-state and polymer chemistry, and vibrational spectroscopy. Some of his more notable achievements were the stepwise protonation of a carbonyl ligand to form methane and the isolation of cluster compounds containing the ketenylidene ligand; both of these systems are relevant to the industrially important Fischer–Tropsch process. He also made significant contributions to the design and synthesis of new polymers for lithium ion batteries and the vibrational signature of metal dioxygen compounds.

Du's book *The Manipulation of Air-Sensitive Compounds* (1969, 1986) is a standard reference in the field of organometallic chemistry, and he coedited *The Chemistry of Metal Cluster Compounds* (1990) with Herbert D. Kaesz (Editor-in-Chief, Volume XXVI, 1989) and Richard D. Adams. His highly successful undergraduate textbook *Inorganic Chemistry* (1990, 1994, 1999, 2006), coauthored with Peter W. Atkins, has been translated into 10 languages and is used to teach this very broad and important subject to students around the world. Du mentored more than 150 students and postdoctoral students who went on to pursue careers in industry and government and at national laboratories, colleges, and universities.

He received many professional awards, including a Guggenheim Fellowship, an Alfred P. Sloan Research Fellowship, the Royal Society of Chemistry Ludwig Mond Medal, the Materials Research Society Medal, and the American Chemical Society Award for Distinguished Service in Inorganic Chemistry. He was a fellow of the American Association for the Advancement of Science.

HERBERT DAVID KAESZ (EDITOR-IN-CHIEF, VOLUME XXVI, 1989)

Herb Kaesz was born on January 4, 1933, in Alexandria, Egypt, to Austrian parents and died on February 26, 2012, in Los Angeles, California, at the age of 79, about a month after he had been diagnosed with cancer. His father, a chemist, had joined his wife's family business, Kurz Optical, to run the Alexandria branch. When Herb was 7, the family emigrated to the United States. He received his A.B. degree from New York University in 1954 with Phi Beta Kappa honors. He earned his M.A. (1956) and Ph.D. degrees (1959) from Harvard University under the supervision of F. Gordon Stone for work on molecular addition compounds of boron. In August 1960, Herb joined the Inorganic Division of the University of California, Los Angeles (UCLA), where he served until his retirement in 2003. He remained an active Professor Emeritus until his death.

Herb's research centered on the synthesis and applications of organometallic compounds, especially metal carbonyls. In 1961, he synthesized $Tc_2(CO)_{10}$, which completed the list of elements, 14 in number, that form isolable binary carbonyls in the zero oxidation state. He discovered many new metal hydride and cluster compounds, and carried out elegant investigations of the nucleophilic activation of coordinated CO ligands under mild conditions. His book, *The Chemistry of Metal Cluster Complexes* (1990), coauthored with Duward F. Shriver (Editor-in-Chief, Volume XIX, 1979) and Richard D. Adams, became the premier reference text on the topic. He also studied main group element compounds and, later in his career, he investigated the development of pyrolytic and photolytic methods of metal film deposition for electronic applications.

He served as Chair of the International Union of Pure and Applied Chemistry (IUPAC) Commission on the Nomenclature of Inorganic Chemistry, was President of *Inorganic Syntheses*, Inc., and served for more than three decades as Associate Editor of the ACS journal, *Inorganic Chemistry* (1969–2001). Herb's honors included the ACS Southern California Section's Tolman Medal (1981), a Fellowship of the American Association for the Advancement of Science (1988), and the ACS Award for Distinguished Service in the Advancement of Inorganic Chemistry (1998). In 2009, the inaugural year of the program, he was elected an ACS Fellow. Herb held two foreign fellowships—a Fellowship from the Japan Society for the Promotion of Science (1978) and a Senior U.S. Scientist Award from the Alexander von Humboldt Foundation in Germany (1988). He also twice held the post of Professeur Invité in France, once in Toulouse (1992) and once in Paris (1995).

GEORGE B. KAUFFMAN
California State University, Fresno, CA

NOTICE TO CONTRIBUTORS
AND CHECKERS

The *Inorganic Syntheses* series (www.inorgsynth.com) publishes detailed and independently checked procedures for making important inorganic and organo-metallic compounds. Thus, the series is the concern of the entire scientific community. The Editorial Board hopes that many chemists will share in the responsibility of producing *Inorganic Syntheses* by offering their advice and assistance in both the formulation and the laboratory evaluation of outstanding syntheses.

The major criterion by which syntheses are judged is their potential value to the scientific community. We hope that the syntheses will be widely used and provide access to a broad range of compounds of importance in current research. The syntheses represent the best available procedures, and new or improved syntheses of well-established compounds are often featured. Syntheses of compounds that are available commercially at reasonable prices are ordinarily not included, however, unless the procedure illustrates some useful technique. *Inorganic Syntheses* is not a repository of primary research data, and therefore submitted syntheses should have already appeared in some form in the primary peer-reviewed literature and, at least to some extent, passed the "test of time." The series offers authors the chance to describe the intricacies of synthesis and purification in greater detail than possible in the original literature, as well as to provide updates of an established synthesis.

Authors wishing to submit syntheses for possible publication should write their manuscripts in a style that conforms with that of previous volumes of *Inorganic Syntheses* (a style guide is available from the Board Secretary). The manuscript should be in English and submitted as an editable electronic document. Nomenclature should be consistent and should follow the recommendations presented in *Nomenclature of Inorganic Chemistry, IUPAC Recommendations 2005*, published for the International Union of Pure and Applied Chemistry by The Royal Society of Chemistry, Cambridge, 2005. This document is available online (as of 2012) at http://www.iupac.org/fileadmin/user_upload/databases/Red_Book_2005.pdf. Abbreviations should conform to those used in publications of the American Chemical Society, particularly *Inorganic Chemistry*.

Submissions should consist of four sections: Introduction, Procedure, Properties, and References. The Introduction should include an indication of the importance and utility of the product(s) in question, and a concise and critical summary of the available procedures for making them and what advantage(s) the chosen method has over the alternatives. The Procedure should present detailed and unambiguous laboratory directions and be written so that it anticipates possible mistakes and misunderstandings on the part of the person who attempts to duplicate the procedure. It should contain an admonition if any potential hazards are associated with the procedure, and what safety precautions should be taken. Sources of unusual starting materials must be given, and, if possible, minimal standards of purity of reagents and solvents should be stated. Ideally, all reagents are readily available commercially or have been described in earlier volumes of *Inorganic Syntheses*. The scale should be reasonable for normal laboratory operation, and problems involved in scaling the procedure either up or down should be discussed if known. Unusual equipment or procedures should be clearly described and, if necessary for clarity, illustrated in line drawings. The yield should be given both in mass and in percentage based on theory. The Procedure section normally will conclude with calculated and found microanalytical data. The Properties section should supply and discuss those physical and chemical characteristics that are relevant to judging the purity of the product and to permitting its handling and use in an intelligent manner. Under References, pertinent literature citations should be listed in the order they appear in the text.

Manuscripts should be submitted electronically to the Secretary of the Editorial Board, Professor Stanton Ching, sschi@conncoll.edu. The Editorial Board determines whether submitted syntheses meet the general specifications outlined above. Every procedure will be checked in an independent laboratory, and publication is contingent on satisfactory duplication of the syntheses. For online access to information and requirements, see www.inorgsynth.com.

Chemists willing to check syntheses should contact the editor of a future volume or make this information known to Professor Ching.

TOXIC SUBSTANCES AND LABORATORY HAZARDS

Chemicals and chemistry are by their very nature hazardous. The obvious hazards in the syntheses reported in this volume are delineated, where appropriate, in the experimental procedure. It is impossible, however, to foresee every eventuality, such as a new biological effect of a common laboratory reagent. As a consequence, all chemicals used and all reactions described in this volume should be viewed as potentially hazardous. Care should be taken to avoid inhalation or other physical contact with reagents and solvents used in this volume. In addition, particular attention should be paid to avoiding sparks, open flames, or other potential sources that could set fire to combustible vapors or gases.

The following sources are especially recommended for guidance:

NIOSH Pocket Guide to Chemical Hazards, U.S. Government Printing Office, Washington, DC, 2005 (ISBN-13: 978-1-59804-052-4), is available free at http://www.cdc.gov/niosh/npg/ and can be purchased in paperback and spiral bound format. It contains information and data for 677 common compounds and classes of compounds.

Organic Syntheses, which is available online at http://www.orgsyn.org, has a concise but useful section "Handling Hazardous Chemicals."

Prudent Practices in the Laboratory: Handling and Disposal of Chemicals, National Academy Press, 1995 (ISBN-13: 978-0-30905-229-0), is available free at http://www.nap.edu/catalog.php?record_id=4911.

W. L. F. Amarego and C. Chai, *Purification of Laboratory Chemicals*, 6th ed., Butterworth-Heinemann, Oxford, 2009 (ISBN-13: 978-1-85617-567-8), is the standard reference for the purification of reagents and solvents. Special attention should be paid to the purification and storage of ethers.

CONTENTS

Chapter One

TRANSITION METAL HALIDE COMPOUNDS

1. OCTAHEDRAL HEXATANTALUM HALIDE CLUSTERS

Submitted by THIRUMALAI DURAISAMY,[*] **DANIEL N. T. HAY,**[*] **and LOUIS MESSERLE**[*]
Checked by ABDESSADEK LACHGAR[†]

Octahedral hexatantalum halide clusters usually exist as extended structures of the form $Ta_6(\mu\text{-}X)_{12}X_2$ with terminal (outer) and bridging (inner) halogen atoms shared between clusters, or as discrete clusters such as $Ta_6(\mu\text{-}X)_{12}X_2 \cdot 8H_2O$ that are better formulated as $Ta_6(\mu\text{-}X)_{12}X_2(OH_2)_4 \cdot 4H_2O$. These clusters consist of six tantalums linked through $Ta-Ta$ bonding to form a Ta_6 octahedron with a halide bridge along each of the 12 octahedral edges and one terminal ligand (halide, water, etc.) located apically on each tantalum.[1] A range of cluster oxidation states have been reported.[2]

Ta_6Cl_{14} was first reported in 1907 from the reduction of Ta_2Cl_{10} (denoted as $TaCl_5$ hereafter) with sodium amalgam,[3] and its structure was determined in 1950.[4] It is prepared typically by high-temperature, solid-state reduction of $TaCl_5$ in vacuum-sealed quartz ampules.[5] Microwave heating has also been employed.[6] Extraction with large volumes of water gives good yields of the discrete cluster[7] $Ta_6(\mu\text{-}Cl)_{12}Cl_2(OH_2)_4 \cdot 4H_2O$ after aqueous reduction of oxidized cluster contaminants with $SnCl_2$. The most commonly used approach is that developed by Koknat et al., involving reduction at 700°C of $TaCl_5$ with a four-fold excess of Ta powder.[8]

[*]Department of Chemistry, The University of Iowa, Iowa City, IA 52242.
[†]Department of Chemistry, Wake Forest University, Winston-Salem, NC 27109.

Inorganic Syntheses, Volume 36, First Edition. Edited by Gregory S. Girolami and Alfred P. Sattelberger.
© 2014 John Wiley & Sons, Inc. Published 2014 by John Wiley & Sons, Inc.

Ta_6Br_{14} was first prepared in 1910 by sodium amalgam reduction of $TaBr_5$.[3b] It has since been prepared by using the reductants aluminum[5a] and excess tantalum[8a] and can be isolated by aqueous extraction as the discrete cluster $Ta_6(\mu\text{-}Br)_{12}Br_2(OH_2)_4 \cdot 4H_2O$.[2a,b] A sample was structurally characterized as $[Ta_6(\mu\text{-}Br)_{12}(OH_2)_6](OH)Br \cdot 4H_2O$,[9] and another structure of the hexaaquo ion $[Ta_6(\mu\text{-}Br)_{12}(OH_2)_6]^{2+}$ was recently reported.[10]

There is considerable interest in the coordination[11] and catalytic[12] chemistries of these discrete clusters. Because of its high electron count, the hexaaquo ion $[Ta_6(\mu\text{-}Br)_{12}(OH_2)_6]^{2+}$ has been used frequently for phase determination[9,13] of isomorphous protein derivatives by SIR, MIR, SIRAS/MIRAS, and SAD/MAD methods in biomacromolecular crystallography. This use is growing as larger biomacromolecular structures and assemblies (e.g., membrane proteins, ribosomes, proteasomes) are studied.

We have found that the main group metal and metalloid reductants mercury, bismuth, and antimony are highly effective[14] in reducing WCl_6 or $MoCl_5$ at surprisingly lower temperatures than commonly used in the solid-state synthesis of early transition metal cluster halides. Borosilicate ampules can be substituted for the more expensive and less easily sealed quartz ampules at these lower temperatures, and the metals and metalloids are not as impacted by oxide coatings that inhibit solid-state reactions with more active metals. These lower temperatures may allow access to kinetic products, such as trinuclear clusters, instead of thermodynamic products.

We report here an extension of this reduction methodology to the convenient preparation[15] of $Ta_6(\mu\text{-}X)_{12}X_2(OH_2)_4 \cdot 4H_2O$ by reduction of TaX_5 with gallium dichloride, $Ga^+GaCl_4^-$ (for $X = Cl$), or gallium (for $X = Br$). Gallium dichloride has not been used as a preparative-scale reductant in transition metal chemistry. Gallium is an effective reductant, but because of its tendency to agglomerate and to adhere to glass, reductions employing Ga need to be agitated several times during the course of the reaction in order to optimize yields by homogenization of reactants. We have not yet tested the use of gallium dibromide as a reductant for $TaBr_5$, but expect that it would eliminate the need to homogenize reactants in gallium-based reductions and might improve the yield. The aquated hexatantalum clusters are liberated from the solid-state products by Soxhlet extraction with water, which greatly simplifies the isolation procedure. We also describe the straightforward preparation of a tetra-alkylammonium derivative of the $[Ta_6(\mu\text{-}Cl)_{12}Cl_6]^{4-}$ anion that has solubility in a broader array of organic solvents than the aquated clusters.

General Procedures

TaX_5 ($X = Cl$, Br; Materion Advanced Chemicals, Milwaukee, WI), hydrochloric acid (12 M, Fisher Scientific), Ga (99.99%, Atlantic Equipment Engineers, Bergenfield, NJ), NaCl (Fisher), KBr (Aldrich Chemical), $SnCl_2 \cdot 2H_2O$ (Fisher), $SnBr_2$ (99.5%, Alfa Aesar), HBr (48%, Fisher), and diethyl ether (anhydrous, Fisher) are used as received. $Ga^+GaCl_4^-$ is purchased and used as received from

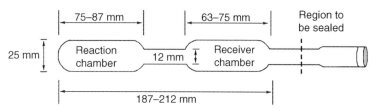

Figure 1. Ampule design for solid-state syntheses.

Alfa Aesar or is prepared by a literature method.[16] Powder X-ray diffraction is performed on samples protected from moisture by a 5 μm polyethylene film. Reactants and solid-state products are handled in a glove box under a dinitrogen atmosphere. A tube furnace with a positionable thermocouple is used in conjunction with a temperature controller in order to maintain and ramp temperatures. Syntheses are performed in dual-chamber, 25 mm OD borosilicate glass ampules with 30–40 mL total chamber volume, a 14/20 ground glass joint at one end, and constrictions between the end reaction chamber and middle receiver chamber (for volatile by-products) and between the middle receiver chamber and ground joint (Fig. 1). Ampules are oven dried at 130°C overnight and brought directly while hot into the glove box and cooled under vacuum or dinitrogen. Reactants are thoroughly mixed (employing a vortex mixer to agitate the reactants in a 20 mL glass scintillation vial) and added to the end reaction chamber via a long-stem funnel that minimizes contamination of the constriction surfaces. A gas inlet adapter[*] is used to seal the ampule, which is then evacuated on a Schlenk line and flame sealed between the joint and receiver chamber.

Tantalum is determined gravimetrically as the metal oxide Ta_2O_5. Samples are decomposed in tared borosilicate test tubes using concentrated nitric acid and hydrogen peroxide. The samples are dried and ignited. Other analyses are performed by Desert Analytics, Tucson, AZ.

A. TETRADECACHLOROHEXATANTALUM OCTAHYDRATE

$$6TaCl_5 + 8Ga^+GaCl_4^- + 20NaCl \rightarrow Na_4[Ta_6(\mu\text{-}Cl)_{12}Cl_6] + 16NaGaCl_4$$

$$Na_4[Ta_6(\mu\text{-}Cl)_{12}Cl_6] + 8H_2O \rightarrow Ta_6(\mu\text{-}Cl)_{12}Cl_2(OH_2)_4 \cdot 4H_2O + 4NaCl$$

Procedure

A vacuum-sealed ampule with $TaCl_5$ (0.96 g, 2.7 mmol), Ga_2Cl_4 (1.00 g, 3.56 mmol), and NaCl (0.52 g, 8.9 mmol) in the end reaction chamber is placed

---

[*]The checkers recommend that a quick-fit be used here to simplify the procedure.

in the center of a 45° inclined tube furnace. The furnace is slowly heated to 500°C over 4 h and kept at 500°C for 24 h. After being cooled, the ampule is opened in air and the dark solid is ground to a green powder with a mortar and pestle. The hygroscopic green powder is transferred to a coarse fritted glass Soxhlet thimble ($25\,mm \times 50\,mm$), containing a layer of borosilicate wool, the end packed with borosilicate wool, and the thimble is placed in a Soxhlet extractor. The apparatus is evacuated and backfilled with argon three times, and the powder is extracted (under argon to minimize air oxidation) for 17 h with argon-degassed distilled water (120 mL). The dark green solution is filtered through a medium-porosity fritted glass funnel in air, in order to remove insoluble white GaO(OH) powder (identified by powder X-ray diffractometry, matching PDF Card Number 06-0180). A solution of $SnCl_2 \cdot 2H_2O$ (1.0 g, 4.5 mmol) in 12 M hydrochloric acid (150 mL) is filtered to remove insoluble material, and a portion of this solution is added to the dark green filtrate, which is stirred and heated to near boiling, then cooled, and the remaining stannous chloride solution added. This step reduces any oxidized cluster contaminants to $Ta_6(\mu\text{-}Cl)_{12}Cl_2(OH_2)_4 \cdot 4H_2O$. The mixture is cooled in an ice bath, and the dark emerald green solid product is collected on a medium-porosity fritted glass funnel. The solid is washed with hydrochloric acid (20 mL), diethyl ether (30 mL), and dried in vacuum. Yield: 0.75 g (95%).

Anal. Calcd. for $H_{16}Cl_{14}O_8Ta_6$: Ta, 62.90; Cl, 28.75; Ga, 0.0. Found: Ta, 62.74; Cl, 28.70; Ga, <0.01.

Properties

$Ta_6(\mu\text{-}Cl)_{12}Cl_2(OH_2)_4 \cdot 4H_2O$ is soluble in water, DMSO, and methanol. The solutions are an intense emerald green. UV–vis (water): 330, 400, 470 (sh), 638, 750 nm. Solid $[Ta_6(\mu\text{-}Cl)_{12}Cl_2(OH_2)_4] \cdot 4H_2O$ and its aqueous solutions are slowly oxidized by air to hexatantalum clusters with higher oxidation states. Stannous chloride is a convenient solution reductant, as it quickly reduces contaminant levels of $[Ta_6(\mu\text{-}Cl)_{12}(OH_2)_6]^{3+}$ and $[Ta_6(\mu\text{-}Cl)_{12}(OH_2)_6]^{4+}$ back to $[Ta_6(\mu\text{-}Cl)_{12}(OH_2)_6]^{2+}$.

B. TETRADECABROMOHEXATANTALUM OCTAHYDRATE

$$18TaBr_5 + 16Ga + 28KBr \rightarrow 3K_4[Ta_6(\mu\text{-}Br)_{12}Br_6] + 16KGaBr_4$$

$$K_4[Ta_6(\mu\text{-}Br)_{12}Br_6] + 8H_2O \rightarrow Ta_6(\mu\text{-}Br)_{12}Br_2(OH_2)_4 \cdot 4H_2O + 4KBr$$

Procedure

A vacuum-sealed ampule with $TaBr_5$ (7.5 g, 12.9 mmol), Ga (0.80 g, 11.4 mmol), and KBr (2.39 g, 19.9 mmol) in the end reaction chamber is placed in the center of a

270°C preheated horizontal tube furnace and heated for 20 min. After being cooled to room temperature, the reactants/products are homogenized by gentle shaking, and this heating and homogenization cycle is repeated two times in order to disperse the molten gallium throughout the reaction mixture. The ampule is then placed in a tube furnace inclined to 45° and heated to 300°C over 1 h and held at that temperature for 12 h. The furnace is turned off and allowed to cool to room temperature. The ampule is removed from the furnace and the products are homogenized by vigorous shaking to give a dark green granular powder. The ampule is returned to the inclined furnace, heated to 400°C in 1 h, and held at that temperature for 24 h. The ampule is removed from the furnace, allowed to cool to room temperature, and opened in air. The dark solid is ground with a mortar and pestle to a dark green powder. The green powder is extracted by Soxhlet extraction with degassed water (170 mL) under argon for 24 h, as described above for $Ta_6(\mu\text{-}Cl)_{12}Cl_2(OH_2)_4 \cdot 4H_2O$. The resulting dark green solution is filtered through a medium-porosity fritted glass funnel in order to remove insoluble white GaO(OH) (as confirmed by powder X-ray diffractometry) powder. The dark green filtrate is treated with a filtered solution of $SnBr_2$ (3.1 g) dissolved in 48% hydrobromic acid (200 mL) to convert any oxidized species to $Ta_6(\mu\text{-}Br)_{12}Br_2(OH_2)_4 \cdot 4H_2O$, as described above for $Ta_6(\mu\text{-}Cl)_{12}Cl_2(OH_2)_4 \cdot 4H_2O$. The mixture is cooled in an ice bath and filtered through a medium-porosity fritted glass funnel. The dark green solid is washed with 48% hydrobromic acid (30 mL) followed by diethyl ether (30 mL). Yield: 4.15 g (86%).

Anal. Calcd. for $H_8Br_{14}O_8Ta_6$: Ta, 46.23; Br, 47.63. Found: Ta, 46.8; Br, 48.22.

Properties

$Ta_6(\mu\text{-}Br)_{12}Br_2(OH_2)_4 \cdot 4H_2O$ has solubility characteristics similar to the chloride analog. Solutions are an intense emerald green. UV–vis (water): 350, 420, 496 (sh), 638, 748 nm.

C. TETRAKIS(BENZYLTRIBUTYLAMMONIUM) OCTADECACHLOROHEXATANTALATE

$$Ta_6(\mu\text{-}Cl)_{12}Cl_2(OH_2)_4 \cdot 4H_2O + 4[N(CH_2Ph)Bu_3]Cl$$

$$\rightarrow [N(CH_2Ph)Bu_3]_4[Ta_6(\mu\text{-}Cl)_{12}Cl_6] + 8H_2O$$

Procedure

The synthesis of this tetraalkylammonium salt of $[Ta_6(\mu\text{-}Cl)_{12}Cl_6]^{4-}$ is adapted from the method developed by McCarley[17] for the synthesis of

$(NMe_4)_4[Nb_6Cl_{18}]$.[*] In a glove box (a glove bag would be sufficient), $Ta_6(\mu\text{-}Cl)_{12}$-$Cl_2(OH_2)_4 \cdot 4H_2O$ (0.50 g, 0.29 mmol) is placed into a coarse porosity fritted glass Soxhlet thimble in a Soxhlet extractor. A condenser with gas inlet and small empty flask are added and the apparatus is attached to a Schlenk line. In a 100 mL Schlenk flask containing a stir bar, a solution of $[N(CH_2Ph)Bu_3]Cl$ (0.36 g, 1.1 mmol) in 100% ethanol (50 mL) is degassed with argon with the help of a gas dispersion tube. The Soxhlet apparatus is joined to the Schlenk flask (with PTFE sleeves) under argon and the $Ta_6(\mu\text{-}Cl)_{12}Cl_2(OH_2)_4 \cdot 4H_2O$ is extracted under argon into the stirring $[N(CH_2Ph)Bu_3]Cl$/ethanol solution for 1 h, resulting in a dark green solution. After the Soxhlet apparatus is detached, the ethanol is removed under vacuum at 30°C. Degassed benzene (70 mL) is added to the dark forest green solid by cannula. A Dean–Stark trap is attached to the Schlenk flask and the solid is dried by means of azeotropic distillation for ~21 h under argon. The remaining benzene is then removed under vacuum. The dark forest green solid is dissolved in a minimum of cold (−40°C) CH_2Cl_2 in the glove box, and the resulting mixture is filtered through Celite® in order to remove a small amount of brown residue. The volume of the filtrate is reduced on a rotary evaporator until a small amount of brown residue forms. The solution is cooled to −40 °C and the brown solid is removed by filtration. The green filtrate is treated dropwise with toluene (~25% of the solution volume) and the volume is reduced by rotary evaporation until a small amount of brown precipitate is noted. The solution is filtered and the volume reduced by rotary evaporation until a clear supernatant is observed over a green solid. The supernatant is decanted and the solid is washed with toluene (3 × 2 mL) and dried in vacuum. Yield: 0.72 g (88%).

Anal. Calcd. for $C_{38}H_{68}N_2Cl_9Ta_3$: C, 32.26; H, 4.84; N, 1.98; Cl, 22.55. Found: C, 32.14; H, 5.04; N, 2.04; Cl, 22.14.

Properties

Green $[N(CH_2Ph)Bu_3]_4[Ta_6(\mu\text{-}Cl)_{12}Cl_6]$ is soluble in dichloromethane and 1,2-dichloroethane, but is insoluble in ether, benzene, and toluene. This tetraalkyl-ammonium cation imparts higher solubility to the salt, compared to the NMe_4^+ salt, and the benzyl group reduces cation crystallographic disorder compared to NBu_4^+ salts. Single-crystal X-ray diffractometry on $[N(CH_2Ph)Bu_3]_4[Ta_6(\mu\text{-}Cl)_{12}Cl_6]$ showed a $[Ta_6(\mu\text{-}Cl)_{12}Cl_6]^{4-}$ core in each of two crystalline forms (one a solvate), with similar metrics to $[Ta_6(\mu\text{-}Cl)_{12}Cl_6]^{4-}$ salts with inorganic cations.[15a]

---

[*]The checkers report that the Me_4N^+ salt of the $Ta_6Cl_{18}^{4-}$ anion is easily prepared, without Soxhlet extraction, by simple addition of NMe_4Cl to the $Ta_6Cl_{18}^{4-}$ solution.

Acknowledgments

This research was supported in part with funds from the National Science Foundation (CHE-0078701), the National Cancer Institute (1 R21 RR14062-01), the Roy J. Carver Charitable Trust (grant 01-93), a pilot grant funded by NIH/NIDDK (P30 DK 54759) and the Cystic Fibrosis Foundation (ENGELH98S0), and the Department of Energy's Energy-Related Laboratory Equipment Program (tube furnace). The Siemens D5000 automated powder diffractometer and Nonius Kappa CCD diffractometer were purchased with support from the Roy J. Carver Charitable Trust (grants 96-36 and 00-192).

References

1. J. Ferguson, *Prep. Inorg. React.* **7**, 93 (1971).
2. (a) B. G. Hughes, J. L. Meyer, P. B. Fleming, and R. E. McCarley, *Inorg. Chem.* **9**, 1343 (1970); (b) B. Spreckelmeyer, *Z. Anorg. Allg. Chem.* **358**, 147 (1968); (c) N. Prokopuk and D. F. Shriver, *Adv. Inorg. Chem.* **46**, 1 (1999); (d) M. Vojnović, N. Brničević, I. Bašic, P. Planinić, and G. Giester, *Z. Anorg. Allg. Chem.* **628**, 401 (2002).
3. (a) M. C. Chabrie, *Compt. Rend.* **144**, 804 (1907); (b) W. H. Chapin, *J. Am. Chem. Soc.* **32**, 323 (1910).
4. P. A. Vaughan, J. H. Sturdivant, and L. Pauling, *J. Am. Chem. Soc.* **72**, 5477 (1950).
5. (a) P. J. Kuhn and R. E. McCarley, *Inorg. Chem.* **4**, 1482 (1965); (b) H. Schäfer, H. Scholz, and R. Gerken, *Z. Anorg. Allg. Chem.* **331**, 154 (1964).
6. A. G. Whittaker and D. M. P. Mingos, *J. Chem. Soc., Dalton Trans.* 2073 (1995).
7. H. Schäfer and D. Bauer, *Z. Anorg. Allg. Chem.* **340**, 62 (1965).
8. (a) F. W. Koknat, J. A. Parsons, and A. Vongvusharintra, *Inorg. Chem.* **13**, 1699 (1974); (b) F. W. Koknat and D. J. Marko, *Inorg. Synth.* **34**, 187 (2004).
9. J. Knäblein, T. Neuefeind, F. Schneider, A. Bergner, A. Messerschmidt, J. Löwe, B. Steipe, and R. Hüber, *J. Mol. Biol.* **270**, 1 (1997).
10. M. Vojnović, I. Bašic, and N. Brničević, *Z. Kristallogr. New Cryst. Struct.* **214**, 435 (1999).
11. T. O. Gray, *Coord. Chem. Rev.* **243**, 213 (2003).
12. (a) D. M. Gardner and R. V. Gutowski, U.S. Patent 4,459,191 (1984); *Chem. Abstr.* **101**, 130218s (1984); (b) P. M. Boorman K. Chong, K. S. Jasim, R. A. Kydd, and J. M. Lewis, *J. Mol. Catal.* **53**, 381 (1989); (c) S. Jin, D. Venkataraman, F. J. DiSalvo, E. C. Peters, F. Svec, and J. M. Fréchet, *J. Polym. Prepr.* **41**, 458 (2000); (d) S. Kamiguchi, M. Noda, Y. Miyagishi, S. Nishida, M. Kodomari, and T. Chihara, *J. Mol. Catal. A* **195**, 159 (2003); (e) S. Kamaguchi, M. Watanabe, K. Kondo, M. Kodomari, and T. Chihara, *J. Mol. Catal. A* **203**, 153 (2003); (f) S. Kamaguchi and T. Chihara, *Catal. Lett.* **85**, 97 (2003); (g) S. Kamaguchi, K. Kondo, M. Kodomari, and T. Chihara, *J. Catal.* **223**, 54 (2004); (h) S. Kamaguchi, S. Iketani, M. Kodomari, and T. Chihara, *J. Cluster Sci.* **15**, 19 (2004).
13. (a) T. A. Ceska and R. Henderson, *J. Mol. Biol.* **213**, 539 (1990); (b) G. Schneider and Y. Lindqvist, *Acta Crystallogr. D* **50** 186 (1994); (c) T. Neuefeind, A. Bergner, F. Schneider, A. Messerschmidt, and J. Knäblein, *Biol. Chem.* **378**, 219 (1997); (d) R. Hüber, J. Knäblein, T. Neuefeind, A. Messerschmidt, and A. Bergner, PCT Int. Appl. WO 9900328 (1999).
14. (a) V. Kolesnichenko and L. Messerle, *Inorg. Chem.* **37**, 3660 (1998); (b) V. Kolesnichenko, D. C. Swenson, and L. Messerle, *Inorg. Chem.* **37**, 3257 (1998); (c) V. Kolesnichenko, J. J. Luci, D. C. Swenson, and L. Messerle, *J. Am. Chem. Soc.* **120**, 13260 (1998); (d) D. N. T. Hay, J. A. Adams, J. Carpenter, S. L. DeVries, J. Domyancich, B. Dumser, S. Goldsmith, M. A. Kruse, A. Leone,

F. Moussavi-Harami, J. A. O'Brien, J. R. Pfaffly, M. Sylves, P. Taravati, J. L. Thomas, B. Tiernan, and L. Messerle, *Inorg. Chim. Acta* **357**, 644 (2004).

15. (a) D. N. T. Hay, D. C. Swenson, and L. Messerle, *Inorg. Chem.* **41**, 4700 (2002); (b) D. N. T. Hay and L. Messerle, *J. Struct. Biol.* **139**, 147 (2002).
16. E. S. Schmidt, A. Schier, N. W. Mitzel, and H. Schmidbaur, *Z. Naturforsch.* **17b**, 337 (2001).
17. F. W. Koknat and R. E. McCarley, *Inorg. Chem.* **11**, 812 (1972).

2. OCTAHEDRAL HEXAMOLYBDENUM HALIDE CLUSTERS

Submitted by CHANG-TONG YANG,[*] DANIEL N. T. HAY,[*] and
LOUIS MESSERLE[*]
Checked by DAVID J. OSBORN III,[†] JEFFREY N. TEMPLETON,[‡] and
LISA F. SZCZEPURA[‡]

Molybdenum halides in lower oxidation states typically adopt dinuclear or polynuclear structures. One of the oldest known (1859) polynuclear clusters is Mo_6X_{12},[1] whose solid-state structure consists of octahedral Mo_6 units bearing eight μ_3-halides (inner chlorides), one over each of the eight octahedral faces, and six terminal (outer) halides in apical positions on the octahedral framework. Mo_6Cl_{12}, also known by its empirical formula $MoCl_2$, has an extended structure in which some of the chlorine atoms are shared between clusters. Although it can be made by direct reduction of higher molybdenum halides, the product obtained is often impure. Much purer material can be obtained by addition of concentrated hydrochloric acid to the crude material, which affords discrete dianionic clusters of stoichiometry $[Mo_6(\mu_3\text{-}Cl)_8Cl_6]^{2-}$. The resulting chloromolybdic acid, $(H_3O)_2[Mo_6(\mu_3\text{-}Cl)_8Cl_6] \cdot 6H_2O$, is sparingly soluble in water and can be recrystallized from hydrochloric acid. This material can be converted in near-quantitative yield to pure Mo_6Cl_{12} by thermolysis in vacuum.[2]

Mo_6Cl_{12}, its discrete halide and chalcogenide derivatives, and its coordination complexes[3] have attracted considerable attention because of their interesting photochemical properties/lifetimes (phosphorescence, luminescence, and electrogenerated chemiluminescence),[4] utility in catalysis,[5] radiochemistry,[6] sensor and conductor research,[7] intercalation chemistry,[8] and possible use as molecular precursors to chalcogenide-based Chevrel-phase materials.[9] There has also been considerable theoretical effort to understand cluster bonding and photochemical lifetimes.[10]

There are a number of published procedures for the synthesis of Mo_6Cl_{12}.[11] The most common are based on (1) comproportionation of $MoCl_5$ and Mo at 650°C to

[*]Department of Chemistry, The University of Iowa, Iowa City, IA 52242.
[†]Department of Chemistry, Michigan State University, East Lansing, MI 48824.
[‡]Department of Chemistry, Campus Box 4160, Illinois State University, Normal, IL 61790.

$MoCl_3$, subsequent disproportionation of $MoCl_3$ to Mo_6Cl_{12} and $MoCl_4$ at the same temperature, and recycling/comproportionation of the $MoCl_4$ with Mo to give additional Mo_6Cl_{12} for an overall 85–91% yield,[11g] (2) reduction of $MoCl_5$ with Al in a chloroaluminate melt at 200–450°C to give a 98% yield of chloromolybdic acid,[11i] and (3) comproportionation at 720°C of $MoCl_5$ and Mo in the presence of NaCl to give a 70–80% yield of Mo_6Cl_{12}.[11j] Two of these approaches require quartz tubing and/or considerable labor.

We have found that nonconventional, mild main group metal and metalloid reductants such as mercury, bismuth, antimony, gallium, and $Ga^+GaCl_4^-$ are highly effective[12] in reducing WCl_6 and TaX_5 (X = Cl, Br) at surprisingly lower temperatures than commonly used in the solid-state synthesis of early transition metal cluster halides. Borosilicate ampules can be substituted for the more expensive and less easily sealed quartz ampules at these lower temperatures, and the metals and metalloids are not as impacted by oxide coatings that inhibit solid-state reactions of more active metals. Bismuth is a particularly ideal reductant because it is nontoxic, inexpensive, not impeded by a surface oxide coating, does not readily reduce Mo_6Cl_{12}, and forms a volatile $BiCl_3$ by-product that does not exert significant pressure within a closed ampule. Bismuth also dissolves in molten $BiCl_3$, and this may improve dispersal of the reductant within the reaction mixture.

We report here an extension of this reduction methodology to the convenient preparation[13] of chloromolybdic acid and Mo_6Cl_{12}.

General Procedures

$MoCl_5$ (Materion Advanced Chemicals, Milwaukee, WI), Bi (325 mesh powder, Materion), hydrochloric acid (12 M, Fisher Scientific), and 8-hydroxyquinoline (Fisher) are used as received. Powder X-ray diffraction is performed on samples protected from moisture via a 5 μm polyethylene film. $MoCl_5$ and solid-state products are handled in either a glove box or a well-purged glove bag under a dinitrogen atmosphere. The molybdenum content of the products is determined gravimetrically as $MoO_2(C_9H_6ON)_2$ (C_9H_6ON = 8-hydroxyquinolinate).[14]

A tube furnace with a positionable thermocouple is used in conjunction with a temperature controller in order to maintain and ramp temperatures. Syntheses are performed in dual-chamber, 25 mm OD borosilicate glass ampules with 30–60 mL total chamber volume, a 14/20 or 19/22 ground glass joint at one end, and constrictions between the end reaction chamber and middle receiver chamber (for volatile by-products) and between the middle receiver chamber and ground joint (Fig. 1).*

---

*The checker used a 2.5 cm OD double-chamber ampule with a 15 cm length reaction chamber, a 10 cm receiver chamber, and a 0.95 cm OD end tube to which a greaseless vacuum valve was attached via a Cajon fitting. The ampule was flame sealed under vacuum between the receiver chamber and the Cajon fitting.

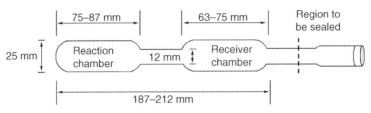

Figure 1. Ampule design for Mo_6Cl_{12} synthesis.

A. TETRADECACHLOROHEXAMOLYBDATE HEXAHYDRATE (CHLOROMOLYBDIC ACID)

$$6MoCl_5 + 6Bi \rightarrow Mo_6Cl_{12} + 6BiCl_3$$

$$Mo_6Cl_{12} + 2HCl(aq) \rightarrow (H_3O)_2[Mo_6(\mu_3\text{-}Cl)_8Cl_6] \cdot 6H_2O$$

Procedure

An ampule as described above is oven dried at 130°C overnight and brought directly while hot into a glove box or glove bag and cooled under vacuum in the antechamber or under dinitrogen, respectively. The reactants, Bi (3.825 g, 18.3 mmol) and ground $MoCl_5$ (5.00 g, 18.3 mmol), are thoroughly mixed and added to the end reaction chamber of the ampule via a long-stem funnel that minimizes contamination of the constriction surfaces. A gas inlet adapter is used to seal the ampule and prevent contamination with air. The ampule is then evacuated on a Schlenk line and flame sealed between the joint and receiver chamber.

The sealed ampule is placed in the center of a horizontal tube furnace. The temperature is ramped up to 230°C over 2 h and then to 350°C over 2 h, and the ampule is heated for 2.5 days (a shorter reaction time leads to yields in the 40–50% range). By sliding the tube, the receiver chamber is then moved out of the furnace and the $BiCl_3$ by-product is removed from the reaction chamber by sublimation over 12 h; the reaction products in the reaction chamber are converted into a free-flowing material by this step. The yellow brown non-volatile material is homogenized by shaking the cooled ampule, and then the ampule is returned to the center of the furnace and heated to 350°C for 3 h.[*] The receiver chamber is again moved out of the furnace, the furnace is reoriented to a slight angle ($\sim$15°) from the horizontal so that the receiver chamber is higher

---

[*]The checker is unsure whether the actions in this sentence are necessary for the success of the preparation.

than the reaction chamber, and the ampule is heated at 350°C for 24 h. The ampule is allowed to cool, the constriction between the receiver and reaction chambers is sealed with a torch, and the reaction chamber is opened in a glove box (a glove bag would be more than adequate for this step). The homogeneous, crystalline, olive green/brown material is collected. Yield of crude material: 4.066 g (theory for Mo_6Cl_{12}, 3.056 g, consistent with appreciable Bi content).

A portion of this solid (2.00 g) is treated in air with HCl (25 mL of 12 M aqueous solution) in a 250 mL Erlenmeyer flask (attached to a water bubbler for trapping HCl vapor) and the mixture heated over a flame with vigorous agitation.[*] Gray-black (presumably Bi-containing) insoluble impurities are removed by hot filtration through a medium-porosity fritted glass funnel. The insoluble material on the frit is washed with hot concentrated aqueous HCl in order to extract chloromolybdic acid that crystallized during filtration. The yellow filtrate is allowed to cool, resulting in the deposition of orange-yellow crystals. The crystalline product is filtered and returned to the Erlenmeyer flask, HCl (25 mL of 12 M aqueous solution) is added, the product redissolved by heating, and then collected by filtration after slow cooling to room temperature to obtain orange-yellow needles of chloromolybdic acid, $(H_3O)_2[Mo_6(\mu_3\text{-}Cl)_8Cl_6] \cdot xH_2O$. Yield: 1.40 g (77%,[†] for $x = 6$, based on $MoCl_5$).[‡]

Analysis of the product is unreliable because of its slow evolution of HCl.[11e] The material can be converted to an air-stable tetraalkylammonium salt if analysis is needed.

Properties

Chloromolybdic acid is insoluble in and slowly reactive with water,[§] but soluble in methanol and ethanol.[¶] The solid gradually oxidizes in air to a white solid with loss

---

[*]The checker reports that this procedure works well as long as the crude material is not exposed to air for more than 1 day.

[†]The checker reports yields for $(H_3O)_2[Mo_6(\mu_3\text{-}Cl)_8Cl_6] \cdot 6H_2O$ of 81–89% at the same scale and $\geq 77\%$ at four times the reaction scale via a 6-day procedure involving (a) heating the sealed ampule from 20 to 230°C over 2 h, (b) heating to 350°C over 2 h, (c) heating at 350°C for 2.5 days with nine ampule rotations separated by a minimum of 2 h during this period, (d) sublimation of $BiCl_3$ over 5 h by positioning the receiver chamber outside the tube furnace, (e) melting of $BiCl_3$ and remixing with nonvolatile material, (f) resublimation of $BiCl_3$ over 5 h, (g) melting of $BiCl_3$ and remixing with nonvolatile material, (h) cooling to 20°C overnight, (i) shaking the ampule contents to remix, (j) heating to 35°C over 3 h, and (k) resubliming $BiCl_3$ to the receiver chamber.

[‡]The checker reports that a singly recrystallized sample of $(H_3O)_2[Mo_6(\mu_3\text{-}Cl)_8Cl_6] \cdot 6H_2O$ contained 750 ppb Bi impurities and ≤ 100 ppb for a triply recrystallized sample. XRD of a triply recrystallized sample matched literature data.

[§]The checker reports that, after being mixed with water, chloromolybdic acid forms an insoluble precipitate that converts to a pale yellow solid upon exposure to air.

[¶]The checker reports that chloromolybdic acid is soluble in acetonitrile.

of HCl. It can be converted into air-stable tetraalkylammonium derivatives by metathesis in hydrochloric acid with tetralkylammonium chlorides.[4c]

B. HEXAMOLYBDENUM DODECACHLORIDE

$$(H_3O)_2[Mo_6(\mu_3\text{-}Cl)_8Cl_6] \cdot 6H_2O \rightarrow Mo_6Cl_{12} + 2HCl + 8H_2O$$

Procedure

A recrystallized sample of chloromolybdic acid (0.809 g, 0.66 mmol) is heated in vacuum using the method of Michel and McCarley.[2] Slowly (2.5 h) raising the temperature to 350°C and heating for 24 h affords an orange-yellow solid. Yield: 0.632 g (95% recovery based on chloromolybdic acid).[*]

Anal. Calcd. for Cl_2Mo: Mo, 57.50. Found (average of two trials): Mo, 57.64.[†]

Properties

Anhydrous Mo_6Cl_{12} changes to a paler yellow solid upon short exposure to air (10–15 min), due to absorption of water. The Mo analysis of an air-exposed sample is consistent with the formation of the neutral dihydrate $Mo_6(\mu_3\text{-}Cl)_8Cl_4(H_2O)_2$.

Anal. Calcd. for $H_4Cl_{12}O_2Mo_6$: Mo, 55.50. Found (average of three trials): 55.28.

Acknowledgments

This research was supported in part with funds from the National Science Foundation (CHE-0078701), the National Cancer Institute (1 R21 RR14062-01), the Roy J. Carver Charitable Trust (grant 01–93), and the Department of Energy's Energy-Related Laboratory Equipment Program (tube furnace). The Siemens D5000 automated powder diffractometer was purchased with support from the Roy J. Carver Charitable Trust.[‡]

---

[*]The checker reports that the method of Mussel and Nocera,[15] heating chloromolybdic acid in vacuum at 150°C for 2 h and then overnight at 210°C, gives consistently yellow Mo_6Cl_{12} in yields of 90–93%.
[†]The checker finds Mo analyses for Mo_6Cl_{12} of 55.46% (average of two trials), 54.07% (three trials), and 54.40% (three trials) for products from three different syntheses.
[‡]D.J.O. acknowledges support (grant #452) from the 21st Century Jobs Trust Fund through the SEIC Board from the State of Michigan.

References

1. (a) W. Blomstrand, *J. Prakt. Chem.* **77**, 88 (1859); (b) C. Brosset, *Ark. Kemi* **1**, 353 (1949).
2. J. B. Michel and R. E. McCarley, *Inorg. Chem.* **21**, 1864 (1982).
3. (a) J. E. Fergusson, B. H. Robinson, and C. J. Wilkins, *J. Chem. Soc. A* 486 (1967); (b) W. M. Carmichael and D. A. Edwards, *J. Inorg. Nucl. Chem.* **29**, 1535 (1967); (c) A. D. Hamer and R. A. Walton, *Inorg. Chem.* **13**, 1446 (1974); (d) P. Lessmeister and H. Schäfer, *Z. Anorg. Allg. Chem.* **417**, 171 (1975); (e) A. D. Hamer, T. J. Smith, and R. A. Walton, *Inorg. Chem.* **15**, 1014 (1976); (f) J. Kraft and H. Schäfer, *Z. Anorg. Allg. Chem.* **524**, 137 (1985); (g) T. Saito, M. Nishida, T. Yamagata, Y. Yamagata, and Y. Yamaguchi, *Inorg. Chem.* **25**, 1111 (1986); (h) G. M. Ehrlich, H. Deng, L. I. Hill, M. L. Steigerwald, P. J. Squattrito, and F. J. DiSalvo, *Inorg. Chem.* **34**, 2480 (1995); (i) D. Bublitz, W. Preetz, and M. K. Simsek, *Z. Anorg. Allg. Chem.* **623**, 1 (1997); (j) N. Prokopuk and D. F. Shriver, *Inorg. Chem.* **36**, 5609 (1997); (k) I. Bašic, N. Brničević, U. Beck, A. Simon, and R. E. McCarley, *Z. Anorg. Allg. Chem.* **624**, 725 (1998); (l) S. M. Malinak, L. K. Madden, H. A. Bullen, J. J. McLeod, and D. C. Gaswick, *Inorg. Chim. Acta* **278**, 241 (1998); (m) L. F. Szczepura, B. A. Ooro, and S. R. Wilson, *J. Chem. Soc., Dalton Trans.* 3112 (2002); (n) N. Brničević, I. Bašic, B. Hoxha, P. Planinić, and R. E. McCarley, *Polyhedron* **22**, 1553 (2003); (o) T. O. Gray, *Coord. Chem. Rev.* **243**, 213 (2003).
4. A. W. Maverick and H. B. Gray, *J. Am. Chem. Soc.* **103**, 1298 (1981); (b) A. W. Maverick, J. S. Najdzionek, D. MacKenzie, D. G. Nocera, and H. B. Gray, *J. Am. Chem. Soc.* **105**, 1878 (1983); (c) Y. Saito, H. K. Tanaka, Y. Sasaki, and T. Azumi, *J. Phys. Chem.* **89**, 4413 (1985); (d) H. K. Tanaka, Y. Sasaki, and K. Saito, *Sci. Pap. Inst. Phys. Chem. Res. (Jpn.)* **78**, 92 (1984); (e) R. D. Mussell and D. G. Nocera, *Polyhedron* **5**, 47 (1986); (f) T. Azumi and Y. Saito, *J. Phys. Chem.* **92**, 1715 (1988); (g) M. D. Newsham, M. K. Cerrata, K. A. Berglund, and D. G. Nocera, *Mater. Res. Soc. Symp. Proc.* **121**, 627 (1988); (h) H. K. Tanaka, Y. Sasaki, M. Ebihara, and K. Saito, *Inorg. Chim. Acta* **161**, 63 (1989); (i) B. Kraut and G. Ferraudi, *Inorg. Chim. Acta* **156**, 7 (1989); (j) B. Kraut and G. Ferraudi, *Inorg. Chem.* **28**, 4578 (1989); (k) J. A. Jackson, C. Turro, M. D. Newsham, and D. G. Nocera, *J. Phys. Chem.* **94**, 4500 (1990); (l) R. D. Mussell and D. G. Nocera, *Inorg. Chem.* **29**, 3711 (1990); (m) M. D. Newsham, R. I. Cukier, and D. G. Nocera, *J. Phys. Chem.* **95**, 9660 (1991); (n) H. Miki, T. Ikeyama, Y. Sasaki, and T. Azumi, *J. Phys. Chem.* **96**, 3236 (1992); (o) L. M. Robinson and D. F. Shriver, *J. Coord. Chem.* **37**, 119 (1996).
5. (a) D. M. Gardner and R. V. Gutowski, U.S. Patent 4,459,191 (1984); *Chem. Abstr.* **101**, 130218s (1984); (b) P. M. Boorman K. Chong, K. S. Jasim, R. A. Kydd, and J. M. Lewis, *J. Mol. Catal.* **53**, 381 (1989); (c) S. Jin, D. Venkataraman, F. J. DiSalvo, E. C. Peters, F. Svec, and J. M. Fréchet, *J. Polym. Prepr.* **41**, 458 (2000); (d) S. Kamiguchi, M. Noda, Y. Miyagishi, S. Nishida, M. Kodomari, and T. Chihara, *J. Mol. Catal. A* **195**, 159 (2003); (e) S. Kamaguchi, S. Watanabe, K. Kondo, M. Kodomari, and T. Chihara, *J. Mol. Catal. A* **203**, 153 (2003); (f) S. Kamaguchi and T. Chihara, *Catal. Lett.* **85**, 97 (2003); (g) S. Kamaguchi, K. Kondo, M. Kodomari, and T. Chihara, *J. Catal.* **223**, 54 (2004); (h) S. Kamaguchi, S. Iketani, M. Kodomari, and T. Chihara, *J. Cluster Sci.* **15**, 19 (2004).
6. L. A. Fucugauchi, S. Millan, A. Mondragon, and M. Solache-Rios, *J. Radioanal. Nucl. Chem.* **178**, 437 (1994).
7. (a) A. Y. Shatalov, *Trudy Inst. Fis. Khim., Akad. Nauk SSSR No. 5, Issledovan. Korrosii Met.* **4**, 237 (1955); *Chem. Abstr.* **50**, 11212e (1956); (b) D. G. Nocera and H. B. Gray, *J. Am. Chem. Soc.* **106**, 824 (1984); (c) H. Fuchs, S. Fuchs, K. Polborn, T. Lehnert, C. P. Heidmann, and H. Mueller, *Synth. Met.* **27**, A271 (1988); (d) P. A. Barnard, I. W. Sun, and C. L. Hussey, *Inorg. Chem.* **29**, 3670 (1990); (e) S. K. D. Strubinger, P. A. Barnard, I. W. Sun, and C. L. Hussey, *Proc. Electrochem. Soc.* **90–17**, 367 (1990); (f) A.-K. Klehe, A. A. House, J. Singleton, W. Hayes, C. J. Kepert, and P. Day, *Synth. Met.* **86**, 2003 (1997); (g) C. J. Kepert, M. Kurmoo, and P. Day, *Proc. R. Soc. Lond. A* **454**, 487 (1998); (h) R. N. Ghosh, G. L. Baker, C. Ruud, and D. G. Nocera, *Appl. Phys. Lett.* **75**, 2885

(1999); (i) A. Deluzet, S. Perruchas, H. Bengel, P. Batail, S. Molas, and J. Fraxedas, *Adv. Funct. Mater.* **12**, 123 (2002); (j) A. Deluzet, P. Batail, Y. Misaki, P. Auban-Senzier, and E. Canadell, *Adv. Mater.* **12**, 436 (2000).

8. S. P. Christiano and T. J. Pinnavaia, *J. Solid State Chem.* **64**, 232 (1986).
9. (a) A. P. Mazhara, A. A. Opalovskii, V. E. Fedorov, and S. D. Kirik, *Zh. Neorg. Khim.* **22**, 1827 (1977); (b) S. J. Hilsenbeck, V. G. Young, and R. E. McCarley, *Inorg. Chem.* **33**, 1822 (1994); (c) M. Ebihara, K. Toriumi, Y. Sasaki, and K. Saito, *Gazz. Chim. Ital.* **125**, 87 (1995); (d) R. E. McCarley, S. J. Hilsenbeck, and X. Xie, *J. Solid State Chem.* **117**, 269 (1995); (e) S. Jin, F. Popp, S. W. Boettcher, M. Yuan, C. M. Oertel, and F. J. DiSalvo, *Dalton Trans.* 3096 (2002).
10. (a) R. J. Gillespie, *Can. J. Chem.* **39**, 2336 (1961); (b) F. A. Cotton and G. G. Stanley, *Chem. Phys. Lett.* **58**, 450 (1978); (c) D. Certain and R. Lissillour, *Ann. Chem. (Paris)* **9**, 953 (1984); (d) R. L. Johnston and D. M. P. Mingos, *Inorg. Chem.* **25**, 1661 (1986); (e) Y. Li, X. Zhang, and J. Wu, *Gaodeng Xuexiao Huaxue Xuebao* **14**, 523 (1993); (f) L. M. Robinson, R. L. Bain, D. F. Shriver, and D. E. Ellis, *Inorg. Chem.* **34**, 5588 (1995); (g) Yu.V. Plekhanov and S. V. Kryuchkov, *Russ. J. Coord. Chem.* **21**, 622 (1995); (h) P.-D. Fan, P. Deglmann, and R. Ahlrichs, *Chem. Eur. J.* **8**, 1059 (2002); (i) H. Honda, T. Noro, K. Tanaka, and E. Miyoshi, *J. Chem. Phys.* **114**, 10791 (2001).
11. (a) A. Rosenheim and E. Dehn, *Ber. Dtsch. Chem. Ges.* **48**, 1167 (1915); (b) K. Lindner, E. Haller, and H. Helwig, *Z. Anorg. Allg. Chem.* **130**, 209 (1923); (c) K. Lindner and H. Frit, *Z. Anorg. Allg. Chem.* **137**, 66 (1924); (d) J. C. Sheldon, *Nature* **184**, 1210 (1959); (e) J. C. Sheldon, *J. Chem. Soc.* 1007 (1960); (f) H. Schäfer, H. G. von Schnering, J. Tillack, F. Kuhnen, H. Woehrle, and H. Baumann, *Z. Anorg. Allg. Chem.* **353**, 281 (1967); (g) P. Nannelli and B. P. Block, *Inorg. Synth.* **12**, 170 (1970); (h) W. C. Dorman, Report IS-T-510 (1972); *Nucl. Sci. Abstr.* **26**, 43612 (1972); *Chem. Abstr.* **78**, 10965t (1973); (i) W. C. Dorman and R. E. McCarley, *Inorg. Chem.* **13**, 491 (1974); (j) F. W. Koknat, T. J. Adaway, S. I. Erzerum, and S. Syed, *Inorg. Nucl. Chem. Lett.* **16**, 307 (1980); (k) W. J. Jolly, *The Synthesis and Characterization of Inorganic Compounds*, Prentice Hall, New York, 1970, p. 456.
12. (a) V. Kolesnichenko and L. Messerle, *Inorg. Chem.* **37**, 3660 (1998); (b) V. Kolesnichenko, D. C. Swenson, and L. Messerle, *Inorg. Chem.* **37**, 3257 (1998); (c) V. Kolesnichenko, J. J. Luci, D. C. Swenson, and L. Messerle, *J. Am. Chem. Soc.* **20**, 13260 (1998); (d) D. N. T. Hay, D. C. Swenson, and L. Messerle, *Inorg. Chem.* **41**, 4700 (2002); (e) D. N. T. Hay and L. Messerle, *J. Struct. Biol.* **139**, 147 (2002).
13. D. N. T. Hay, J. A. Adams, J. Carpenter, S. L. DeVries, J. Domyancich, B. Dumser, S. Goldsmith, M. A. Kruse, A. Leone, F. Moussavi-Harami, J. A. O'Brien, J. R. Pfaffly, M. Sylves, P. Taravati, J. L. Thomas, B. Tiernan, and L. Messerle, *Inorg. Chim. Acta* **357**, 644 (2004).
14. A. I. Busev, translated by J. Schmorak, *Analytical Chemistry of Molybdenum*, Ann Arbor-Humphrey Scientific Publishers, Ann Arbor, MI, 1969, pp. 22–23, 144–145.
15. R. D. Mussel, Ph.D. dissertation, Michigan State University, 1988, p. 26.

3. ETHER COMPLEXES OF MOLYBDENUM(III) AND MOLYBDENUM(IV) CHLORIDES

Submitted by SÉBASTIEN MARIA[*] and RINALDO POLI[*]
Checked by KIMBERLEY J. GALLAGHER,[†] ADAM S. HOCK,[†] and
MARC J. A. JOHNSON[†]

Trichlorotris(tetrahydrofuran)molybdenum(III), $MoCl_3(THF)_3$, is a widely used starting material for the synthesis of a broad range of molybdenum compounds. It was first prepared by a carbonyl route,[1] but later the compound was obtained by a reductive procedure in three steps starting from $MoCl_5$, as initially reported by the Chatt group.[2] This synthetic method, which was later improved by several minor modifications,[3–6] has since remained the most widely employed route to this compound. In the first step in the three-step procedure, $MoCl_5$ is dissolved in acetonitrile, which causes a spontaneous reduction to $MoCl_4(MeCN)_2$. Tetrahydrofuran cannot be used in this step, because the powerful Lewis acidic properties of $MoCl_5$ lead to the ring-opening polymerization of this solvent. In two subsequent steps, the nitrile ligands in $MoCl_4(MeCN)_2$ are replaced with THF to give $MoCl_4(THF)_2$, and the synthesis is then completed by a chemical reduction with tin. In spite of the various incremental improvements, this three-step procedure remains rather long and requires, in our hands, isolation of both the $MoCl_4(MeCN)_2$ and $MoCl_4(THF)_2$ intermediates in order to obtain product of sufficient quality. With individual step yields of 63–86%, the overall yield from $MoCl_5$ is 46% at best, with a time investment of ≥2 days.

Unlike THF, diethyl ether is not susceptible to acid-catalyzed polymerization and, unlike acetonitrile, it is not susceptible to oxidation. Ethereal solutions of $MoCl_5$ are stable and their conversion to $MoCl_4(OEt_2)_2$ by reducing agents such as norbornene[7] or allyltrimethylsilane[8] has been previously reported. As it turns out, this reduction can also be conveniently carried out by metallic tin. The reduction reaction of $MoCl_5$ by tin in diethyl ether is fast but controlled at room temperature, without an excessive exothermicity. Therefore, this reaction does not require any special precautions except exclusion of air and moisture. In addition, the tin chloride coproduct is ether soluble, whereas $MoCl_4(OEt_2)_2$ is only sparingly soluble. Therefore, the Mo(IV) product is easily purified by ether washings with only minimal product loss. When $MoCl_4(OEt_2)_2$ is dissolved in THF in the presence of metallic tin, smooth conversion to $MoCl_3(THF)_3$ occurs. This

[*]Laboratoire de Chimie de Coordination, UPR CNRS 8241, 205 Route de Narbonne, 31077 Toulouse Cedex, France.
[†]Chemical Sciences and Engineering Division, Argonne National Laboratory, 9700 South Cass Avenue, Argonne, IL 60439.

transformation is the combined result of two reactions: ligand exchange to give $MoCl_4(THF)_2$ and reduction of the latter to $MoCl_3(THF)_3$ by tin as previously reported.[5] This reaction gives excellent results without the need to isolate the $MoCl_4(OEt_2)_2$ product after its generation from $MoCl_5$, nor the need to isolate the second $MoCl_4(THF)_2$ intermediate. Thus, it suffices to use enough tin reagent to reduce Mo(V) to Mo(III) and to replace the solvent after completion of the first reduction step, in order to achieve a convenient high-yield synthesis of $MoCl_3(THF)_3$ in one half day from $MoCl_5$. This contribution describes both the synthesis and isolation of the $MoCl_4(OEt_2)_2$ complex and the direct, two-step, single-flask synthesis of $MoCl_3(THF)_3$ from $MoCl_5$.

General Procedures

Solvents (diethyl ether, tetrahydrofuran) are distilled from sodium benzophenone ketyl under argon.[*] The tin granules were activated by heating in an oven at 120° C for several hours.[†]

A. TETRACHLOROBIS(DIETHYL ETHER)MOLYBDENUM(IV)

$$2MoCl_5 + 4Et_2O + Sn \rightarrow 2MoCl_4(OEt_2)_2 + SnCl_2$$

Procedure

A 100 mL Schlenk tube equipped with a magnetic stir bar is connected to a vacuum/argon line through the sidearm with ground glass stopcock.[‡] After the tube is evacuated and filled with argon three times, pentachloromolybdenum(V) (1.20 g, 4.39 mmol) and activated tin granules (20 mesh;[§] 1.04 g, 8.76 mmol) are added. Freshly distilled ether (30 mL) is added, and the suspension is then stirred at room temperature until it yields a suspension of an orange solid in an orange solution (about 30 min).[¶] The orange solid is mechanically separated from the excess tin by taking advantage of the large difference in density between the two solids, as follows. Under gentle magnetic stirring that stirs up the finely divided orange product but

---

[*]The checkers report that one can also use solvents purified with the column method described in Ref. 9.
[†]The checkers activated the tin granules by drying them under vacuum on a Schlenk line overnight. Using tin that is not activated may result in longer reaction times, but can give satisfactory yields of the product.
[‡]The checkers carried out the synthesis in an inert atmosphere glove box, and note that dinitrogen can be used instead of argon as an inert gas.
[§]Subsequent separation of the tin is made much easier by using granules or shot (20 mesh) rather than powder.
[¶]The checkers note that a more typical time for this step is 60 min. The formation of an orange solid suspended in an orange solution indicates that consumption of the $MoCl_5$ is complete.

leaves the tin at the bottom of the Schlenk tube, the supernatant suspension of the $MoCl_4(Et_2O)_2$ product is transferred into a new Schlenk tube through a medium-sized (G15) cannula, taking care not to transfer any of the metallic tin. The product is then collected by filtration on a glass frit,[*] washed with Et_2O (5×5 mL), and finally dried under vacuum at $-20°$ C, the low temperature being used in order to avoid the loss of diethyl ether. Yield: 1.36 g (80%).

Anal. Calcd. for $C_8H_{20}Cl_4MoO_2$: C, 24.89; H, 5.22. Found: C, 25.31; H, 5.23.

Properties

Tetrachlorobis(diethyl ether)molybdenum(IV) is an orange, air-sensitive crystalline powder. It is paramagnetic, with a triplet ground state ($\mu_{eff} = 2.33\mu_B$). The IR spectrum is reported.[8] The ^1H NMR spectrum in C_6D_6 shows two broad resonances for the methyl and the methylene protons of the ether ligands, at positions that are not significantly contact shifted from those of the free ether (δ 1.1, 3.5). These resonances are not due to the coordinated ether molecules, but instead are due to *free* ether molecules resulting from dissociation equilibria. The compound has a marked tendency to lose diethyl ether, and samples dried without special precautions systematically give low carbon and hydrogen microanalyses. Heating the compound at $80°$ C under vacuum affords a reactive form of $MoCl_4$, which can be transformed easily and in high yields into a variety of other useful compounds such as $MoCl_4(THF)_2$, $MoCl_3(THF)_3$, $MoCl_3(PMe_3)_3$, and $Mo(O$-t-$Bu)_4$.[10] It is often convenient to generate $MoCl_4(OEt_2)_2$ in situ, and to use it in subsequent reactions. This strategy is employed for the synthesis of $MoCl_3(THF)_3$ in the next section.

B. TRICHLOROTRIS(TETRAHYDROFURAN)MOLYBDENUM(III)

$$2MoCl_5 + 4Et_2O + Sn \rightarrow 2MoCl_4(OEt_2)_2 + SnCl_2$$

$$2MoCl_4(OEt_2)_2 + 6THF + Sn \rightarrow 2MoCl_3(THF)_3 + 4Et_2O + SnCl_2$$

Procedure

An orange suspension of $MoCl_4(OEt_2)_2$ and excess tin in diethyl ether is prepared according to the procedure detailed in the previous section, starting from

---

[*]The additional cannula transfer step to an intermediate container makes it possible to evaluate whether any metallic tin has inadvertently been transferred together with the product. The checkers note that the reaction solution can be transferred directly to the glass frit (and the intermediate transfer step omitted) by using a smaller gauge cannula, through which the desired product but not the tin powder can pass.

MoCl$_5$ (3.18 g, 11.64 mmol) and Sn granules (5–20 mesh; 2.77 g, 23.33 mmol) in diethyl ether (30 mL). Once the orange solid has settled well at the bottom of the Schlenk tube, the supernatant liquid is removed by using a thin cannula,[*] and freshly distilled THF (30 mL) is added to the remaining solid. The resulting mixture is then stirred at room temperature until a beige solid forms in a dark purple solution (~4 h).[†] The product is separated mechanically from the excess tin by taking advantage of the large difference in density between the two solids. Under gentle magnetic stirring that stirs up the finely divided product but leaves the tin at the bottom of the Schlenk tube, the supernatant suspension of MoCl$_3$(THF)$_3$ is transferred into a new Schlenk tube through a medium-sized (G15) cannula, taking care not to transfer any of the metallic tin. The product is then collected by filtration on a glass frit, washed with Et$_2$O (3 × 15 mL),[‡] and finally dried under vacuum. The complex is stored under dry argon. Yield: 3.99 g (82%).[§]

Anal. Calcd. for C$_{12}$H$_{24}$Cl$_3$MoO$_3$: C, 34.43; H, 5.78. Found: C, 34.04; H, 5.51.

Properties

Trichlorotris(tetrahydrofuran)molybdenum(III) is a beige[¶] crystalline product that deteriorates upon exposure to the laboratory atmosphere. When dissolved or suspended in solvents other than THF, it progressively loses THF and transforms to darker dinuclear complexes such as Mo$_2$Cl$_6$(THF)$_4$ or Mo$_2$Cl$_6$(THF)$_3$.[11] It also decomposes slowly in THF, especially upon warming, with formation of 1,4-dichlorobutane and more soluble molybdenum oxo derivatives whose precise nature is unknown. Therefore, when MoCl$_3$(THF)$_3$ is used as a starting material, it is important to use its solutions in THF promptly.

References

1. A. D. Westland and N. Muriithi, *Inorg. Chem.* **11**, 2971 (1972).
2. M. W. Anker, J. Chatt, G. J. Leigh, and A. G. Wedd, *J. Chem. Soc., Dalton Trans.* 2639 (1975).
3. J. R. Dilworth and R. L. Richards, *Inorg. Synth.* **20**, 119 (1980).
4. J. R. Dilworth and J. Zubieta, *Inorg. Synth.* **24**, 193 (1986).
5. J. R. Dilworth and R. L. Richards, *Inorg. Synth.* **28**, 33 (1990).
6. D. Zeng and M. J. Hampden-Smith, *Polyhedron* **11**, 2585 (1992).

[*]The checkers removed the solution with a cannula filter. As mentioned above, subsequent separation of the tin is made much easier by using granules or shot (20 mesh) rather than powder.
[†]In the checker's hands, this step is typically more rapid and is usually complete in about 2.5 h. The appearance of the purple color in the supernatant signals the proper time for filtration.
[‡]The checkers prefer to wash first with small quantities of THF until the washings are orange with no hint of red or purple, and then with Et$_2$O as described.
[§]The checkers obtained a yield of 72%.
[¶]The color of the solid may have orange or pink overtones.

7. L. Castellani and M. C. Gallazzi, *Transit. Met. Chem.* **10**, 194 (1985).
8. C. Persson and C. Andersson, *Inorg. Chim. Acta* **203**, 235 (1993).
9. P. J. Alaimo, D. W. Peters, J. Arnold, and R. G. Bergman, *J. Chem. Educ.* **78**, 64 (2001).
10. F. Stoffelbach, D. Saurenz, and R. Poli, *Eur. J. Inorg. Chem.* 2699 (2001).
11. R. Poli and H. D. Mui, *J. Am. Chem. Soc.* **112**, 2446 (1990).

4. OCTAHEDRAL HEXATUNGSTEN HALIDE CLUSTERS

Submitted by DONALD D. NOLTING[*] and LOUIS MESSERLE[*]
Checked by MIN YUAN[†] and FRANCIS J. DISALVO[†]

Tungsten dichloride or, more correctly, hexatungsten dodecachloride, W_6Cl_{12}, has an octahedral hexanuclear cluster structure with μ_3-chlorines over each of the eight octahedral faces and six terminal chlorines, some of which are shared between clusters in an extended array. Treatment of W_6Cl_{12} with concentrated hydrochloric acid yields the discrete cluster dianion $[W_6(\mu_3\text{-}Cl)_8Cl_6]^{2-}$ as the H_3O^+ salt. The $[W_6(\mu_3\text{-}Cl)_8Cl_6]^{2-}$ cluster, with a variety of cations, has attracted considerable interest because of its long photo-excited state lifetimes and photochemical, electrochemical, and spectroscopic properties.[1]

W_6Cl_{12} and $[W_6(\mu_3\text{-}Cl)_8Cl_6]^{2-}$ have been used in the synthesis of hexatungsten chalcogenide molecular precursors for tungsten analogs of Chevrel phases,[2] organotungsten compounds,[3] alkene polymerization and metathesis catalysts,[4] and tungsten thiochlorides.[5]

The principal routes to W_6Cl_{12} have involved reduction of WCl_6, with yields of 35–60%. Reductants have included Na/Hg,[6] Al,[1f,7] Al in molten $NaAlCl_4$,[8] Mg,[7b] Zn,[7b] Pb,[7b] W,[2c,9] and WO_2.[10] Tungsten tetrachloride has also been used as a precursor to W_6Cl_{12} by either thermal disproportionation[11] or reduction with Fe.[2b] Disadvantages of many of these preparative routes include low yields, slow reactions, high temperatures necessitating the use of quartz vessels, contamination by reductant or intermediate reduction products, and violent explosions during the opening of sealed tubes.[1f] Some procedures require (1) small scales, (2) careful control of ampule position within a furnace, (3) careful control of thermal gradients, (4) tungsten powder pre-purification in H_2 at 1000°C, or (4) rocking furnaces. W_6Cl_{12} is usually purified by crystallization from hot concentrated hydrochloric acid, forming the chloro acid $(H_3O)_2[W_6(\mu_3\text{-}Cl)_8Cl_6] \cdot 7H_2O$, followed by thermolysis to brown to pale yellow W_6Cl_{12}.

[*]Department of Chemistry, The University of Iowa, Iowa City, IA 52242.
[†]Department of Chemistry and Chemical Biology, Cornell University, Ithaca, NY 14853.

We have reported that mild main group metal reductants such as mercury, bismuth, antimony, gallium, and $Ga^+GaCl_4^-$ are highly effective solid-state reductants[12] for reducing WCl_6, $MoCl_5$, and TaX_5 (X = Cl, Br) at considerably lower temperatures than those used in other syntheses of early transition metal cluster halides. These lower temperatures allow the use of borosilicate glass ampules in place of more expensive, harder-to-seal quartz ampules.

Bismuth is an ideal reductant for WCl_6 because it is nontoxic, inexpensive, not impeded by a surface oxide, does not readily reduce W_6Cl_{12}, and forms volatile but low vapor pressure $BiCl_3$ for greatly improved safety in sealed tube reactions. Bismuth is also easily dispersed in reaction mixtures because it dissolves in molten $BiCl_3$.

We report here an extension of this approach to the convenient preparation of $(H_3O)_2[W_6(\mu_3\text{-}Cl)_8Cl_6] \cdot 6H_2O$ and thus W_6Cl_{12} at significantly lower temperatures and greater yield than other published procedures.[13] The procedure is simple enough for use in undergraduate inorganic synthesis laboratory courses, particularly when combined with instruction in basic glassblowing.

General Procedures

WCl_6 (Materion Advanced Chemicals, Milwaukee, WI), Bi (325 mesh powder, Materion), and hydrochloric acid (12 M, Fisher Scientific) are used as received. Powder X-ray diffraction is performed on samples protected from moisture by a 5 µm polyethylene film. WCl_6 and solid-state products are handled in either a glove box or a well-purged glove bag under a dinitrogen atmosphere. A tube furnace with a positionable thermocouple and temperature controller for ramping temperatures is employed for the solid-state reduction. Syntheses are performed in dual-chamber, 25 mm OD borosilicate glass ampules with 60–70 mL total chamber volume, a 19/22 ground glass joint at one end, and constrictions between the end reaction chamber and middle receiver chamber (for volatile by-products) and between the middle receiver chamber and ground joint (Fig. 1). The ampule is oven dried at 130°C overnight and brought directly while hot into the glove box or glove bag and cooled under vacuum (in the antechamber) or dinitrogen, respectively. Reactants are thoroughly mixed and added to the end reaction chamber via a long-stem funnel that minimizes contamination of the constriction surfaces. A gas inlet adapter is used to seal the ampule, which is then evacuated on a Schlenk line and flame sealed between the joint and receiver chamber. Recrystallizations involving concentrated aqueous HCl are performed in stoppered filter flasks with the outlet connection attached by rubber tubing to a water trap in order to absorb HCl gas released from the heated HCl solution. Tungsten analyses are based on gravimetric determination of WO_3 after oxidation with HNO_3 and ignition.

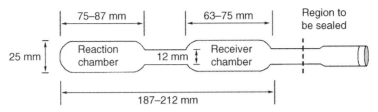

Figure 1. Ampule employed in the synthesis of W_6Cl_{12}.

A. BIS(HYDROXONIUM) TETRADECACHLOROHEXATUNGSTATE HEPTAHYDRATE (CHLOROTUNGSTIC ACID)

$$6WCl_6 + 8Bi \rightarrow W_6Cl_{12} + 8BiCl_3$$

$$W_6Cl_{12} + 2HCl(aq) \rightarrow (H_3O)_2[W_6(\mu_3\text{-}Cl)_8Cl_6] \cdot 7H_2O$$

Procedure

WCl_6 (15.00 g, 37.8 mmol) and Bi (10.54 g, 50.4 mmol) are mixed in the glove box and sealed in an ampule under vacuum. The ampule is positioned at the tube furnace center and the temperature slowly raised to 230°C over 2 h,[*] and then to 355°C over 2 h. The ampule is then positioned with the receiver chamber in a cooler area of the furnace and heated at the reaction chamber end at 350–355°C for 24 h. After being cooled to room temperature, the ampule is opened in the glove box, taking care not to mix any nonvolatile material and $BiCl_3$ from their respective chambers. Alternatively, before opening the ampule, the two chambers can be separated with a torch in order to separate/store the sublimed $BiCl_3$. The chocolate brown nonvolatile material weighs 18.28 g, consistent with appreciable bismuth content because the theoretical yield of W_6Cl_{12} is 9.64 g.

A portion of the chocolate brown product (3.00 g) is treated with concentrated aqueous HCl (5 mL) while agitating for 10–15 min at room temperature. This step results in the formation of a fine, yellow-brown precipitate. The precipitate is separated by decanting off the liquid, and then the precipitate is washed with concentrated HCl (5 mL) and dissolved in boiling concentrated HCl (20 mL). When the solution cools to room temperature, greenish yellow needles form. The needles are collected by filtration through a coarse porosity fritted glass funnel. The

---

[*]The checkers recommend removing the ampule with heat-resistant gloves after it reaches 230°C, thoroughly shaking the tube in order to mix the reactants, and then placing the tube in the furnace with the receiver area in a cooler region of the furnace so that $BiCl_3$ can sublime away while heating to 350°C.

product is recrystallized in the same manner three times to afford the pure product. Yield: 1.65 g (91% based on WCl_6, for workup of 3 g of the 18.28 g total impure nonvolatile product).[*]

Anal. Calcd. for $H_{20}O_9Cl_{14}W_6$ $\{(H_3O)_2[W_6(\mu_3\text{-}Cl)_8Cl_6] \cdot 7H_2O\}$: W, 62.55. Found: W, 62.41.

Properties

Yellow, crystalline $(H_3O)_2[W_6(\mu_3\text{-}Cl)_8Cl_6] \cdot 7H_2O$ is insoluble in water, with which it slowly reacts, but is soluble in methanol. The crystals slowly lose HCl and water upon prolonged storage, but recrystallization from concentrated hydrochloric acid restores the material to crystalline form with a small loss of material from oxidation. For long-term storage, it can be converted to the air-stable, crystalline tetrabutylammonium salt by metathesis with Bu_4NCl in halo-genated solvents such as $CHCl_3$ or CH_2Cl_2.

B. HEXATUNGSTEN DODECACHLORIDE

$$(H_3O)_2[W_6(\mu_3\text{-}Cl)_8Cl_6] \cdot 7H_2O \rightarrow W_6Cl_{12} + 2HCl + 9H_2O$$

Procedure

A triply recrystallized chlorotungstic acid sample is thermolyzed in vacuum by slowly (over 3–3.5 h) raising the temperature under nitrogen to 325° C and heating for 1 h. Yield: 98%.

Anal. Calcd. for Cl_2W: W, 72.2. Found: W, 72.0.

Properties

Pale canary yellow W_6Cl_{12} absorbs water from the air to form the very pale yellow neutral dihydrate. The material is slowly oxidized or hydrolyzed in air to a white solid, presumed to be WO_3. W_6Cl_{12} is best stored under anhydrous conditions and under a dioxygen-free atmosphere or in vacuum.

Acknowledgments

This research was supported in part with funds from the National Science Foundation (CHE-0078701), the National Cancer Institute (1 R21 RR14062-01),

---

[*]The checkers report consistent yields of ~90%.

the Roy J. Carver Charitable Trust (grant 01-93), Nycomed Inc., and the Department of Energy's Energy-Related Laboratory Equipment Program (tube furnace). The Siemens D5000 automated powder diffractometer was purchased with support from the Roy J. Carver Charitable Trust.

References

1. (a) A. W. Maverick, J. S. Najdzionek, D. MacKenzie, D. G. Nocera, and H. B. Gray, *J. Am. Chem. Soc.* **105**, 1878 (1983); (b) D. G. Nocera and H. B. Gray, *J. Am. Chem. Soc.* **106**, 824 (1984); (c) T. C. Zietlow, M. D. Hopkins, and H. B. Gray, *J. Solid State Chem.* **57**, 112 (1985); (d) T. C. Zietlow, W. P. Schaefer, B. Sadeghi, N. Hua, and H. B. Gray, *Inorg. Chem.* **25**, 2195 (1986); (e) J. A. Jackson, C. Turro, M. D. Newsham, and D. G. Nocera, *J. Phys. Chem.* **94**, 4500 (1990); (f) R. D. Mussell and D. G. Nocera, *Inorg. Chem.* **29**, 3711 (1990); (g) J. R. Schoonover, T. C. Zietlow, D. L. Clark, J. A. Heppert, M. H. Chisholm, H. B. Gray, A. P. Sattelberger, and W. H. Woodruff, *Inorg. Chem.* **35**, 6606 (1996), and references therein.

2. (a) T. Saito, A. Yoshikawa, T. Yamagata, H. Imoto, and K. Unoura, *Inorg. Chem.* **28**, 3588 (1989); (b) X. Zhang and R. E. McCarley, *Inorg. Chem.* **34**, 2678 (1995); (c) G. M. Ehrlich, C. J. Warren, D. A. Vennos, D. M. Ho, R. C. Haushalter, and F. J. DiSalvo, *Inorg. Chem.* **34**, 4454 (1995); (d) X. Xie and R. E. McCarley, *Inorg. Chem.* **34**, 6124 (1995); (e) X. Xie and R. E. McCarley, *Inorg. Chem.* **35**, 2713 (1996); (f) R. E. McCarley, S. J. Hilsenbeck, and X. Xie, *J. Solid State Chem.* **117**, 269 (1995); (g) X. Xie and R. E. McCarley *Inorg. Chem.* **36**, 4011 (1997); (h) S. Ihmaine, C. Perrin, and M. Sergent, *Eur. J. Solid State Inorg. Chem.* **34**, 169 (1997); (i) X. Xie and R. E. McCarley, *Inorg. Chem.* **36**, 4665 (1997).

3. T. Saito, H. Manabe, T. Yamagata, and H. Imoto, *Inorg. Chem.* **26**, 1362 (1987).

4. J.-L. Herisson, Y. Chauvin, N. H. Phung, and G. Lefebvre, *C. R. Acad. Sci. Ser. C* **269**, 661 (1969).

5. P. E. Rauch, F. J. DiSalvo, W. Zhou, D. Tang, and P. P. Edwards, *J. Alloys Compd.* **182**, 253 (1992).

6. J. B. Hill, *J. Am. Chem. Soc.* **38**, 2383 (1916).

7. (a) K. Lindner and A. Köhler, *Ber. Dtsch. Chem. Ges.* **55**, 1458 (1922); (b) K. Lindner and A. Köhler, *Z. Anorg. Allg. Chem.* **140**, 357 (1924); (c) R. D. Hogue and R. E. McCarley, *Inorg. Chem.* **9**, 1354 (1970).

8. W. C. Dorman and R. E. McCarley, *Inorg. Chem.* **13**, 491 (1974).

9. (a) H. Schäfer, M. Trenkel, and C. Brendel, *Monatsh. Chem.* **102**, 1293 (1971); (b) G. M. Ehrlich, P. E. Rauch, and F. J. DiSalvo, *Inorg. Synth.* **30**, 1 (1995).

10. S. S. Eliseev, L. E. Malysheva, E. E. Vozhdaeva, and N. V. Gaidaenko, *Molybdenum and Tungsten Chlorides and Oxychlorides*, Donish, Dushanbe (USSR), 1989, pp. 190–191; *Chem. Abstr.* **112**, 110790w (1990).

11. R. E. McCarley and T. M. Brown, *Inorg. Chem.* **3**, 1232 (1974), and references therein.

12. (a) V. Kolesnichenko, D. C. Swenson, and L. Messerle, *Inorg. Chem.* **37**, 3257 (1998); (b) V. Kolesnichenko, J. J. Luci, D. C. Swenson, and L. Messerle, *J. Am. Chem. Soc.* **120**, 13260 (1998); (c) D. N. T. Hay, D. C. Swenson, and L. Messerle, *Inorg. Chem.* **41**, 4700 (2002); (d) D. N. T. Hay and L. Messerle, *J. Struct. Biol.* **139**, 147 (2002); (e) D. N. T. Hay, J. A. Adams, J. Carpenter, S. L. DeVries, J. Domyancich, B. Dumser, S. Goldsmith, M. A. Kruse, A. Leone, F. Moussavi-Harami, J. A. O'Brien, J. R. Pfaffly, M. Sylves, P. Taravati, J. L. Thomas, B. Tiernan, and L. Messerle, *Inorg. Chim. Acta* **357**, 644 (2004).

13. V. Kolesnichenko and L. Messerle, *Inorg. Chem.* **37**, 3660 (1998).

5. TRINUCLEAR TUNGSTEN HALIDE CLUSTERS

Submitted by JOHN H. THURSTON,[*] VLADIMIR KOLESNICHENKO,[*] and
LOUIS MESSERLE[*]
Checked by SUSAN E. LATTURNER[†] and WILLIAM AINSWORTH[†]

Triangular tritungsten compounds with one or two face-bridging chalcogenides are well known.[1] Monocapped clusters[2] with three edge-bridging ligands μ-L (e.g., μ-O) and three terminal ligands per metal can be described as three octahedra that share a common vertex (the capping μ3-L) and edges in a pairwise fashion.

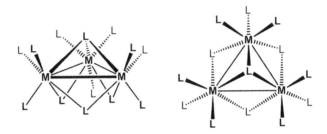

These clusters have also been described as incomplete metallacubanes, lacking only a fourth metal that would bond to the three bridging ligands μ-L. Tritungsten clusters with μ3-halides are rare,[3] and monocapped group 6 clusters with *exclusively* halides were unknown before our research. In contrast, perhalo trinuclear clusters are more common for the nearby elements Re,[4] Nb,[5] and Ta.[6]

Mild main group metal reductants such as mercury, bismuth, antimony, gallium, and $Ga^+GaCl_4^-$ are highly effective[7] solid-state reductants for reducing WCl_6, $MoCl_5$, and TaX_5 (X = Cl, Br) at considerably lower temperatures than those used in traditional syntheses of group 5 and 6 metal cluster halides. Borosilicate glass ampules can thus replace more expensive, harder-to-seal quartz ampules. We have reported convenient routes to tungsten dichloride, W_6Cl_{12},[7b] and to crystalline[7a] and reactive[8] forms of tungsten tetrachloride, $(WCl_4)_x$, from reduction of WCl_6 with Hg, Sb, or Bi. Antimony is oxidized to the volatile (moderate vapor pressure) liquid $SbCl_3$, which can act as a solid-state flux. Nontoxic, inexpensive bismuth is another ideal reductant for WCl_6 because it is not impeded by a surface oxide, it forms the volatile by-product $BiCl_3$ (which has a low vapor pressure for improved safety in heated sealed tube reactions), and it is easily dispersed through dissolution in molten $BiCl_3$.

[*]Department of Chemistry, The University of Iowa, Iowa City, IA 52242.
[†]Department of Chemistry, Florida State University, Tallahassee, FL 32306.

In an attempt to extend this methodology to the preparation of tungsten trichloride, $W_6(\mu\text{-Cl})_{12}Cl_6$,[9] we instead found a new monocapped, eight-electron binary tungsten chloride, W_3Cl_{10}, and the perchlorotritungstates, $Na_3[W_3(\mu_3\text{-Cl}) (\mu\text{-Cl})_3Cl_9]$ and $[N(CH_2Ph)Bu_3]_3[W_3(\mu_3\text{-Cl})(\mu\text{-Cl})_3Cl_9]$.[10] The latter trianionic clusters can be oxidized in solution to salts of the seven-electron isostructural dianion, $[N(CH_2Ph)Bu_3]_2[W_3(\mu_3\text{-Cl})(\mu\text{-Cl})_3Cl_9]$. These perchlorotritungsten clusters constitute new synthons, readily accessed from WCl_6, for the development of tritungsten chemistry. They have potential use in the synthesis of heterometalla-cubanes and higher nuclearity[11] clusters. Meyer has recently reported an alternative synthesis of $Na_3[W_3(\mu_3\text{-Cl})(\mu\text{-Cl})_3Cl_9]$ from WCl_4 and its solid-state structure.[12]

General Procedures

WCl_6 (Materion Advanced Chemicals, Milwaukee, WI), Bi (325 mesh powder, Materion), Sb (100 mesh powder, Aldrich Chemical, Milwaukee, WI), and tetrahydrofuran (anhydrous, inhibitor-free, Aldrich) are used as received. Acetonitrile (Aldrich) is degassed with argon and distilled from granular P_2O_5. Hydrochloric acid (12 M, Fisher Scientific) is used as received. Benzyltrialkylammonium chloride salts (alkyl = ethyl, butyl; Aldrich) are dried by azeotropic distillation with benzene. NaCl (Fisher) is dried overnight in an oven at 130°C and then evacuated overnight in the glove box antechamber under vacuum while cooling. Powder X-ray diffractometry is performed on samples protected from moisture by a 5 μm polyethylene film. WCl_6 and solid-state products are handled in a glove box under a dinitrogen atmosphere. A tube furnace with a positionable thermocouple and temperature controller for ramping temperatures is employed for the solid-state reductions. Syntheses at scales described below are performed in dual-chamber, 25 mm OD borosilicate glass ampules with ~60 mL total chamber volume, a 19/22 ground glass joint at one end, and constrictions between the end reaction chamber and middle receiver chamber (for volatile by-products) and between the middle receiver chamber and ground joint (Fig. 1). The ampule size should be adjusted for different reaction scales.

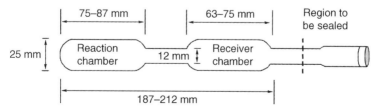

Figure 1. Ampule employed in the synthesis of trinuclear tungsten clusters.

Ampules are oven dried at $130°C$ overnight and cooled under vacuum in the glove box antechamber. Reactants are thoroughly mixed and ground and then added to the end reaction chamber by means of a long-stem funnel that minimizes contamination of the constriction surfaces. A gas inlet adapter is used to seal the ampule, which is then evacuated on a Schlenk line and flame sealed between the joint and receiver chamber. Tungsten analyses are based on gravimetric determination of WO_3 after oxidation with HNO_3 and ignition.

A. TRITUNGSTEN DECACHLORIDE

$$9WCl_6 + 8Bi \rightarrow 3W_3Cl_{10} + 8BiCl_3$$

Procedure

In the glove box, WCl_6 (10.00 g, 25.2 mmol) and Bi (4.68 g, 22.4 mmol) are thoroughly mixed, ground in a polished agate mortar, and added to a dual-chamber borosilicate glass ampule. The ampule is evacuated and flame sealed and then placed into a horizontal tube furnace and the temperature ramped to $220°C$ over 2 h. After 1 day, the ampule is withdrawn from the furnace and the purple-brown solid is homogenized by gentle shaking for $\sim$1 min. This and subsequent homogenization steps increase the overall yield. The ampule is returned to the tube furnace, angled at $\sim$45° to the vertical in order to facilitate refluxing of the $BiCl_3$ by-product, and the temperature is raised to $350°C$ over 8 h. After 1 day, the furnace is repositioned to horizontal and the ampule is oriented so that the empty receiving chamber is located in a cooler end of the furnace. The by-product $BiCl_3$ is removed by sublimation in order to make the nonvolatile materials free flowing. After 24 h, the ampule is removed and the nonvolatile material is homogenized by gentle shaking for 1 min. The ampule is heated at $350°C$ at a furnace angle of 45° for 2 days, the furnace is reoriented to horizontal, and the $BiCl_3$ is removed by sublimation at $350°C$ for 2 days. Once sublimation of $BiCl_3$ is judged to be complete, the ampule is cooled to room temperature, and the chamber containing the $BiCl_3$ is flame sealed at the constriction in order to separate it from the W_3Cl_{10} chamber. The W_3Cl_{10} chamber is opened under inert atmosphere. Yield: 9.0 g (92%).[*] According to X-ray powder diffractometry, the product is a single phase free from Bi, $BiCl_3$, WCl_4, and W_6Cl_{12} impurities. The purity can be further improved by washing the solid with aqueous HCl (6–12 M), which removes any residual $BiCl_3$ and trace amounts of WCl_4 and W_6Cl_{12}.

---

[*]The checkers reported a yield of 6.88 g (70%) with powder XRD showing mostly single-phase material with the dominant impurity being WCl_4. The lower yield may be the result of employing different ampule dimensions.

Anal. Calcd. for $Cl_{10}W_3$: W, 60.87; Cl, 39.13. Found: W, 60.3; Cl, 39.81.

Properties

Brown, nonvolatile, microcrystalline W_3Cl_{10} appears to be indefinitely stable when stored under an inert atmosphere. The compound is insoluble in all common organic solvents and does not react with concentrated aqueous HCl or HNO_3 (facilitating its separation from WCl_4 and W_6Cl_{12}). The compound as obtained by this procedure is crystalline by powder X-ray diffraction. W_3Cl_{10} is converted into $Na_3W_3Cl_{13}$ when heated to $350°C$ with NaCl and $SbCl_3$ (which acts as solvent). The structure of W_3Cl_{10} is believed to be analogous to that of Nb_3Cl_8 [5a,b] and $Na_2Ti_3Cl_8$,[13] with an extended structural net of nondiscrete $W_3(\mu_3\text{-}Cl)(\mu\text{-}Cl)_3Cl_9$ units that share chloride ions.

B. TRISODIUM TRIDECACHLOROTRITUNGSTATE

$$9WCl_6 + 8Sb + 9NaCl \rightarrow 3Na_3W_3Cl_{13} + 8SbCl_3$$

Procedure

WCl_6 (10.9 g, 25.4 mmol), Sb (3.0 g, 22.4 mmol), and NaCl (1.6 g, 25.4 mmol) are mixed and ground in an agate mortar in a glove box and then loaded into a double-chamber borosilicate ampule. After the ampule is evacuated and flame sealed, it is placed in a tube furnace oriented at $45°$ to the vertical, in order to facilitate $SbCl_3$ reflux, and the temperature is ramped to $200°C$ over 2 h. After 1 day at $200°C$, the tube furnace is pivoted to the horizontal position and the ampule repositioned so that the receiver chamber is in a cool region of the furnace. After 1 day, the sublimation of $SbCl_3$ is complete. The nonvolatile product is homogenized through gentle ampule shaking for ~1 min (this step improves the overall yield). The ampule is returned to the tube furnace, at a $45°$ orientation, and the reaction mixture is heated at $240°C$ for 5 days. The furnace is rotated to the horizontal and the ampule is repositioned so that the receiver chamber is in a cool region of the furnace. $SbCl_3$ is removed by sublimation at $240°C$ for 24 h. The ampule is removed from the furnace, allowed to cool to room temperature, and opened under an inert atmosphere (taking care to prevent the sublimed by-products from falling back into the bottom chamber of the reaction vessel). Yield: 9.8 g (98%).[*]

Anal. Calcd. for $Cl_{13}Na_3W_3$: W, 51.00; Cl, 42.62. Found: W, 50.9; Cl, 42.85.

---

[*]The checkers obtained 9.563 g of a dark purple powder, with powder XRD showing mostly single-phase material with weak NaCl reflections and possibly trace amounts of W_6Cl_{12}.

Properties

$Na_3W_3Cl_{13}$ is a free-flowing purple, nonvolatile, microcrystalline solid that is stable to air. It is homogeneous by powder X-ray diffractometry. The compound is insoluble in common organic solvents, including dichloromethane, and modestly soluble in degassed water and CH_3CN, producing deep purple solutions. Aqueous solutions in air slowly oxidize to an unknown, deep blue complex over the course of several months. The compound is stable in vacuum to $\leq 370°$ C.

C. TRIS(BENZYLTRIBUTYLAMMONIUM) TRIDECACHLOROTRITUNGSTATE

$$Na_3W_3Cl_{13} + 3[N(CH_2Ph)Bu_3]Cl \rightarrow [N(CH_2Ph)Bu_3]_3[W_3(\mu_3\text{-}Cl)(\mu\text{-}Cl)_3Cl_9] + 3NaCl$$

Procedure

$Na_3W_3Cl_{13}$ (2.00 g, 1.85 mmol) and $[N(CH_2Ph)Bu_3]Cl$ (1.73 g, 5.55 mmol) are stirred in anhydrous acetonitrile (20 mL). After 3 days, the NaCl precipitate is removed by filtration and the solvent is removed by rotary evaporation to give a viscous oil. The oil is layered with tetrahydrofuran (15 mL) and allowed to stand undisturbed. After 4 days, the mother liquor is decanted, and the solid is washed with tetrahydrofuran (3 × 2 mL) and dried in vacuum. The dark purple crystals are washed again with tetrahydrofuran (3 mL, 10 min), the solution is decanted, and the solid is dried in vacuum for 3 h. Yield: 3.34 g (98%).[*]

Anal. Calcd. for $C_{57}H_{102}N_3Cl_{13}W_3$: W, 29.95; Cl, 25.02. Found: W, 29.6; Cl, 25.18.

Properties

Dark purple $[N(CH_2Ph)Bu_3]_3[W_3Cl_{13}]$ is diamagnetic (or weakly paramagnetic) at 25° C, with a low effective moment of 0.73–0.79μ_B. This tetraalkylammonium cation imparts higher solubility for the salt, compared to the NMe_4^+ salt, and the benzyl group reduces cation crystallographic disorder compared to NBu_4^+ salts. The UV–vis spectrum (λ in nm, ϵ in $M^{-1} cm^{-1}$) of $[N(CH_2Ph)Bu_3]_3[W_3Cl_{13}]$ in dichloromethane exhibits maxima at 555 (1010), 420 (shoulder; 1600), 350 (6630), 270 (shoulder; 25,100), and 240 (35,400). Mass spectrometry (FAB, NPOE matrix, negative ion mode, m/e, base peak of isotopic grouping): 1290 $[N(CH_2Ph)Bu_3][W_3Cl_{13}]$, 1253 (loss of Cl). The $N(CH_2Ph)Et_3$ salt can be isolated

---

[*]The checkers stirred the initial mixture for 5 days, and after filtration, rotary evaporation, and crystallization, they obtained 0.962 g of pure phase material as shown by powder XRD.

in similar fashion, with the Et groups being less disordered than the Bu groups. Single-crystal X-ray diffractometry on $[N(CH_2Ph)Et_3]_3[W_3(\mu_3\text{-}Cl)(\mu\text{-}Cl)_3Cl_9]$ shows a solid-state structure with an average $W-W$ distance of 2.778(5) Å,[10] an average $W-Cl-W$ angle of 70.4(1)°, and a $W-Cl$(terminal) distance of 2.415(3) Å. Chemical oxidation of $[N(CH_2Ph)Bu_3]_3[W_3Cl_{13}]$ with 2,4-dichlorophenyliodine dichloride in dichloromethane gives the seven-electron cluster compound $[N(CH_2Ph)Bu_3]_2[W_3(\mu_3\text{-}Cl)(\mu\text{-}Cl)_3Cl_9]$ with longer $W-W$ distances.[10]

Acknowledgments

This research was supported in part with funds from the National Science Foundation (CHE-0078701), the National Cancer Institute (1 R21 RR14062-01), the Roy J. Carver Charitable Trust (grant 01-93), Nycomed Inc., and the Department of Energy's Energy-Related Laboratory Equipment Program (tube furnace). The Siemens D5000 automated powder diffractometer was purchased with support from the Roy J. Carver Charitable Trust.

References

1. (a) F. A. Cotton and G. Wilkinson, *Advanced Inorganic Chemistry*, 5th ed., Wiley, New York, 1988, pp. 825–827; (b) A. Müller, R. Jostes, and F. A. Cotton, *Angew. Chem., Int. Ed. Engl.* **19**, 875 (1980); (c) T. Shibahara, *Coord. Chem. Rev.* **123**, 73 (1993).
2. Y. Jiang, A. Tang, R. Hoffmann, J. Huang, and J. Lu, *Organometallics* **4**, 27 (1985).
3. (a) F. A. Cotton, T. R. Felthouse, and D. G. Lay, *J. Am. Chem. Soc.* **102**, 1431 (1980); (b) F. A. Cotton, T. R. Felthouse, and D. G. Lay, *Inorg. Chem.* **20**, 2219 (1981).
4. F. A. Cotton and R. A. Walton, *Multiple Bonds Between Metal Atoms*, 2nd ed., Clarendon Press, Oxford, 1993, p. 533.
5. (a) H.-G. von Schnering, H. Wohrle, and H. Schäfer, *Naturwissenschaften* **48**, 159 (1961); (b) A. Simon and H.-G.von Schnering, *J. Less Common Met.* **11**, 31 (1966); (c) F. A. Cotton, M. P. Diebold, X. Feng, and W. J. Roth, *Inorg. Chem.* **27**, 3413 (1988).
6. (a) E. Babaian-Kibala, F. A. Cotton, and M. Shang, *Inorg. Chem.* **29**, 5148 (1990); (b) M. D. Smith and G. J. Miller, *J. Am. Chem. Soc.* **118**, 12238 (1996).
7. (a) V. Kolesnichenko, D. C. Swenson, and L. Messerle, *Inorg. Chem.* **37**, 3257 (1998); (b) V. Kolesnichenko and L. Messerle, *Inorg. Chem.* **37**, 3660 (1998); (c) D.N.T. Hay, D.C. Swenson, and L. Messerle, *Inorg. Chem.* **41**, 4700 (2002); (d) D. N. T. Hay and L. Messerle, *J. Struct. Biol.* **139**, 147 (2002); (e) D. N. T. Hay, J. A. Adams, J. Carpenter, S. L. DeVries, J. Domyancich, B. Dumser, S. Goldsmith, M. A. Kruse, A. Leone, F. Moussavi-Harami, J. A. O'Brien, J. R. Pfaffly, M. Sylves, P. Taravati, J. L. Thomas, B. Tiernan, and L. Messerle, *Inorg. Chim. Acta* **357**, 644 (2004).
8. V. Kolesnichenko, D. C. Swenson, and L. Messerle, *Chem. Commun.* 2137 (1998).
9. (a) R. Siepmann, H.-G.von Schnering, and H. Schäfer, *Angew. Chem., Int. Ed. Engl.* **6**, 637 (1967); (b) D. L. Kepert, R. E. Marshall, and D. Taylor, *J. Chem. Soc., Dalton Trans.* 506 (1974); (c) Y. Q. Zheng, E. Jonas, J. Nuss, and H.-G.von Schnering, *Z. Anorg. Allg. Chem.* **624**, 1400 (1998); (d) A. Nägele, J. Glaser, and H.-J. Meyer, *Z. Anorg. Allg. Chem.* **627**, 244 (2001); (e) S. Dill, J. Glaser, M. Ströbele, S. Tragl, and H.-J. Meyer, *Z. Anorg. Allg. Chem.* **630**, 987 (2004).
10. V. Kolesnichenko, J. J. Luci, D. C. Swenson, and L. Messerle, *J. Am. Chem. Soc.* **120**, 13260 (1998).

11. T. Saito, N. Yamamoto, T. Yamagata, and H. Imoto, *J. Am. Chem. Soc.* **110**, 1646 (1988).
12. M. Weisser, S. Tragl, H.-J. Meyer, *Z. Anorg. Allg. Chem.* **632**, 1885 (2006).
13. D. J. Hinz, G. Meyer, T. Dedecke, and W. Urland, *Angew. Chem., Int. Ed. Engl.* **34**, 71 (1995).

6. CRYSTALLINE AND AMORPHOUS FORMS OF TUNGSTEN TETRACHLORIDE

Submitted by YIBO ZHOU,[*] VLADIMIR KOLESNICHENKO,[*] and
LOUIS MESSERLE[*]
Checked by SELIM ALAYOGLU[†] and BRYAN EICHHORN[†]

Tungsten tetrachloride is the most synthetically useful binary tungsten chloride after WCl_6 and is an important precursor in W^{IV} coordination[1] and organometallic chemistries,[2] metal–metal bonded ditungsten chemistry,[3] materials synthesis,[4] hexatungsten chloride cluster synthesis,[5] and catalysis.[6]

Of the many published procedures for the synthesis of WCl_4, the two most commonly employed methods are reduction of WCl_6 with red phosphorus in a sealed tube,[7] or reduction of WCl_6 with $W(CO)_6$ in refluxing chlorobenzene.[8] The suitability of WCl_4 for some chemical uses is dependent on its preparative route because of associated impurities. For example, tungsten phosphides and phosphorus are potential contaminants in WCl_4 that is derived from reduction with red phosphorus. Tungsten hexacarbonyl, $W(CO)_6$, is an expensive reductant, toxic CO is liberated, and chlorobenzene can be difficult to remove.

Main group metal and metalloid reductants such as mercury, bismuth, antimony, gallium, and $Ga^+GaCl_4^-$ are highly effective[9] in reducing WCl_6, $MoCl_5$, and TaX_5 (X = Cl, Br) at surprisingly lower temperatures (allowing the use of borosilicate ampules) than commonly used in the solid-state synthesis of early transition metal polynuclear halides. These metals and metalloids are not as impacted by oxide coatings that inhibit solid-state reactions of more active metal reductants such as aluminum or the alkali metals.

We report here an extension of this reduction methodology to the convenient preparations[10] of WCl_4 from WCl_6 in a crystalline, edge-sharing bioctahedral polymer form via solid-state methods (employing Sb as reductant) and a reactive

[*]Department of Chemistry, University of Iowa, Iowa City, IA 52242.
[†]Department of Chemistry, University of Maryland, College Park, MD 20742.

(or more readily extracted) amorphous powder form of unknown structure via solution methods (employing Sn).

General Procedures

WCl_6 (Materion Advanced Chemicals, Milwaukee, WI), Sb (100 mesh powder, Aldrich Chemical, Milwaukee, WI), and Sn (325 mesh, Aldrich) are used as received. Methylene chloride (Fisher Scientific) and 1,2-dichloroethane (Aldrich) are degassed with argon and distilled from granular P_2O_5. Powder X-ray diffraction is performed on samples protected from moisture by a 5 μm polyethylene film. WCl_6 and solid-state products are handled in either a glove box or a well-purged glove bag under a dinitrogen atmosphere. A tube furnace with a positionable thermocouple is used in combination with a temperature controller in order to maintain and ramp temperatures. Syntheses are performed in dual-chamber, 25 mm OD borosilicate glass ampules with 30–60 mL total chamber volume, a 14/20 or 19/22 ground glass joint at one end, and constrictions between the end reaction chamber and middle receiver chamber (for volatile by-products) and between the middle receiver chamber and ground joint (Fig. 1). In order to improve temperature homogeneity and preclude cooler zones where WCl_5 could collect, the ratio of lengths of tube furnace and ampule is kept at ≥1.6. The ampule and all glassware are oven dried at 130°C overnight and brought directly while hot into the glove box and cooled under vacuum in the antechamber or, in the glove bag, under dinitrogen. Reactants are thoroughly mixed and added to the end reaction chamber via a long-stem funnel that minimizes contamination of the ampule constrictions. A gas inlet adapter is used to seal the ampule, which is then evacuated on a Schlenk line and flame sealed between the joint and receiver chamber.

A literature procedure is followed for gravimetric determination of tungsten as WO_3 after treatment with HNO_3 and ignition.[11] Chlorine content is determined gravimetrically (average of two determinations) as AgCl after removal of tungsten. An accurately weighed sample of WCl_4 (0.17–0.19 g) is decomposed at 20°C with aqueous 1 M KOH (5 mL). 30% Hydrogen peroxide (two drops) is added to the suspension and, after 1 h the solution is diluted to

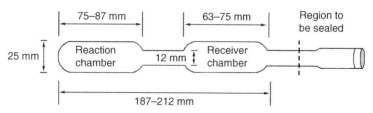

Figure 1. Ampule design for crystalline $(WCl_4)_x$ synthesis.

50 mL with water and heated at 100°C for 20 min. To the hot solution is added dropwise $Ba(NO_3)_2$ (7 mL of a 0.1 M aqueous solution) and the mixture heated at 100°C for 20 min. After settling overnight at room temperature, the $BaWO_4$ precipitate is removed by filtration through a medium-porosity fritted glass funnel and the solid is washed with distilled water. The combined filtrate and wash is diluted to 75 mL and neutralized with concentrated HNO_3 (~12 drops). $AgNO_3$ (5.5 mL of a 0.5 M aqueous solution) is added dropwise. The mixture is heated at 100°C for 20 min and then allowed to settle overnight, protected from light, at room temperature. The AgCl precipitate is recovered by filtration, washed with distilled water, oven dried for 2 h at 150°C, and weighed.

A. CRYSTALLINE TUNGSTEN TETRACHLORIDE BY SOLID-STATE REDUCTION

$$3WCl_6 + 2Sb \rightarrow 3(WCl_4)_x + 2SbCl_3$$

Procedure

WCl_6 (10.0 g, 25.2 mmol) and Sb (2.04 g, 16.7 mmol) are sealed in a glass ampule under vacuum. After the contents are mixed, the ampule is placed in the center of the tube furnace and heated at 75°C for 12 h, 105°C for 12 h, and 130°C for 12 h (the calculated vapor pressure of $SbCl_3$ at 130°C is 46 mmHg).[*] The ampule's receiver chamber is moved out of the furnace and the $SbCl_3$ is removed by distillation at 130°C.[†] Residual WCl_5 and $WOCl_4$ are removed by sublimation at 160°C for 1 day and 215°C for 1 day. The ampule is cooled and opened in a glove box. Black, nonvolatile, crystalline $(WCl_4)_x$ is removed from the end chamber. Yield: 7.99 g (97% based on WCl_6).

Anal. Calcd. for Cl_4W: W, 56.45; Cl, 43.55. Found: W, 56.5; Cl, 43.39. No Sb was detected by qualitative microanalysis.

Properties

$(WCl_4)_x$ is an air- and moisture-sensitive solid. It can be stored indefinitely under dry, oxygen-free dinitrogen or argon. This crystalline form of $(WCl_4)_x$ has solubility properties that are similar (i.e., insoluble in noncoordinating solvents)

---

[*]If all of the above heating is done at 200°C, the yield of crystalline $(WCl_4)_x$ drops to 75%.

[†]To remove the $SbCl_3$, the checkers used an L-shaped ampule, similar to that used in the synthesis[7] of $(WCl_4)_x$ from red phosphorus and WCl_6. They reported a lower yield of $(WCl_4)_x$ starting from $WOCl_4$-contaminated WCl_6.

and reactivity that is similar to that of WCl_4 prepared by red phosphorus or mercury reduction.[10] The solid-state structure of $(WCl_4)_x$ from reduction of WCl_6 with mercury is an edge-shared bioctahedral linear polymer, as predicted by McCarley[11] by analogy to $(NbCl_4)_x$, with alternating short $W=W$ (2.688(2) Å) and long (3.787(3) Å) nonbonded $W-W$ distances. Other structural features support this bonding description.[10]

B. AMORPHOUS TUNGSTEN TETRACHLORIDE BY SOLUTION-PHASE REDUCTION

$$2WCl_6 + Sn \rightarrow 2(WCl_4)_x + SnCl_4$$

Procedure

A 50 mL Schlenk flask is loaded with WCl_6 (2.00 g, 5.04 mmol), Sn powder (0.299 g, 2.52 mmol), and 1,2-dichloroethane (10 mL) in the glove box. The mixture is stirred at room temperature for 20 h, yielding a gray suspension. A reflux condenser is attached to the flask, and the mixture is heated and stirred at 85°C for 2 days under argon. The appearance of the shiny dark suspension is unchanged during heating. After the flask is cooled and transferred to a glove box, the suspension is removed and centrifuged. The supernatant is discarded, the solid is resuspended in CH_2Cl_2 (10 mL), and the mixture is centrifuged again. The supernatant is discarded and the resuspension in CH_2Cl_2/centrifugation process is repeated four times in order to remove $SnCl_4$.[*] The resulting fine powder is rinsed into a 20 mL flask with CH_2Cl_2, the supernatant is decanted, and the precipitate is dried under dynamic vacuum at room temperature. Yield: 1.576 g (96% based on WCl_6).

Anal. Calcd. for Cl_4W: W, 56.45; Cl, 43.55. Found: W, 56.6; Cl, 43.27. No Sn was detected by qualitative microanalysis.

Properties

The $(WCl_4)_x$ produced by Sn reduction of WCl_6 resembles graphite in appearance. It is amorphous as shown by X-ray diffraction. It is more extractable (suspensions in organic solvents have slight color) and more reactive than WCl_4 prepared by other methods.[10] It reacts readily with tetrahydrofuran to give the known $WCl_4(THF)_2$, with acetonitrile to yield the known $WCl_4(MeCN)_2$ in 81% yield, and with methanol to yield the known $W_2Cl_4(OMe)_4(HOMe)_2$ in 73% yield.[10] It reacts with tetraalkylammonium chlorides to give the face- and edge-bridging

---

[*]The presence of $SnCl_4$ in the washings can be detected by addition of pyridine, which causes precipitation of $SnCl_4$ in virtually quantitative yield as $SnCl_4(py)_2$.

ditungsten(IV) complexes $W_2(\mu\text{-Cl})_3Cl_6^-$ and $W_2(\mu\text{-Cl})_2Cl_8^{2-}$ with formal W=W double bonds.[3h]

Acknowledgments

This research was supported in part with funds from the National Science Foundation (CHE-0078701) and Nycomed Inc. The Siemens D5000 automated powder diffractometer was purchased with support from the Roy J. Carver Charitable Trust.

References

1. (a) F. M. Bickelhaupt, B. Neumüller, M. Plate, and K. Dehnicke, *Z. Anorg. Allg. Chem.* **624**, 1455 (1998); (b) P. R. Sharp, *Organometallics* **3**, 1217 (1984); (c) J. C. Bryan, S. J. Geib, A. L. Rheingold, and J. M. Mayer, *J. Am. Chem. Soc.* **109**, 2826 (1987).
2. (a) C. Persson and C. Andersson, *Organometallics* **12**, 2370 (1993); (b) S. G. Bott, D. L. Clark, M. L. H. Green, and P. Mountford, *J. Chem. Soc., Dalton Trans.* 471 (1991).
3. (a) P. M. Boorman, N. L. Langdon, V. J. Mozol, M. Parvez, and G. P. A. Yap, *Inorg. Chem.* **37**, 6023 (1998); (b) F. A. Cotton, E. V. Dikarev, and S. Herrero, *Inorg. Chem. Commun.* **2**, 98 (1999); (c) J. C. Bollinger, M. H. Chisholm, D. R. Click, K. Folting, C. M. Hadad, D. B. Tiedtke, and P. J. Wilson, *Dalton Trans.* 2074 (2001); (d) A. P. Sattelberger, K. W. MacLaughlin, and J. C. Huffman, *J. Am. Chem. Soc.* **103**, 2880 (1981); (e) D. J. Santure, J. C. Huffman, and A. P. Sattelberger, *Inorg. Chem.* **24**, 371 (1985); (f) D. J. Santure, K. W. McLaughlin, J. C. Huffman, and A. P. Sattelberger, *Inorg. Chem.* **22**, 1877 (1983); (g) M. H. Chisholm, B. W. Eichhorn, K. Folting, J. C. Huffman, C. D. Ontiveros, W. E. Streib, and W. E. Van Der Sluys, *Inorg. Chem.* **26**, 3182 (1987); (h) V. Kolesnichenko, D. C. Swenson, and L. Messerle, *Chem. Commun.* 2137 (1998).
4. (a) D. Zeng and M. J. Hampden-Smith, *Chem. Mater.* **5**, 681 (1993); (b) A. M. Nartowski, I. P. Parkin, M. MacKenzie, A. J. Craven, and I. MacLeod, *J. Mater. Chem.* **9**, 1275 (1999).
5. X. Zhang and R. E. McCarley, *Inorg. Chem.* **34**, 2678 (1995).
6. H. Balcar and M. Pacovska, *J. Mol. Catal. A: Chem.* **115**, 101 (1997).
7. (a) G. I. Novikov, N. V. Andreeva, and O. G. Polyachenok, *Russ. J. Inorg. Chem.* **6**, 1019 (1961); (b) M. H. Chisholm and J. D. Martin, *Inorg. Synth.* **29**, 137 (1992).
8. (a) M. A. Schaefer-King and R. E. McCarley, *Inorg. Chem.* **12**, 1972 (1973); (b) R. R. Schrock, L. G. Sturgeoff, and P. R. Sharp, *Inorg. Chem.* **22**, 2801 (1983); (c) D. J. Santure and A. P. Sattelberger, *Inorg. Synth.* **26**, 219 (1989).
9. (a) V. Kolesnichenko and L. Messerle, *Inorg. Chem.* **37**, 3660 (1998); (b) V. Kolesnichenko, J. J. Luci, D. C. Swenson, and L. Messerle, *J. Am. Chem. Soc.* **120**, 13260 (1998); (c) D. N. T. Hay, D. C. Swenson, and L. Messerle, *Inorg. Chem.* **41**, 4700 (2002); (d) D. N. T. Hay and L. Messerle, *J. Struct. Biol.* **139**, 147 (2002); (e) D. N. T. Hay, J. A. Adams, J. Carpenter, S. L. DeVries, J. Domyancich, B. Dumser, S. Goldsmith, M. A. Kruse, A. Leone, F. Moussavi-Harami, J. A. O'Brien, J. R. Pfaffly, M. Sylves, P. Taravati, J. L. Thomas, B. Tiernan, and L. Messerle, *Inorg. Chim. Acta* **357**, 644 (2004).
10. V. Kolesnichenko, D. C. Swenson, and L. Messerle, *Inorg. Chem.* **37**, 3257 (1998).
11. R. E. McCarley and T. M. Brown, *Inorg. Chem.* **3**, 1232 (1964).

Chapter Two

CYCLOPENTADIENYL COMPOUNDS

7. SODIUM AND POTASSIUM CYCLOPENTADIENIDE

Submitted by TARUN K. PANDA,[*]
MICHAEL T. GAMER,[*] and PETER W. ROESKY[*]
Checked by HYOJONG YOO[†‡] and DONALD H. BERRY[†]

Alkali metal cyclopentadienide salts are important reagents in organometallic chemistry, having been used to prepare innumerable cyclopentadienyl complexes. Potassium cyclopentadienide, KCp, is usually made by deprotonation of cyclopentadiene with potassium metal either in organic solvents such as THF and benzene or in liquid ammonia,[1–3] or by deprotonation with KH[4] or KOH.[5] The analogous sodium compound, NaCp, was reported by the groups of Fischer[2] and

[*]Institut für Chemie und Biochemie, Freie Universität Berlin, Fabeckstraße 34–36, 14195 Berlin, Germany; Institut für Anorganische Chemie, Karlsruher Institut für Technologie (KIT), Engesserstr. 15, Geb. 30.45, 76131 Karlsruhe, Germany.
[†]Department of Chemistry, University of Pennsylvania, 231 S. 34 Street, Philadelphia, PA 19104.
[‡]Present address: Department of Chemistry, Hallym University, 1 Hallymdaehak-gil, Chuncheon, Gangwon-do 200-702, South Korea.

Inorganic Syntheses, Volume 36, First Edition. Edited by Gregory S. Girolami and Alfred P. Sattelberger.
© 2014 John Wiley & Sons, Inc. Published 2014 by John Wiley & Sons, Inc.

Ziegler,[6] and NaCp can be obtained by a similar deprotonation of cyclopentadiene with sodium metal, NaH, NaO-*t*-Bu, or NaOH.[3,5-8] The disadvantage of all the reactions described above is that monomeric cyclopentadiene must be used as the starting material. Because monomeric cyclopentadiene is not stable at room temperature, it is usually freshly prepared by a thermal retro-Diels–Alder reaction from the commercially available dicyclopentadiene.

As reported earlier,[9] both NaCp and KCp can be obtained in a one-pot synthesis directly by the reaction of dicyclopentadiene with the respective alkali metal without using any other solvent. This procedure is much more convenient, does not require dry solvents such as THF or decahydronaphthalene, and produces a product free of colored impurities.

General Procedures

All manipulations were performed under nitrogen. *n*-Pentane was distilled under nitrogen from $LiAlH_4$. All other chemicals used as starting materials were obtained commercially (Aldrich) and used without further purification.

A. SODIUM CYCLOPENTADIENIDE

Procedure

■ **Caution.** *Sodium reacts violently with water, liberating flammable hydrogen gas. NaCp reacts violently with water.*

A 1 L Schlenk flask, which is attached to a pressure-release valve, is flushed with nitrogen and charged with dicyclopentadiene (400 mL). Freshly cut sodium metal (10.00 g, 0.435 mol) is added. The mixture is heated for 6 h at 160°C. The solution initially turns blue (around 35°C), but then slowly discolors. At 150°C, a white solid begins to precipitate. When the alkali metal is quantitatively consumed, the dihydrogen evolution stops. To ensure a quantitative conversion, the heating is continued for another 30 min after the dihydrogen evolution ends. The reaction mixture is cooled, and the white residue is collected by filtration, washed with *n*-pentane (3 × 50 mL), and dried in vacuum. The unreacted dicyclopentadiene (260 mL) can be reused later for the same reaction. Yield: 38 g (99%).[*]

Properties

Sodium cyclopentadienide is white when rigorously pure, but often darkens with use as it is exposed to increasing amounts of oxygen. As long as the color is not too

---

[*]The checkers obtained a yield of 93%.

dark, the material is synthetically useful. ^{1}H NMR (d_8-THF, 400 MHz, 25°C): δ 5.60 (s). ^{13}C{^{1}H} NMR (d_8-THF, 100.4 MHz, 25°C): δ 103.3 (s).

B. POTASSIUM CYCLOPENTADIENIDE

Procedure

■ **Caution.** *Potassium reacts violently with water, liberating flammable hydrogen gas. KCp reacts violently with water.*

Potassium cyclopentadienide is made from potassium (1.32 g, 33.8 mmol) and dicyclopentadiene (50 mL) using the same procedure as described above. The reaction time is 4 h at 160°C. Yield: 3.34 g (95%).[*]

Properties

The properties of potassium cyclopentadienide are similar to those of the sodium salt discussed above. ^{1}H NMR (d_8-THF, 400 MHz, 25°C): δ 5.60 (s).[†]

Acknowledgments

This work was supported by the Deutsche Forschungsgemeinschaft and the Fonds der Chemischen Industrie. Reprinted with permission from Ref. 9, © 2003, American Chemical Society.

References

1. J. Thiele, *Chem. Ber.* **34**, 68 (1901).
2. (a) E. O. Fischer, R. Jira, and K. Hafner, *Z. Naturforsch. B* **8**, 327 (1953); (b) E. O. Fischer, W. Hafner, and H. O. Stahl, *Z. Anorg. Allg. Chem.* **282**, 47 (1955).
3. W. P. Fehlhammer, W. A. Herrmann, and K. Öfele, in *Synthetic Methods of Organometallic and Inorganic Chemistry*, W. A. Herrmann and G. Brauer, eds., Thieme, Stuttgart, 1997, Vol. 3, p. 50.
4. F. H. Köhler, W. A. Geike, and N. Hertkorn, *J. Organomet. Chem.* **334**, 359 (1987).
5. A. P. Borisov and V. D. Makhaev, *Metalloorg. Khim.* **2**, 680 (1989); *Organomet. Chem. USSR* **2**, 680 (1989).
6. K. Ziegler, H. Froitzheim-Kühlbom, and K. Hafner, *Chem. Ber.* **80**, 434 (1956).
7. J. C. Smart and C. J. Curtis, *Inorg. Chem.* **16**, 1788, (1977).
8. C. Janiak, *Z. Anorg. Allg. Chem.* **636**, 2387 (2010).
9. T. K. Panda, M. T. Gamer, and P. W. Roesky, *Organometallics* **22**, 877 (2003).

---

[*]The checkers obtained a yield of 90%.
[†]The checkers report that KCp is only very slightly soluble in THF, and caution that ^{1}H NMR spectroscopy may not be a useful check of purity.

8. (PENTAFLUOROPHENYL)CYCLOPENTADIENE AND ITS SODIUM SALT

Submitted by PAUL A. DECK[*]
Checked by KIMBERLY C. JANTUNEN,[†] FELICIA L. TAW,[†]
and JAQUELINE L. KIPLINGER[†]

Pentafluorophenyl-substituted cyclopentadienide anions are useful ligands for transition metals because the C_6F_5 substituent is not only highly electron withdrawing but also relatively stable.[1–6] Up to four C_6F_5 substituents may be attached to cyclopentadiene (or up to two to indene[7]) by one-pot nucleophilic aromatic substitution reactions of cyclopentadienyl or indenyl anions with hexafluorobenzene. The title diene was also reported as the major organic product of the photolysis of $Cp_2Ti(C_6F_5)_2$.[8]

We now present the synthesis of the monoarylated cyclopentadiene $C_5H_5C_6F_5$ on a useful but manageable scale (50 mmol). The procedure is similar to one that we first published[1] and subsequently revised.[2] The sequence begins with a nucleophilic substitution of cyclopentadienide for one fluorine substituent in hexafluorobenzene. Sodium hydride traps the product diene as the corresponding sodium salt $Na(C_5H_4C_6F_5)$ in situ so that the diene can neither dimerize[1] nor transfer its acidic proton to unreacted cyclopentadienyl anion. The resulting arylcyclopentadienyl anion (and the excess sodium hydride) is subsequently hydrolyzed in a cold aqueous workup. The resulting diene $C_5H_5C_6F_5$ is not thermally stable, but is easily converted back to the corresponding sodium salt. The procedure uses convenient, commercially available reagents, and product workup has been optimized.

General Procedures

Except where noted, the reactions should be carried out under an inert atmosphere either in a fume hood using a Schlenk manifold (N_2 or Ar) equipped with a bubbler to release hydrogen formed during the reactions or in a dry box (He or N_2).[‡] Sodium cyclopentadienide (2.0–3.0 M solution in THF, Strem), sodium hydride (60% mineral oil dispersion, Strem), hexafluorobenzene (Aldrich), hydrochloric

[*]Department of Chemistry, Virginia Tech, Blacksburg, VA 24061.
[†]Chemistry Division, Los Alamos National Laboratory, Los Alamos, NM 87545.
[‡]The checkers conducted the inert atmosphere steps under He or N_2 in a Vacuum Atmospheres Model HE-533-2 dry box with a MO-40-2 Dri-Train, using either a cold well or a freezer to maintain the target reaction temperatures for the reactions.

acid (6 M, Fisher), anhydrous THF (Aldrich), anhydrous hexanes (Aldrich), and anhydrous pentane (Aldrich) are used as received.

A. (PENTAFLUOROPHENYL)CYCLOPENTADIENE

$$Na(C_5H_5) + C_6F_6 + NaH \rightarrow Na(C_5H_4C_6F_5) + NaF + H_2$$

$$Na(C_5H_4C_6F_5) + H^+_{(aq)} \rightarrow Na^+_{(aq)} + C_5H_5C_6F_5$$

■ **Caution.** *Sodium hydride reacts violently with water to release hydrogen. Oil-free sodium hydride may burst into flame spontaneously upon exposure to air.*

Procedure

A dry 250 mL Schlenk flask is fitted with a large egg-shaped stir bar and charged with sodium hydride (3.2 g, 60% dispersion, 80 mmol). Under a nitrogen counterstream, the NaH dispersion is washed with hexanes (3 × 40 mL) to remove the oil. The flask is then sealed with a rubber septum. Anhydrous THF (100 mL) and sodium cyclopentadienide solution (25.0 mL of a 2.0 M solution in THF, 50.0 mmol) are added via syringe. The mixture is stirred and cooled using an ice water bath. With continued stirring and ice cooling, hexafluorobenzene (11.2 g, 6.89 mL, 60.0 mmol) is added in small portions (about 0.5 mL min⁻¹) using a syringe. The mixture changes color immediately.* After the addition is complete, the ice bath is removed, and the mixture is stirred under nitrogen for about 36 h at room temperature. The solvent is then removed in vacuum using a room-temperature water bath† and a liquid nitrogen trap. Complete removal of the solvent is important; therefore, the pumping should be continued for at least 12 h (overnight). Pentane (100 mL, *not* hexane) is then added, and against a nitrogen counterstream, a spatula is used carefully to break up the lumps of residue until the stirrer can be restarted. The mixture is then cooled in an ice water bath, and with vigorous stirring, 50 mL of water is *cautiously* added (the first 5 mL dropwise), through a nitrogen counterstream, to hydrolyze the unreacted sodium hydride. After the addition of water is complete, the flask may be disconnected from the

---

*The author and checkers found that, depending on the source and purity of the sodium cyclopentadienide and hexafluorobenzene, the color of the reaction can range from yellow to purple or brown. The author notes that freshly prepared solutions of sodium cyclopentadienide give paler colors. The checkers did not observe this to be the case.

†The checkers found that complete removal of the solvent can also be carried out without using a water bath.

nitrogen manifold, and the remainder of the procedure may be carried out in air. Ice-cold hydrochloric acid (50 mL of a 6 M aqueous solution) is added, and the resulting mixture is stirred for 10 min with continued ice cooling, after which the hydrolysis should be complete. The biphasic mixture is then filtered through a large coarse frit padded with Celite®. Additional pentane (100 mL) is used to rinse the flask and the filter cake. The biphasic filtrate is transferred to a separatory funnel, and the lower aqueous layer is removed and discarded. The organic layer is washed with ice-cold water (2 × 25 mL), and dried over $MgSO_4$. The $MgSO_4$ slurry is gravity filtered, using pentane (50 mL) to rinse the drying agent. The filtrate is evaporated on a rotary evaporator equipped with aspirator suction, with the evaporator bath filled with ice water to minimize loss of the volatile product. Hydrolyzed, unreacted starting material (cyclopentadiene, bp 41°C) is removed in this step. The crude product is purified by flash chromatography on silica gel using a column that is about 80 mm in diameter and packed to a height of about 15 cm.[*] Pentane is used to pack, load, and run the column. A head pressure of 5–8 psi is sufficient. Because the desired product is eluted first, the onset of elution is easily detected by spotting the eluent periodically on a fluorescent-indicating TLC plate and observing a bright blue spot under a UV lamp. The next 500 mL of eluent is then run into a 1 L recovery flask. Rotary evaporation to remove the pentane gives the product as a white solid. Yield: 7.3–9.4 g (63–80%).

Properties

The isolated product is either a colorless oil or solid, is volatile, and has a characteristic sweet odor. It is ~90–95% pure as judged by NMR spectroscopy. The diene may be stored in a freezer for several days without significant decomposition. The primary mode of decomposition is the formation of a Diels–Alder dimer. Solutions stored at room temperature show evidence of dimerization (NMR) within several hours. ^{1}H NMR (300 MHz, $CDCl_3$): major isomer, δ 7.22 (s, 1H), 6.71 (s, 2H), 3.53 (s, 2H), assigned to 1-(pentafluorophenyl)-cyclopentadiene; minor isomer, δ 6.88 (m, 1H), 6.82 (m, 1H), 6.61 (m, 1H), 3.20 (s, 2H), assigned to 2-(pentafluorophenyl)cyclopentadiene. ^{19}F NMR (282 MHz, $CDCl_3$): major isomer, δ −142.17 (m, 2F), −159.96 (t, 1F), −164.88 (m, 2F); minor isomer, δ −142.48 (m, 2F), −158.44 (t, 1F), −164.48 (m, 2F). Small amounts (about 5–10 mol%) of diarylated cyclopentadienes are observed at δ −155.46 and −155.93 (1,3-isomer) and δ −158.52 (1,4-isomer) in the ^{19}F NMR spectrum.

---

[*]The checkers found that, in the absence of such a large column, purification using a large coarse frit (150 mL, 5 cm × 7 cm) with a silica plug (3.5 cm) and elution with pentane (500 mL) gives identical results.

B. SODIUM (PENTAFLUOROPHENYL)CYCLOPENTADIENIDE

$$C_5H_5C_6F_5 + NaH \rightarrow Na(C_5H_4C_6F_5) + H_2$$

Procedure

■   **Caution.**   *Sodium hydride reacts violently with water to release hydrogen. Oil-free sodium hydride may burst into flame spontaneously upon exposure to air.*

A dry 200 mL Schlenk flask is charged with sodium hydride (1.6 g, 60% dispersion, 40 mmol). After the flask is flushed with nitrogen, the dispersion is washed with hexanes (3×25 mL) to remove the oil. A stir bar is added, the flask fitted with a rubber septum, and THF (75 mL) is added. The mixture is stirred and cooled in an ice water bath. A separate, dry 100 mL Schlenk flask is charged with the entire product from the previous reaction (7.3–9.4 g, 32–40 mmol), which has been dissolved in THF (50 mL) to form a colorless to yellow solution. The diene solution is transferred by cannula or syringe into the cold, stirring NaH suspension. Hydrogen is evolved, resulting in a yellow to orange reaction mixture. After the addition is complete, the mixture is stirred for 2 h at room temperature. The mixture is then filtered through a Celite® padded frit. The filter cake is rinsed with additional anhydrous THF (50 mL).

■   **Caution.**   *(The filter cake may still contain small portions of unreacted sodium hydride and should be handled with care.)*

The dark, air-sensitive filtrate is evaporated under reduced pressure. After all visible liquid has been removed, the residue is dried at 80°C for several hours (e.g., overnight) using a vacuum pump. The resulting solid is washed with pentane (2×50 mL) and dried briefly under reduced pressure. This washing step removes any dimerized diene that may have been present and any traces of oil that may be left from the sodium hydride dispersion. Yield: 6.8–7.4 g (84–92% based on 7.3 g of $C_5H_5C_6F_5$).

Properties

The product is a yellow powder that is stable indefinitely in the absence of air. Samples exposed to laboratory light will darken over several weeks, but the NMR spectra remain unchanged. Samples stored in the dry box freezer do not discolor. ^{1}H NMR (300 MHz, THF-d_8): δ 6.41 (m, 2H), 5.90 (m, 2H). ^{19}F NMR (282 MHz, THF-d_8): δ −144.0 (d, 2F), −166.5 (t, 2F), −172.4 (t, 1F). Small amounts (about 5 mol%) of residual THF are typically present.

Acknowledgments

We acknowledge LANL (G.T. Seaborg Institute Fellowship to K.C.J. and Director's Postdoctoral Fellowship to F.L.T.), the Division of Chemical Sciences, Office of Basic Energy Sciences (J.L.K.), the Los Alamos National Laboratory LDRD Program (J.L.K.), and the Petroleum Research Fund (P.A.D.) for financial support.

References

1. P. A. Deck, W. F. Jackson, and F. R. Fronczek, *Organometallics* **15**, 5287 (1996).
2. M. P. Thornberry, C. Slebodnick, P. A. Deck, and F. R. Fronczek, *Organometallics* **19**, 5352 (2000).
3. M. P. Thornberry, C. Slebodnick, P. A. Deck, and F. R. Fronczek, *Organometallics* **20**, 920 (2001).
4. M. D. Blanchard, R. P. Hughes, T. E. Concolino, and A. L. Rheingold, *Chem. Mater.* **12**, 1604 (2000).
5. R. J. Maldanis, J. C. W. Chien, and M. D. Rausch, *J. Organomet. Chem.* **599**, 107 (2000).
6. P. A. Deck and F. R. Fronczek, *Organometallics* **19**, 327 (2000).
7. B. Klingert, A. Roloff, B. Urwyler, and J. Wirz, *Helv. Chim. Acta* **71**, 1858 (1988).
8. A. Roloff, K. Meier, and M. Reidiker, *Pure Appl. Chem.* **58**, 1267 (1986). (b) B. Klingert, A. Roloff, B. Urwyler, and J. Wirz, *Helv. Chim. Acta* **71**, 1858 (1988).

9. BIS(η^5-PENTAMETHYLCYCLOPENTADIENYL) COMPLEXES OF SCANDIUM

Submitted by ENDY Y.-J. MIN[*] and JOHN E. BERCAW[*]
Checked by ALYSON L. KENWARD[†] and WARREN PIERS[†]

Use of pentamethylcyclopentadienyl ligands (C_5Me_5) in place of cyclopentadienyl (Cp) provides reactive, but thermally robust monomeric 14-electron scandocene derivatives. A useful starting material for the synthesis of a variety of derivatives of permethylscandocene is the chloride complex (η^5-C_5Me_5)$_2$ScCl (**1**).[1] Compound **1** is obtained, together with its THF adduct (THF = tetrahydrofuran), when solvent is removed from a reaction mixture consisting of $ScCl_3(THF)_3$ and 2 equiv of LiC_5Me_5, which has been heated at reflux in xylenes. The coordinated THF is removed on sublimation of the residue under dynamic vacuum, affording pale yellow, crystalline **1**. Compound **1** is monomeric; in contrast, $[(\eta^5$-$C_5H_5)_2ScCl]_2$ is a robust dimer with the two chloride ligands bridging scandium centers. Moreover, (η^5-$C_5H_5)_2$ScCl(THF) does not dissociate its THF ligand on sublimation. Both of these differences are undoubtedly attributable to the much greater steric bulk of the pentamethylcyclopentadienyl ligand in **1**. (η^5-$C_5Me_5)_2$ScCl readily undergoes straightforward metathetical reactions with main group organometallic compounds. Thus, (η^5-$C_5Me_5)_2$ScCH$_3$ (**2**), (η^5-$C_5Me_5)_2$ScC$_6$H$_5$ (**3**), (η^5-$C_5Me_5)_2$-Sc(*o*-$C_6H_4CH_3$) (*o*-**4**), (η^5-$C_5Me_5)_2$Sc(*m*-$C_6H_4CH_3$) (*m*-**4**), and (η^5-$C_5Me_5)_2$Sc-(*p*-$C_6H_4CH_3$) (*p*-**4**) are obtained by treatment of **1** with the appropriate organolithium compounds.[1]

[*]Department of Chemistry, California Institute of Technology, Pasadena, CA 91125.
[†]Department of Chemistry, University of Calgary, Calgary, Alberta, Canada T2N 1N4.

General Procedures

All experiments are conducted under argon in a glove box or a high-vacuum line.[2] Solvents are purified by passing through standard purification columns and distilled at room temperature from titanocene,[3*] on the vacuum line at $<10^{-4}$ mmHg. The reagents $ScCl_3(THF)_3$,[4] LiC_5Me_5,[5] and the isomers of tolyllithium[6] are prepared by literature routes. Methyllithium and phenyllithium are purchased from Aldrich.

A. BIS(η^5-PENTAMETHYLCYCLOPENTADIENYL)-CHLOROSCANDIUM

$$ScCl_3(THF)_3 + 2LiC_5Me_5 \rightarrow (\eta^5\text{-}C_5Me_5)_2ScCl(THF) + 2LiCl + 2THF$$

$$(\eta^5\text{-}C_5Me_5)_2ScCl(THF) \rightarrow (\eta^5\text{-}C_5Me_5)_2ScCl + THF$$

Procedure

■ **Caution.** *Both lithium pentamethylcyclopentadienide and 1 are extremely air sensitive and should be handled using high-vacuum line[2] and inert atmosphere glove box techniques.*

In a dinitrogen-filled glove box, $ScCl_3(THF)_3$ (10.0 g, 27 mmol) and LiC_5Me_5 (9.0 g, 63 mmol) are transferred to a 500 mL round-bottomed flask fitted with a Teflon®-coated magnetic stir bar, reflux condenser, and a Teflon needle valve assembly.[1] The flask is sealed, removed from the glove box, and attached to a high-vacuum line. Mixed xylenes (250 mL) are transferred onto the solid in vacuum,[3] the mixture is warmed to room temperature with stirring, an atmosphere of argon is introduced, and the orange mixture is heated to reflux for 3 days in a silicone oil bath. Periodically (about once per day), heating is discontinued, ~30 mL of solvent and evolved THF are removed under vacuum, argon is readmitted, and refluxing is resumed. After 3 days, all volatile materials are removed in vacuum to give an orange solid. The solid residue is transferred to a sublimation apparatus. Sublimation over 4 days at 120°C (dynamic vacuum, $<10^{-5}$ mmHg) affords bright yellow crystals, which are scraped from the sublimation probe and stored in a glove box. Yield: 8.0 g (80%).[†]

Anal. Calcd. for $C_{20}H_{30}ClSc$: C, 68.47; H, 8.77; Cl, 10.10. Found: C, 68.05; H, 8.77; Cl, 10.12. Molecular weight by ebulliometry in benzene:[1] 365 (calcd. 350).

---

[*]The checkers used sodium/benzophenone to dry their solvents.
[†]The checkers carried out the preparation beginning with 5 g of $ScCl_3(THF)_3$ and obtained a yield of 68%.

Properties

The yellow crystalline product is very air and moisture sensitive. It is soluble in hydrocarbons and diethyl ether, but complexes THF tightly. Its infrared spectrum in Nujol displays prominent bands at 2715, 1030, 800, 430, and 360 cm^{-1}. ^{1}H NMR (benzene-d_6): δ 1.83 (s).

B. BIS(η^5-PENTAMETHYLCYCLOPENTADIENYL)-METHYLSCANDIUM

$$(\eta^5\text{-}C_5Me_5)_2ScCl + LiCH_3 \rightarrow (\eta^5\text{-}C_5Me_5)_2ScCH_3 + LiCl$$

Procedure

■ **Caution.** *Both methyllithium and **1** are extremely air sensitive and should be handled using high-vacuum line[2] and inert atmosphere glove box techniques.*

In a dinitrogen-filled glove box, **1** (5.0 g, 14 mmol) is transferred to a 250 mL round-bottomed flask fitted with a swivel frit assembly.[2] The flask is sealed, removed from the glove box, and attached to a high-vacuum line. Toluene (75 mL) is transferred in vacuum onto the solid.[2] While the mixture is stirred in a −78°C dry ice/acetone bath, methyllithium (8.0 mL of a 1.78 M in diethyl ether, 14.2 mmol) is added slowly by a syringe to the solution via the septum-capped sidearm of the swivel frit. The reaction mixture is stirred for 30 min at room temperature, and the volatile materials are removed under reduced pressure. Petroleum ether (30 mL) is transferred onto the solid in vacuum, and the resulting solution is filtered to remove LiCl. The residue is washed with additional pentane (2 × 30 mL) using standard swivel frit techniques. The filtered washings are combined, concentrated to 10 mL, and cooled to −78°C. The resulting pale yellow precipitate is isolated by cold filtration. Yield: 2.6 g (55%).[*]

Anal. Calcd. for C$_{21}$H$_{33}$Sc: C, 76.33; H, 10.07. Found: C, 76.02; H, 9.98.

Properties

The product is a pale yellow crystalline solid that is very air and moisture sensitive. Solid samples of **2** should be stored at 10°C to prevent decomposition. It is soluble

---

[*]The checkers carried out the preparation beginning with 1.0 g of **1** and obtained a yield of 48%.

in hydrocarbons. Its infrared spectrum in Nujol displays prominent bands at 2730, 1490, 1114, 1028, 800, 605, 580, and 420 cm^{-1}. ^{1}H NMR (benzene-d_6): δ 1.87 (s, C$_5$Me$_5$), 0.07 (s, Sc-CH$_3$).

C. BIS(η^5-PENTAMETHYLCYCLOPENTADIENYL)-PHENYLSCANDIUM

$$(\eta^5\text{-}C_5Me_5)_2ScCl + LiC_6H_5 \rightarrow (\eta^5\text{-}C_5Me_5)_2ScC_6H_5 + LiCl$$

Procedure

■ **Caution.** *Both phenyllithium and 1 are extremely air sensitive and should be handled using high-vacuum line[2] and inert atmosphere glove box techniques.*

In a dinitrogen-filled glove box, **1** (2.5 g, 7.6 mmol) is transferred to a 50 mL round-bottomed flask fitted with a swivel frit assembly.[2] The flask is sealed, removed from the glove box, and attached to a high-vacuum line. Petroleum ether (20 mL) is transferred onto the solid in vacuum. While the mixture is stirred in a −78°C dry ice/acetone bath, phenyllithium (3.11 mL of a 2.4 M solution in cyclohexane–diethyl ether, 7.5 mmol) is added slowly by a syringe to the solution via the septum-capped sidearm of the swivel frit. The reaction mixture is stirred for 30 min at room temperature, and the volatile materials are removed under reduced pressure. Petroleum ether (20 mL) is transferred onto the solid in vacuum, and the resulting solution is filtered to remove LiCl. The residue is washed with additional pentane (2 × 20 mL) using standard swivel frit techniques. The filtered washings are combined, concentrated to 5 mL, and cooled to −78°C. The resulting pale yellow precipitate is isolated by cold filtration. Yield: 1.71 g (60%).

Anal. Calcd. for C$_{26}$H$_{35}$Sc: C, 79.56; H, 8.99. Found: C, 79.58; H, 8.77.

Properties

The product is a pale yellow crystalline solid that is very air and moisture sensitive. Solid samples of **3** should be stored at 10°C to prevent decomposition. It is soluble in hydrocarbons. Its infrared spectrum in Nujol displays prominent bands at 3430 (w), 2730 (w), 1590 (w), 1563 (w), 1485 (w), 1312 (w), 1233 (w), 1165 (w), 1067 (m), 1025 (m), 983 (w), 738 (w), 722 (s), 708 (s), 462 (s), and 418 (s) cm^{-1}. ^{1}H NMR (benzene-d_6): δ 1.73 (s, C$_5$Me$_5$), 7.0–7.4 (m, Ph).

D. BIS(η^5-PENTAMETHYLCYCLOPENTADIENYL)(o-TOLYL) SCANDIUM

$$(\eta^5\text{-}C_5Me_5)_2ScCl + Li(o\text{-}C_6H_4CH_3) \rightarrow (\eta^5\text{-}C_5Me_5)_2Sc(o\text{-}C_6H_4CH_3) + LiCl$$

Procedure

■   **Caution.**   *Both ortho-tolyllithium and* **1** *are extremely air sensitive and should be handled using high-vacuum line[2] and inert atmosphere glove box techniques.*

In a dinitrogen-filled glove box, **1** (0.420 g, 1.2 mmol) and *ortho*-tolyllithium (0.18 g, 1.8 mmol) are transferred to a 50 mL round-bottomed flask fitted with a swivel frit assembly.[2] The reaction mixture is stirred for 12 h at room temperature, and the volatile materials are removed under reduce pressure. Petroleum ether (10 mL) is transferred onto the solid in vacuum, and the resulting solution is filtered to remove LiCl. The residue is washed with additional pentane (2 × 10 mL) using standard swivel frit techniques. The filtered washings are combined, concentrated to 3 mL, and cooled to −78°C. The resulting off-white precipitate is isolated by cold filtration. Yield: 0.285 g (59%).

Anal. Calcd. for $C_{27}H_{37}Sc$: C, 79.77; H, 9.17. Found: C, 79.79; H, 9.08.

Properties

The off-white crystalline product is very air and moisture sensitive. Solid samples of **4** should be stored at 10°C to prevent decomposition. It is soluble in hydrocarbons. Its infrared spectrum in Nujol displays prominent bands at 2715, 1578, 1300, 1252, 1222, 1208, 1160, 1140, 1079, 1053, 1030, 1016, 793, 774, 720, 549, 480, and 417 cm^{-1}. ^{1}H NMR (benzene-d_6): δ 1.77 (s, C_5Me_5), 7.26 (d, J_{HH} = 6.0 Hz, C_3H), 7.20 (m, $C_{4,5}H$), 6.32 (t, J_{HH} = 5.3 Hz, C_6H), 2.12 (s, o-CH$_3$).

The same procedure can be used to prepare $(\eta^5\text{-}C_5Me_5)_2Sc(m\text{-}C_6H_4CH_3)$ and $(\eta^5\text{-}C_5Me_5)_2Sc(p\text{-}C_6H_4CH_3)$. ^{1}H NMR of *meta*-isomer (benzene-d_6): δ 1.77 (s, C_5Me_5), 6.78 (d, J_{HH} = 5.7 Hz, C_2H), 7.07 (d, J_{HH} = 5.7 Hz, C_4H), 7.32 (t, J_{HH} = 5.7 Hz, C_5H), 6.85 (d, J_{HH} = 5.7 Hz, C_6H), 2.43 (s, m-CH$_3$). ^{1}H NMR of *para*-isomer (benzene-d_6): δ 1.76 (s, C_5Me_5), 7.26 (d, J_{HH} = 5.4 Hz, C_2H), 6.92 (d, J_{HH} = 5.4 Hz, C_3H), 2.33 (s, p-CH$_3$).

References

1. M. E. Thompson, S. M. Baxter, A. R. Bulls, B. J. Burger, M. C. Nolan, B. D. Santarsiero, W. P. Schaefer, and J. E. Bercaw, *J. Am. Chem. Soc.* **109**, 203 (1987).

2. B. J. Burger and J. E. Bercaw, in *New Developments in the Synthesis, Manipulation and Characterization of Organometallic Compounds*, A. Wayda and M. Y. Darensbourg, eds., American Chemical Society, Washington, DC, 1987, Vol. 357, Chapter 4.
3. R. H. Marvich and H. H. Brintzinger, *J. Am. Chem. Soc.* **93**, 2046 (1971).
4. P. G. Hayes and W. E. Piers, *Inorg. Synth.* **35**, 20 (2010).
5. J. M. Manriquez, E. E. Bunel, and B. Oelkers, *Inorg. Synth.* **31**, 214 (1997).
6. M. Schlosser and V. Ladenberger, *J. Organomet. Chem.* **8**, **193**, (1967).

10. BIS(η^5-PENTAMETHYLCYCLOPENTADIENYL) COMPLEXES OF TITANIUM, ZIRCONIUM, AND HAFNIUM

Submitted by ENDY Y.-J. MIN* and JOHN E. BERCAW*
Checked by WESLEY BERNSKOETTER† and PAUL J. CHIRIK‡

The chemistry of organometallic derivatives of titanium and zirconium bearing two pentamethylcyclopentadienyl groups as ancillary ligands has proven to be rich and varied, whereas that for hafnium is still in its formative stages. Derivatives of permethyltitanocene, permethylzirconocene, and permethylhafnocene offer a number of advantages over their (η^5-C$_5$H$_5$) analogs. Generally, they exhibit higher thermal stability, solubility, and crystallinity. Furthermore, the steric bulk of the pentamethylcyclopentadienyl group discourages oligomerization through single-atom bridges (e.g., H, O, N), thus rendering permethylmetallocene derivatives monomeric. Such properties have facilitated the development of a rich and diverse reaction chemistry for compounds of the group 4 transition metals with penta-methylcyclopentadienyl ancillary ligands.

A. BIS(η^5-PENTAMETHYLCYCLOPENTADIENYL)-DICHLOROTITANIUM(IV)

$$Na + NH_3 \rightarrow NaNH_2 + \tfrac{1}{2}H_2$$

$$NaNH_2 + C_5Me_5H \rightarrow NaC_5Me_5 + NH_3$$

$$TiCl_3 + 2NaC_5Me_5 \rightarrow (\eta^5\text{-}C_5Me_5)_2TiCl + 2NaCl$$

$$(\eta^5\text{-}C_5Me_5)_2TiCl + HCl + \tfrac{1}{2}O_2 \rightarrow (\eta^5\text{-}C_5Me_5)_2TiCl_2 + \tfrac{1}{2}H_2O$$

*Department of Chemistry, California Institute of Technology, Pasadena, CA 91125.
†Department of Chemistry and Chemical Biology, Cornell University, Ithaca, NY 14853.
‡Department of Chemistry, Princeton University, Princeton, NJ 08544.

Procedure[1]

■ **Caution.** *Sodium metal is air and (extremely) water sensitive and should be handled under an inert atmosphere or only briefly in dry air. Ammonia has a sharp, penetrating odor and can be fatal if inhaled in excess.*

A 1 L three-necked flask is fitted with a dry ice/acetone-filled cold finger for ammonia reflux, and glass stoppers for the remaining two necks, and placed in a fume hood with excellent ventilation. A gas inlet at the top of the cold finger is connected to a mercury bubbler and the flask is flushed with argon. Sodium (2.3 g, 100 mmol) is rinsed with petroleum ether to remove the mineral oil in which it is usually stored. The metal is cut into pea-sized pieces with a razor blade, and the pieces of sodium and a couple of small crystals of $Fe(NO_3)_3 \cdot 9H_2O$ are placed in the flask. Liquid ammonia ($\sim$500 mL) is transferred through rubber tubing into a neck of the 1 L flask from a steel cylinder that is inverted to allow the liquid to flow through the valve. After restoppering the neck used for addition of liquid ammonia, the solution is stirred at $-33°C$ until the initial blue color changes to light gray. Then C_5Me_5H (13.5 g, 100 mmol) is syringed into the stirred sodium amide–liquid ammonia suspension against a counterflow of Ar, and the mixture is allowed to stir at $-33°C$ for 2 h. The ammonia is then allowed to evaporate through the mercury bubbler, by allowing the dry ice in the reflux condenser to evaporate and the cold finger to warm. Pale yellow crystals of NaC_5Me_5 form in the residue.[*] Residual ammonia is further removed from the crystals by passing pure Ar through the flask and out an open neck of the flask at room temperature for a few minutes. THF (200 mL), which has been freshly vacuum transferred from benzophenone ketyl,[2] is added via cannula under Ar to dissolve the NaC_5Me_5. A separate 500 mL two-necked round-bottomed flask is charged with anhydrous, aluminum-free $TiCl_3$ (4.38 g, 28.4 mmol, Aldrich) in a nitrogen-filled glove box. The flask is equipped with a Teflon®-coated stir bar, a Teflon needle valve adapter for attachment to the Schlenk line, and a rubber septum in the other neck. The NaC_5Me_5 solution is carefully decanted (to avoid transferring insoluble solids) and cannula transferred through a septum in the second neck of flask into the round-bottomed flask containing $TiCl_3$ while cooling at $-30°C$ under Ar. The mixture is allowed to warm slowly and stirred for 24 h at room temperature. HCl (50 mL of a 12 M aqueous solution, 60 mmol) is added to the green-brown suspension at $-20°C$ via a syringe, and the flask is opened to the air, whereupon red-brown crystals precipitate from the mixture. Chloroform (250 mL) is added and the dark red-brown chloroform/ THF layer is separated from the aqueous layer. The organic layer is dried over anhydrous Na_2SO_4, and the solvents are removed under reduced pressure on a rotary evaporator. The residue is transferred to an extraction thimble and placed in

---

[*]The checkers point out that the use of NaC_5Me_5 in this preparation is essential; the more common lithium salt gives little to no yield of the desired product. See Ref. 3.

a Soxhlet extractor. The (η^5-C$_5$Me$_5$)TiCl$_3$ contaminant is removed by extracting this red-orange compound from the residue with petroleum ether (bp 30–60°C) saturated with HCl(g); the product (η^5-C$_5$Me$_5$)$_2$TiCl$_2$ is only slightly soluble in petroleum ether. The collection flask is then replaced, and the residue is extracted with CCl$_4$ saturated with HCl. The purple-brown (η^5-C$_5$Me$_5$)$_2$TiCl$_2$ crystallizes from the cooled CCl$_4$ solution as long needles. Yield: 3.4 g (31%). Large, well-formed dark purple crystals may be obtained by recrystallization from a small amount of chloroform saturated with HCl(g).

Anal. Calcd. for C$_{20}$H$_{30}$Cl$_2$Ti: C, 61.7; H, 7.77; Cl, 18.2; Ti, 12.3. Found: C, 61.8; H, 7.92; Cl 18.09; Ti, 12.16.

Properties

The product is a dark purple crystalline solid that is air and moisture stable. It is soluble in CCl$_4$ or CHCl$_3$. ^{1}H NMR (chloroform-d_1): δ 2.00 (s). Its infrared spectrum displays prominent bands at 2880, 1601, 1450, 1420, 1370, 1020, 410, 360, and 340 cm^{-1} (Nujol). Melting point: 273°C (dec.).

B. BIS(η^5-PENTAMETHYLCYCLOPENTADIENYL) DICHLOROZIRCONIUM(IV)

$$\text{Li}(n\text{-Bu}) + \text{C}_5\text{Me}_5\text{H} \rightarrow \text{LiC}_5\text{Me}_5 + n\text{-BuH}$$

$$\text{ZrCl}_4 + 2\text{LiC}_5\text{Me}_5 \rightarrow (\eta^5\text{-C}_5\text{Me}_5)_2\text{ZrCl}_2 + 2\text{LiCl}$$

Procedure[4]

■ **Caution.** *n-Butyllithium is extremely air sensitive and should be handled using Schlenk, high-vacuum line, and inert atmosphere glove box techniques.*

A 500 mL three-necked round-bottomed flask is equipped with a Teflon-coated stirring bar, a Vigreux column topped with a Teflon valve adapter for attachment to a vacuum line, and two glass stoppers. 1,2-Dimethoxyethane (300 mL) is vacuum transferred from LiAlH$_4$ into the flask.[2] The flask is cooled to −78°C with a dry ice/acetone bath. Under an Ar counterflow, one glass stopper is replaced with a septum, and C$_5$Me$_5$H (20.4 g, 150 mmol) followed by n-butyllithium (70 mL, 150 mmol) are slowly added via a syringe through the septum. The septum is replaced with a glass stopper. This mixture, while open to a bubbler, is warmed slowly to room temperature and stirred for 30 min. The mixture is cooled again to −78°C, and freshly sublimed ZrCl$_4$ (15.3 g, 63 mmol) is added slowly using a solid addition ampule.[*] (An ampule containing the $ZrCl_4$ having an 18/9 ball joint sealed at a slight bend with clamped 18/9 socket as cap is prepared in the glove box. A 24/40 standard taper-to-18/9 socket joint adapter is then inserted into one neck of the flask in place of a glass stopper against an Ar counterflow. The cap of the ampule is removed and the ampule is quickly attached to the 18/9 socket joint and clamped in place. By slowly rotating the ampule around the lightly clamped socket joint and tapping with a wooden rod, the $ZrCl_4$ solid may be added slowly to the cooled solution.) This mixture is allowed to warm to 25°C and then heated to mild reflux for 3 days. The solvent is removed under reduced pressure to leave a thick, pale brown residue, which is taken up into a mixture of $CHCl_3$ (250 mL) and HCl (100 mL of 6 M aqueous solution). The aqueous layer is separated and washed with $CHCl_3$. The combined chloroform layers are washed once with distilled water, dried over Na_2SO_4, and then concentrated to ~50 mL by rotary evaporation. Petroleum ether (200 mL, bp 90–100°C) is added, and the solvent is slowly removed by rotary evaporation. The residual solution (~50 mL) is cooled, and the product is collected by filtration and washed with cold petroleum ether to afford pale yellow crystals. Yield: 15 g (55%).

Anal. Calcd. for $C_{20}H_{30}Cl_2Zr$: C, 55.56; H, 6.94; Cl, 16.40; Zr, 21.10. Found: C, 55.57; H, 6.99; Cl, 16.36; Zr, 21.21.

Properties

The product is a yellow crystalline solid that is air and moisture stable. It is soluble in hydrocarbons. ^1H NMR (chloroform-d_1): δ 2.0 (s).

C. BIS(η^5-PENTAMETHYLCYCLOPENTADIENYL)-DICHLOROHAFNIUM(IV)

$$Li(n\text{-}Bu) + C_5Me_5H \rightarrow LiC_5Me_5 + n\text{-}BuH$$

$$HfCl_4 + 2LiC_5Me_5 \rightarrow (\eta^5\text{-}C_5Me_5)_2HfCl_2 + 2LiCl$$

Procedure[5]

■ **Caution.** *n-Butyllithium is extremely air sensitive and should be handled using inert atmosphere techniques.*

---

[*]The checkers note that they have carried out this procedure and obtained product in similar yield and purity by mixing isolated LiC_5Me_5 with $ZrCl_4$, adding toluene, heating to reflux for 2 days, and carrying out the workup as described. This approach avoids the necessity of using a solid addition ampule.

An oven-dried 1 L three-necked round-bottomed flask is equipped with a Teflon-coated stir bar, a Vigreux column topped with a Teflon valve adapter for attachment to a vacuum line,[2] and two glass stoppers. The flask is purged with argon, and *m*-xylene (400 mL) is vacuum transferred[2] (slowly!) into the flask. Under an Ar counterflow, a sample of C_5Me_5H (24.7 g, 0.181 mmol) is added via a syringe to the solution through a septum that has been inserted into one neck. The flask is cooled in a −10°C bath, and *n*-BuLi (115 mL of a 1.6 M solution in hexane, 0.184 mmol) is then added via a syringe into the reaction flask. The solution turns yellow immediately and, after several hours at room temperature, appears orange and gelatinous. After 24 h, $HfCl_4$ (23.1 g, 0.072 mmol) is added by addition ampule (see the procedure for adding $ZrCl_4$).[*] The mixture is then heated to mild xylene reflux for 3 days. After the flask is cooled to room temperature, the *m*-xylene is removed under vacuum, and the residue is suspended in CH_2Cl_2 (200 mL) and HCl (200 mL of a 4 M aqueous solution). The mixture is filtered, and the organic and aqueous layers are separated and both layers kept. The aqueous layer is extracted with CH_2Cl_2 (3 × 50 mL). The combined organic fractions are dried over Na_2SO_4 and filtered. The solvent is then removed under reduced pressure. The remaining crystalline solid is transferred to a thimble and Soxhlet extracted with CCl_4. The receiving flask is cooled, and off-white crystals form. The crystals are isolated by filtration and dried. Yield: 26.4 g (70%).

Anal. Calcd. for $C_{20}H_{30}Cl_2Hf$: C, 46.21; H, 5.82; Cl, 13.64; Hf, 34.33. Found: C, 46.26; H, 5.72; Cl, 13.48; Hf, 33.98.

Properties

The product is an off-white crystalline solid that is air and moisture stable. It is soluble in hydrocarbons. ^1H NMR (benzene-d_6): δ 1.88 (s).

References

1. J. E. Bercaw, R. H. Marvich, L. G. Bell, and H. H. Brintzinger, *J. Am. Chem. Soc.* **94**, 1219 (1972).
2. B. J. Burger and J. E. Bercaw, in *New Developments in the Synthesis, Manipulation and Characterization of Organometallic Compounds*, A. Wayda and M. Y. Darensbourg, eds., American Chemical Society, Washington, DC, 1987, Chapter 4.
3. J. W. Pattiasina, H. J. Heeres, F. Van Bolhuis, A. Meetsma, J. H. Teuben, and A. L. Spek, *Organometallics* **6**, 1004 (1987).
4. J. M. Manriquez, D. R. McAlister, E. Rosenberg, A. M. Shiller, K. L. Williamson, S. I. Chan, and J. E. Bercaw, *J. Am. Chem. Soc.* **100**, 3078 (1978).
5. D. M. Roddick, M. D. Fryzuk, P. F. Seidler, G. L. Hillhouse, and J. E. Bercaw, *Organometallics* **4**, 97 (1985); **4**, 1694 (1985).

[*]The checkers note that they have carried out this procedure and obtained product in similar yield and purity by mixing isolated LiC_5Me_5 with $HfCl_4$, adding toluene, heating to reflux for 2 days, and carrying out the workup as described. This approach avoids the necessity of using a solid addition ampule.

11. BIS(η⁵-PENTAMETHYLCYCLOPENTADIENYL) COMPLEXES OF NIOBIUM AND TANTALUM

Submitted by ENDY Y.-J. MIN[*] and JOHN E. BERCAW[*]
Checked by SHUJIAN CHEN[†] and ZI-LING XUE[†]

Bent metallocene derivatives having η^5-C_5Me_5 ligands ("permethylmetallocene" derivatives) offer advantages over their η^5-C_5H_5 analogs in that they generally display greater thermal stability and are less prone to dimerize through single-atom bridges. These properties have allowed the development of a rich chemistry of group 3 and 4 d-block and f-block transition metal compounds with pentamethylcyclopentadienyl ancillary ligands. Although permethylniobocene and permethyltantalocene derivatives display reactivities similar to the parent niobocene and tantalocene analogs, their increased electron density at Nb and Ta, their greater steric crowding, and their higher thermal stabilities offer some advantages in probing fundamental transformations such as α- and β-migratory insertion and elimination processes.[1]

Convenient synthetic entry points to these compounds are $(\eta^5$-$C_5Me_5)_2NbCl_2$ and $(\eta^5$-$C_5Me_5)_2TaCl_2$. These compounds can be reduced under dihydrogen to yield $(\eta^5$-$C_5Me_5)_2MH_3$ (M = Nb, Ta), which catalyzes the exchange of molecular hydrogen with benzene-d_6.[2] Olefin hydride derivatives may be obtained by treatment of these hydrides with excess olefin, or by treatment of the dichlorides with main group alkyls having β-hydrogens.[3] Reduction of the dichloride with Na/Hg in tetrahydrofuran, followed by reaction with alkyl or alkenyl or aryl lithium reagents, affords alkylidene, vinylidene, or phenylene hydride complexes.[4] The very reactive niobocene hydride is generated by the thermally induced reductive elimination of H_2 from the trihydride; in the presence of carbon monoxide, $(\eta^5$-$C_5Me_5)_2Nb(H)CO$ may be generated.

General Procedures

THF and 1,2-dimethoxyethane (glyme) are stored over sodium benzophenone ketyl and are transferred at room temperature on the vacuum line at $<10^{-4}$ mmHg. Pentane is dried over titanocene. Trimethylphosphine (PMe_3) and pentamethylcyclopentadiene (C_5Me_5H) are commercially available (Aldrich)

[*]Department of Chemistry, California Institute of Technology, Pasadena, CA 91125.
[†]Department of Chemistry, The University of Tennessee, Knoxville, TN 37996.

and are handled and stored under inert atmosphere. PMe$_3$ is dried over CaH$_2$. Tantalum pentachloride (Alfa) is sublimed before use. Similarly, NbCl$_5$ is purified by two sublimations of the commercially available material (Aldrich or Alfa). The first sublimation at 110°C separates NbCl$_5$ from most of the oxychloride impurities. The second sublimation through a plug of glass wool (to prevent mechanical transfer of oxychloride to the sublimation probe) at 100°C yields pure NbCl$_5$, which shows no niobium oxide stretches in the IR spectrum. Removal of all oxychloride appears to be essential for the success of the procedure below.

A. BIS(η^5-PENTAMETHYLCYCLOPENTADIENYL)-DICHLOROTANTALUM(IV)

$$KH + C_5Me_5H \rightarrow KC_5Me_5 + H_2$$

$$KC_5Me_5 + SnCl(n\text{-}Bu)_3 \rightarrow (\eta^1\text{-}C_5Me_5)Sn(n\text{-}Bu)_3 + KCl$$

$$TaCl_5 + (\eta^1\text{-}C_5Me_5)Sn(n\text{-}Bu)_3 \rightarrow (\eta^5\text{-}C_5Me_5)TaCl_4 + SnCl(n\text{-}Bu)_3$$

$$(\eta^5\text{-}C_5Me_5)TaCl_4 + PMe_3 + \tfrac{1}{2}\,Mg \rightarrow (\eta^5\text{-}C_5Me_5)TaCl_3(PMe_3) + \tfrac{1}{2}MgCl_2$$

$$(\eta^5\text{-}C_5Me_5)TaCl_3(PMe_3) + KC_5Me_5 \rightarrow (\eta^5\text{-}C_5Me_5)_2TaCl_2 + KCl + PMe_3$$

Procedure[5,6]

■   **Caution.**   *Potassium hydride is air sensitive and should be handled using Schlenk, high-vacuum line,*[7] *and inert atmosphere glove box techniques.*

KC$_5$Me$_5$ is prepared by weighing KH (6.0 g, 0.15 mol) in a 500 mL round-bottomed flask equipped with a Teflon®-coated magnetic stir bar, a swivel frit, a 500 mL round-bottomed collection flask, and a Teflon valve adapter. THF (250 mL) is vacuum transferred onto the KH.[*] While the mixture is stirred in a −78°C bath, C$_5$Me$_5$H (20.44 g, 0.150 mol) is added through a sidearm by a syringe under an Ar counterflow. The suspension is allowed to stir in the −78°C bath and slowly warm to room temperature, after which it is stirred for an additional 2 h. The suspension is filtered and washed with THF. The collected solid is dried in vacuum.

---

[*]The checkers used a cannula to transfer the THF.

In the dry box, KC_5Me_5 (26.1 g, 150 mmol), $(n\text{-}Bu)_3SnCl$ (48.9 g, 150 mmol), and toluene (600 mL, dried over sieves) are added to a 1 L round-bottomed flask. The mixture is stirred at room temperature for 4 days followed by filtration to yield a light yellow-orange filtrate. The collected solid is washed with three 150 mL portions of toluene, and all filtrates are combined in a round-bottomed flask equipped with a Teflon valve adapter and Teflon magnetic stir bar. The toluene is removed on the vacuum line to yield the crude $(\eta^1\text{-}C_5Me_5)Sn(n\text{-}Bu)_3$ as a slightly viscous yellow-orange liquid. A distillation apparatus is connected to the Schlenk line via a Teflon valve adapter, and another Teflon valve adapter and round-bottomed flask are connected to the collection end of the distillation apparatus. The solution is distilled at 0.005 mmHg. The fraction collected between 115 and 120°C affords the light yellow-orange air-sensitive product, $(\eta^1\text{-}C_5Me_5)Sn(n\text{-}Bu)_3$. Yield: 49.9–56.1 g (78–88%).

TaCl$_5$ (35.0 g, 0.098 mol) is added to a 1 L round-bottomed flask equipped with Teflon-coated stir bar and a Teflon Y-valve adapter for attachment to the vacuum line.[1] Toluene (400 mL) is vacuum transferred onto the TaCl$_5$. On the Schlenk line, under strong Ar purge, the Teflon stopcock is replaced with a septum. $(\eta^1\text{-}C_5Me_5)$Sn$(n\text{-}Bu)_3$ (49.9 g, 0.116 mol) is cannula transferred slowly to the stirred suspension of TaCl$_5$ over a period of 30 min; mild exothermicity ensues. Stirring is continued for a further 20 h to give a copious amount of a yellow suspension. The yellow suspension is filtered through a Schlenk frit via cannula transfer and is washed with dry petroleum ether (3 × 50 mL). The residue is dried in vacuum. Yield of $(\eta^5\text{-}C_5Me_5)TaCl_4$: 43.5 g (97%).

In the dry box, $(\eta^5\text{-}C_5Me_5)TaCl_4$ (43.5 g, 0.095 mol) and magnesium turnings (1.15 g, 0.047 mol) are weighed in a 500 mL round-bottomed flask equipped with Teflon-coated stir bar and a Teflon valve adapter for attachment to the vacuum line. THF (300 mL) is vacuum transferred into the mixture at −78°C, followed by trimethylphosphine (7.60 g, 0.10 mol). The mixture is allowed to warm to room temperature with stirring. The solution rapidly takes on a red coloration, and stirring is continued until all the magnesium turnings are consumed (∼6 h). The volatile material is then removed under reduced pressure, and the crude product is dried in vacuum. Yield of crude $(\eta^5\text{-}C_5Me_5)\text{-}TaCl_3(PMe_3)$: 52 g.[*] $(\eta^5\text{-}C_5Me_5)TaCl_3(PMe_3)$ is obtained as red crystals.

---

[*]For preparation of $(\eta^5\text{-}C_5Me_5)_2TaCl_2$, there is no need to purify the crude $(\eta^5\text{-}C_5Me_5)TaCl_3(PMe_3)$. If an analytically pure sample of $(\eta^5\text{-}C_5Me_5)TaCl_3(PMe_3)$ is desired, however, the crude material in the round-bottomed flask should be dissolved in toluene (200 mL) and a small amount of PMe$_3$ (∼1 mL) added to inhibit irreversible PMe$_3$ dissociation. The suspension is filtered, and the red filtrate is concentrated and cooled to −78°C to afford red crystals of $(\eta^5\text{-}C_5Me_5)TaCl_3(PMe_3)$. *Anal.* Calcd. for $C_{13}H_{24}C_{13}PTa$: C, 31.48; H, 4.82. Found: C, 31.29; H, 4.81.

Its infrared spectrum as a Nujol mull displays prominent bands at 2725 (w), 1307 (w), 1290 (m), 1260 (w), 1025 (m, br), 965 (s), 950 (m), 740 (w), and 732 (w) cm^{-1}.

KC_5Me_5 is prepared by weighing KH (3.0 g, 0.075 mol) in a 250 mL round-bottomed flask equipped with a Teflon-coated magnetic stir bar, a swivel frit, a 250 mL round-bottomed collection flask, and a Teflon valve adapter. THF (125 mL) is vacuum transferred onto the KH. While the mixture is stirred in a −78°C bath, C_5Me_5H (10.22 g, 0.075 mol) is added by syringe through a sidearm under a strong Ar purge. The suspension is allowed to stir in the −78°C bath and slowly warm to room temperature, after which it is stirred for an additional 2 h. The suspension is filtered and washed with THF. The collected solid is dried in vacuum. In the dry box, KC_5Me_5 (13.0 g, 0.075 mol) is weighed in a 500 mL round-bottomed flask along with 40 g of the 52 g of the crude $(\eta^5\text{-}C_5Me_5)TaCl_3(PMe_3)$ intermediate. The flask is equipped with a Teflon valve adapter. PMe_3 (0.5 mL) and toluene (250 mL) are condensed onto the mixture at −78°C. After being allowed to warm to room temperature, the flask is equipped with a condenser, and the mixture is stirred and heated slowly to 100°C. This temperature is maintained for 24 h, after which large quantities of brown crystals are seen. The volatile materials are removed under reduced pressure, and the residue is dried in vacuum. Purification is achieved by toluene Soxhlet extraction under Ar, followed by very slow cooling to afford large dark brown needle crystals of $(\eta^5\text{-}C_5Me_5)_2TaCl_2$. Yield of $(\eta^5\text{-}C_5Me_5)_2TaCl_2$: 25.4 g (50%).

Anal. Calcd. for $C_{20}H_{30}Cl_2Ta$: C, 46.10; H, 5.80. Found: C, 45.98; H, 5.75.

Properties

$(\eta^5\text{-}C_5Me_5)_2TaCl_2$ is obtained as paramagnetic brown crystals. Mass spectrum: *m/z* 521.

B. BIS(η⁵-PENTAMETHYLCYCLOPENTADIENYL)-DICHLORONIOBIUM(IV)

$$Li(n\text{-}Bu) + C_5Me_5H \rightarrow LiC_5Me_5 + n\text{-}BuH$$

$$NbCl_5 + 2LiC_5Me_5 + 3NaBH_4 \rightarrow (\eta^5\text{-}C_5Me_5)_2NbBH_4 + 3NaCl$$

$$+2LiCl + H_2 + B_2H_6$$

$$(\eta^5\text{-}C_5Me_5)_2NbBH_4 + 2HCl + 3H_2O \rightarrow (\eta^5\text{-}C_5Me_5)_2NbCl_2 + \tfrac{9}{2}H_2 + B(OH)_3$$

Procedure[8,9]

■ **Caution.** *n-BuLi and NaBH₄ are air sensitive and should be handled using Schlenk, high-vacuum line,*[7] *and inert atmosphere glove box techniques.*

A two-necked 500 mL round-bottomed flask is equipped with a Teflon valve adapter and a glass stopper. 1,2-Dimethoxyethane (300 mL) is vacuum or cannula transferred into the flask. At −78°C under an argon counterflow, C_5Me_5H (18.15 g, 0.133 mol) and then *n*-BuLi (0.133 mol) are syringe added to the reaction flask and the solution allowed to warm slowly to room temperature. The reaction is stirred at 25°C for 6 h and then recooled to −78°C. $NaBH_4$ (9.0 g, 0.238 mol) is added slowly at −78°C using a solid addition ampule. (An ampule containing the $NaBH_4$ having an 18/9 ball joint sealed at a slight bend with clamped 18/9 socket as cap is prepared in the glove box. A 24/40 standard taper-to-18/9 socket joint adapter is then inserted into one neck of the flask in place of a glass stopper against an Ar counterflow. The cap of the ampule is removed and the ampule is quickly attached to the 18/9 socket joint and clamped in place. By slowly rotating the ampule around the lightly clamped socket joint and tapping with a wooden rod, the $NaBH_4$ solid may be added slowly to the solution.) The solution is cooled to −78°C, and freshly sublimed $NbCl_5$ (16 g, 0.059 mol) is added slowly using a solid addition ampule and with vigorous stirring. After the addition is complete, the addition ampule is replaced against an argon counterflow with a condenser topped with a stopcock and gas inlet attached to a mercury bubbler. The solution is warmed to room temperature and then heated to reflux for 3 days. The condenser is replaced with a glass stopper against an argon counterflow. The volatile materials are removed in vacuum. In the dry box, the glass stopper is replaced with a swivel frit and 250 mL round-bottomed flask. The suspension is filtered and extracted with vacuum-transferred pentane (4 × 100 mL). The maroon or dark blue pentane extracts are combined and concentrated to small volume to yield green crystals. Further purification is achieved by recrystallization from octane or sublimation. Yield of $(\eta^5\text{-}C_5Me_5)_2NbBH_4$: 8.5–11.2 g (37–50%).

Anal. Calcd. for $C_{20}H_{34}BNb$: C, 63.52; H, 9.06; Nb, 24.56; B, 2.86. Found: C, 63.36; H, 8.96; Nb, 25.04; B, 2.92. High-resolution mass spectrum: *m/e* 378.178. ¹H NMR (benzene-d_6): δ 1.67 (s, 30H), 5.17 (d, 2H), −18.2 (s, 2H). IR (Nujol, cm⁻¹): 2716 (w), 2452 (s), 2428 (s), 1728 (w), 1620 (m), 1516 (m), 1483 (s), 1417 (m), 1390 (w), 1171 (s), 1028 (s), 864 (s), 798 (w), 420 (m). Melting point: 239–241°C (dec.).

In a nitrogen-filled glove box, $(\eta^5\text{-}C_5Me_5)_2NbBH_4$ (5.3 g, 0.014 mol) is weighed in a 250 mL round-bottomed flask fitted with a Teflon needle valve. The flask is transferred to the high-vacuum line, and benzene (100 mL) is added by vacuum transfer from a source of benzene stored over sodium benzophenone ketyl. Into a 25 mL round-bottomed flask fitted with a Teflon needle valve, HCl (15 mL of a 3 M aqueous solution) is transferred by a syringe. The flask is transferred to the high-vacuum line. The HCl solution is freeze–pump–thawed three times to remove all oxygen gas. While being kept at 0°C, the HCl solution is *slowly* cannula transferred onto the (η^5-C$_5$Me$_5$)$_2$NbBH$_4$ solution. The solution evolves gas vigorously and turns brown while a yellow precipitate forms. After being stirred for 20 min, all volatile material is removed under reduced pressure. The brown residue is transferred to a Soxhlet extraction apparatus and is Soxhlet extracted with benzene. The solvent is removed under reduced pressure. The brown microcrystalline residue is dissolved in CH$_2$Cl$_2$ (120 mL). The solution is added to degassed grade III alumina (500 mg) to separate (η^5-C$_5$Me$_5$)$_2$NbCl$_2$ from all hydroxy niobium compounds. The solid is recrystallized from toluene. Yield of (η^5-C$_5$Me$_5$)$_2$NbCl$_2$: 4.0 g (66%).

Anal. Calcd. for C$_{20}$H$_{30}$Cl$_2$Nb: C, 55.31; H, 6.96; Nb, 21.39; Cl, 16.33. Found: C, 55.41; H, 7.03; Nb, 21.52; Cl, 16.29.

Properties

(η^5-C$_5$Me$_5$)$_2$NbCl$_2$ is paramagnetic and has an ESR spectrum in benzene at 25°C that consists of a decet due to hyperfine coupling with the $I = 9/2$ niobium nucleus, $g = 2.01$. Its magnetic moment at room temperature is 1.91μ_B.

References

1. G. Parkin, E. Bunel, B. J. Burger, M. S. Trimmer, A. Van Asselt, and J. E. Bercaw, *J. Mol. Catal.* **1–2**, 21 (1987).
2. R. A. Bell, S. A. Cohen, N. M. Doherty, R. S. Threlkel, and J. E. Bercaw, *Organometallics* **5**, 972 (1986).
3. (a) B. J. Burger, B. D. Santarsiero, M. S. Trimmer, and J. E. Bercaw, *J. Am. Chem. Soc.* **110**, 3134 (1988); (b) N. M. Doherty and J. E. Bercaw, *J. Am. Chem. Soc.* **107**, 2670 (1985).
4. A. Van Asselt, B. J. Burger, V. C. Gibson, and J. E. Bercaw, *J. Am. Chem. Soc.* **108**, 5347 (1986).
5. V. C. Gibson, J. E. Bercaw, W. J. Bruton Jr., and R. D. Sanner, *Organometallics* **5**, 976 (1986).
6. R. D. Sanner, S. T. Carter, and W. J. Bruton Jr., *J. Organomet. Chem.* **240**, 157 (1982).
7. B. J. Burger and J. E. Bercaw, in *New Developments in the Synthesis, Manipulation and Characterization of Organometallic Compounds*, A. Wayda and M.Y. Darensbourg, eds., American Chemical Society, Washington DC, 1987, Chapter 4
8. R. S. Threlkel, Ph.D. thesis, California Institute of Technology, 1980.
9. R. A. Bell, S. A. Cohen, N. M. Doherty, R. S. Threlkel, and J. E. Bercaw, *Organometallics* **5**, 972 (1986).

12. BIS(η^5-PENTAMETHYLCYCLOPENTADIENYL) COMPLEXES OF MOLYBDENUM

Submitted by JUN HO SHIN,[*†] **DAVID G. CHURCHILL,**[*‡]
and GERARD PARKIN[*]
Checked by STEFAN ROGGAN,[§] **CHRISTEL JANKOWSKI,**[§]
CHRISTIAN LIMBERG,[§] **IAN TONKS,**[**] **and JOHN BERCAW**[**]

Molybdenocene and tungstenocene complexes have played a prominent role in the development of organometallic chemistry. By comparison to the parent molybdenocene and tungstenocene systems, however, the chemistry of permethylmolybdenocene and permethyltungstenocene complexes has been little explored. With respect to the permethylmolybdenocene system, the first report of the dihydride (η^5-C$_5$Me$_5$)$_2$MoH$_2$ and several other derivatives appeared in 1973,[1] but there were no subsequent reports until 1991 when Cloke synthesized (η^5-C$_5$Me$_5$)$_2$MoH$_2$ by using metal vapor synthesis techniques.[2] The paucity of studies pertaining to (η^5-C$_5$Me$_5$)$_2$MoH$_2$ may be attributed to the inability of other researchers to substantiate the 1973 synthesis of (η^5-C$_5$Me$_5$)$_2$MoH$_2$.[2] In addition, a modified synthetic method for (η^5-C$_5$Me$_5$)$_2$MoH$_2$ was reported to proceed with insufficient yield for it to be useful as a starting material for subsequent derivatization.[3] More recently, reproducible syntheses of (η^5-C$_5$Me$_5$)$_2$MoCl$_2$, (η^5-C$_5$Me$_5$)$_2$MoH$_2$, and a variety of other derivatives have been reported.[4–6] Of these, synthetic details for (η^5-C$_5$Me$_5$)$_2$MoCl$_2$ and (η^5-C$_5$Me$_5$)$_2$MoH$_2$ are described here.

(η^5-C$_5$Me$_5$)$_2$MoCl$_2$ is obtained by a two-step sequence involving (i) the reaction of MoCl$_5$ with a mixture of KC$_5$Me$_5$ and NaBH$_4$ to give crude (η^5-C$_5$Me$_5$)$_2$MoH$_2$, followed by (ii) addition of CHCl$_3$. The latter step is important because the lower solubility of (η^5-C$_5$Me$_5$)$_2$MoCl$_2$ in pentane compared to that of (η^5-C$_5$Me$_5$)$_2$MoH$_2$ facilitates isolation of the organomolybdenum products. Specifically, the greater solubility of (η^5-C$_5$Me$_5$)$_2$MoH$_2$ makes it difficult to separate the complex from oily by-products such as C$_5$Me$_5$H and (C$_5$Me$_5$)$_2$.

(η^5-C$_5$Me$_5$)$_2$MoCl$_2$ is a useful precursor for a variety of other permethylmolybdenocene derivatives, which include hydride, alkyl, carbonyl, and oxo complexes, as illustrated in Scheme 1.[4] For example, the dihydride (η^5-C$_5$Me$_5$)$_2$MoH$_2$ is readily obtained by reaction of (η^5-C$_5$Me$_5$)$_2$MoCl$_2$ with LiAlH$_4$.

[*]Department of Chemistry, Columbia University, New York, NY 10027.
[†]Present address: Department of Chemistry, Queensborough Community College, Bayside, NY 11364.
[‡]Present address: Molecular Logic Gate Laboratory, Department of Chemistry, KAIST, Daejeon 305-701, South Korea.
[§]Institut für Chemie, Humboldt-Universität zu Berlin, D-12489 Berlin, Germany.
[**]Department of Chemistry, California Institute of Technology, Pasadena, CA 91125.

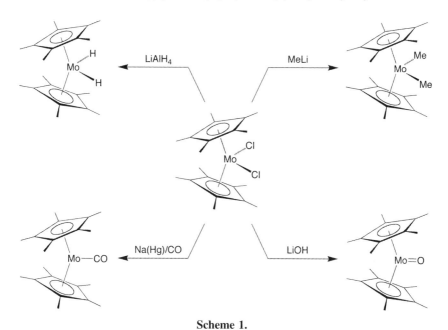

Scheme 1.

General Procedures

All manipulations are performed using a combination of inert atmosphere glove box and Schlenk techniques. Solvents are purified and degassed by standard procedures. All commercially available reagents are used as received without any further purification. Potassium pentamethylcyclopentadienide (KC_5Me_5) is synthesized by a literature route.[7] ^1H and ^{13}C NMR chemical shifts are reported in ppm relative to $SiMe_4$ (δ 0) and are referenced internally to the residual ^1H NMR resonance (δ 7.15 for C_6D_5H) or ^{13}C NMR resonance (δ 128.0 for C_6D_6) of the solvent.

A. BIS(PENTAMETHYLCYCLOPENTADIENYL)- DICHLOROMOLYBDENUM(IV)

$$MoCl_5 \xrightarrow[\text{NaBH}_4]{\text{K(C}_5\text{Me}_5)} \quad \xrightarrow{\text{CHCl}_3}$$

Procedure

■ **Caution.** *Tetrahydrofuran (THF) is flammable and forms explosive peroxides upon prolonged exposure to air. Sodium borohydride (NaBH₄) reacts violently with acids and acidic solutions to form hydrogen, which can form explosive mixtures with air. Chloroform is a suspected carcinogen.*

A sample of $MoCl_5$ (4.02 g, 14.7 mmol) is placed in a Schlenk tube (see, for example, Chemglass AF-0538-23), cooled to −78°C, and treated sequentially with toluene (10 mL) and THF (50 mL) that had been precooled to −78°C. The dark suspension is then added through a wide-bore cannula to a mixture of KC_5Me_5 (15.70 g, 90.06 mmol), $NaBH_4$ (1.49 g, 39.4 mmol), and THF (150 mL) in a 350 mL glass vessel (see, for example, Chemglass AF-0096-04) equipped with a Teflon stopcock at −78°C. The resulting mixture is allowed to warm to room temperature and stirred for 3 h. After this period, the mixture is heated for 24 h at 65°C. The volatile components are removed from the mixture and the residue is dried in vacuum overnight. The residue is extracted into pentane (200 mL) and filtered. The filtrate is concentrated to 100 mL and treated with $CHCl_3$ (5 mL). The mixture is stirred overnight at room temperature and the resulting precipitate is isolated by filtration, washed with pentane (2 × 30 mL), and dried in vacuum to give $(\eta^5\text{-}C_5Me_5)_2MoCl_2$ as a brown solid. Yield: 4.13 g (64%).[*]

Anal. Calcd. for $C_{20}H_{30}Cl_2Mo$: C, 54.9; H, 6.9. Found: C, 54.3; H, 7.2.

Properties

The IR spectrum of $(\eta^5\text{-}C_5Me_5)_2MoCl_2$ as a KBr disk shows peaks at 2959 (s), 2904 (vs), 1456 (vs), 1374 (vs), 1069 (m), 1019 (vs), 804 (w), 679 (w), 609 (w), and 415 (w) cm^{-1}. Its ^1H NMR spectrum in C_6D_6 contains a singlet at δ 1.46, and its ^{13}C NMR spectrum in C_6D_6 shows resonances at δ 11.3 (q, $^1J_{C-H} = 128$, C_5Me_5) and 108.9 (s, C_5Me_5).

---

[*]The checkers obtained yields of 55–61%. They also emphasize the need for using a wide-bore cannula to avoid losses during transfer of the $MoCl_5$/THF suspension, which tends to clump.

B. BIS(PENTAMETHYLCYCLOPENTADIENYL)-DIHYDRIDOMOLYBDENUM(IV)

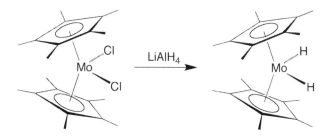

Procedure

■   **Caution.**   *Diethyl ether is flammable. Lithium aluminum hydride (LiAlH$_4$) reacts violently with water, acids, and acidic solutions, and can decompose violently upon heating to produce hydrogen gas, which can form explosive mixtures with air.*

To a stirred suspension of (η^5-C$_5$Me$_5$)$_2$MoCl$_2$ (1.00 g, 2.29 mmol) in Et$_2$O (30 mL) is added LiAlH$_4$ (14 mL of a 1.0 M solution in Et$_2$O, 14 mmol) at −78°C. The mixture is allowed to warm slowly to room temperature and stirred overnight. After this period, the mixture is cooled to 0°C and treated dropwise with degassed water (2 mL). The mixture is allowed to warm to room temperature and stirred for 1 h. The volatile components are removed in vacuum, the residue is extracted into pentane (~100 mL), and the mixture is filtered. The filtrate is concentrated to ~2 mL and cooled to −78°C for ~1 h, thereby depositing (η^5-C$_5$Me$_5$)$_2$MoH$_2$ as a green-brown solid that is isolated by filtration and dried in vacuum. Yield: 0.52 g (62%).[*]

Anal. Calcd. for C$_{20}$H$_{32}$Mo: C, 65.2; H, 8.8. Found: C, 65.3; H, 9.2.

Properties

The IR spectrum of (η^5-C$_5$Me$_5$)$_2$MoH$_2$ as a KBr disk shows bands at 2975 (vs), 2898 (vs), 2712 (m), 1864 (s, ν_{Mo-H}), 1478 (s), 1452 (s), 1426 (s), 1375 (vs), 1069 (m), 1028 (vs), 908 (w), 884 (w), 792 (w), 670 (w), 590 (w), and 414 (w) cm^{-1}. Its ^{1}H NMR spectrum in C$_6$D$_6$ features resonances at δ −8.35 (s, MoH$_2$) and 1.85

---

[*]The checkers obtained a yield of 46–57% in the first crop, but were able to isolate an additional 20% by concentrating and cooling the mother liquors. They also report that, although the green-brown product is analytically and spectroscopically pure, the dihydride can be sublimed at 100°C and 10^{-3} mmHg with minimal loss to afford a yellow solid, which they propose is the true color of this compound.

(s, C_5Me_5), and its ^{13}C NMR in C_6D_6 exhibits peaks at δ 12.2 (q, $^1J_{C-H} = 126$, C_5Me_5) and 92.9 (s, C_5Me_5).

Acknowledgment

The authors thank the U.S. Department of Energy, Office of Basic Energy Sciences (DE-FG02-93ER14339) for support of this research.

References

1. J. L. Thomas, *J. Am. Chem. Soc.* **95**, 1838 (1973).
2. F. G. N. Cloke, J. P. Day, J. C. Green, C. P. Morley, and A. C. Swain, *J. Chem. Soc., Dalton Trans.* 789 (1991).
3. T. Ito and T. Yoden, *Bull. Chem. Soc. Jpn.* **66**, 2365 (1993).
4. J. H. Shin, D. G. Churchill, B. M. Bridgewater, K. L. Pang, and G. Parkin, *Inorg. Chim. Acta* **359**, 2942 (2006).
5. J. H. Shin, B. M. Bridgewater, D. G. Churchill, M.-H. Baik, R. A. Friesner, and G. Parkin, *J. Am. Chem. Soc.* **123**, 10111 (2001).
6. K. E. Janak, J. H. Shin, and G. Parkin, *J. Am. Chem. Soc.* **126**, 13054 (2004).
7. V. C. Gibson, J. E. Bercaw, W. J. Bruton Jr., and R. D. Sanner, *Organometallics* **5**, 976 (1986).

13. (η^5-CYCLOPENTADIENYL)TRICARBONYLMANGANESE(I) COMPLEXES

Submitted by PATRICK N. KIRK[*] and MICHAEL P. CASTELLANI[*]
Checked by MICHAEL RIZZO[†] and JIM D. ATWOOD[†]

Since the initial preparation of (η^5-C_5H_5)Mn(CO)$_3$ by Fischer and Jira in 1954,[1] many methods have been developed for the synthesis of this compound (often referred to as "cymantrene") and its C_5 substituted derivatives.[2] Interestingly, each of these preparations has at least one significant drawback from a preparative perspective (e.g., thallium reagents, high pressures, poor yields, or reactants that require significant preparative investment).

Surprisingly missing from this collection of preparative methods is the direct reaction between commercially available Mn(CO)$_5$Br and NaC$_5$H$_5$. This omission suggests that the reaction proceeds in poor yields, which is consistent with our experience. In 1988, however, Smart and coworkers found MnBr(CO)$_3$(pyridine)$_2$[3] to be an effective reagent in making a substituted

[*]Department of Chemistry, Marshall University, Huntington, WV 25755.
[†]Department of Chemistry, University of Buffalo, Buffalo, NY 14260.

indacenylmanganese tricarbonyl complex.[4] Here we take advantage of the fact that this pyridine complex also serves as an effective manganese starting material for the high-yield syntheses of cymantrene and its derivatives such as $(\eta^5\text{-}C_5Me_5)Mn(CO)_3$.

A. (η[5]-CYCLOPENTADIENYL)TRICARBONYLMANGANESE(I)

$$MnBr(CO)_3(py)_2 + Na(C_5H_5) \cdot DME \rightarrow (\eta^5\text{-}C_5H_5)Mn(CO)_3 + 2py$$

$$+ NaBr + DME$$

Procedure

A Schlenk flask equipped with a magnetic stir bar is charged under an inert atmosphere, with $MnBr(CO)_3(pyridine)_2$[5] (0.400 g, 1.1 mmol) and $Na(C_5H_5) \cdot$ DME[6] (0.230 g, 1.3 mmol), where DME = 1,2-dimethoxyethane. Dry, deoxygenated THF (10 mL) is added and the flask is stoppered. After the mixture has been stirred overnight, the THF is removed in vacuum, being careful to avoid loss of the volatile product by placing the residue under a nitrogen atmosphere as soon as the mixture is dry. Against a counterstream of nitrogen, a water-cooled cold finger is inserted into the flask, which is then sealed and evacuated. The flask is warmed in a 50°C water bath for 2 h. The flask is refilled with nitrogen, and the $CpMn(CO)_3$ product can be collected from the cold finger in the air. Yield: 0.165 g (73%).[*]

Properties

$(\eta^5\text{-}Cyclopentadienyl)tricarbonylmanganese(I)$ is a yellow, air-stable solid that sublimes at room temperature. It dissolves in many nonpolar and moderately polar solvents (e.g., hexane, toluene, CH_2Cl_2, and acetone). The solution IR spectrum displays $\nu(C\equiv O)$ bands at 2024 (m, sh) and 1937 (s, br) cm[-1] in dichloromethane or 2029 (m, sh) and 1947 (s, br) cm[-1] in hexane.[7] The ^{1}H NMR spectrum in $CDCl_3$ displays a singlet at δ 4.75.

B. (η[5]-PENTAMETHYLCYCLOPENTADIENYL)- TRICARBONYLMANGANESE(I)

$$MnBr(CO)_3(py)_2 + LiC_5Me_5 \rightarrow (\eta^5\text{-}C_5Me_5)Mn(CO)_3 + 2py + LiBr$$

---

[*]The checkers carried out the sublimation by transferring the dried reaction residue to a separate sublimator. This extra step probably was responsible for the checkers obtaining a lower yield of 51%.

Procedure

A Schlenk flask equipped with a magnetic stir bar is charged under an inert atmosphere with $MnBr(CO)_3(pyridine)_2$[5] (0.400 g, 1.1 mmol) and LiC_5Me_5[8] (0.18 g, 1.3 mmol). Dry, deoxygenated THF (10 mL) is added and the flask is stoppered. After the mixture has been stirred overnight, the THF is removed in vacuum. Against a counterstream of nitrogen, a water-cooled cold finger is inserted into the flask, which is then sealed and evacuated. The flask is warmed in a 90–95°C water bath for 3 h. The flask is refilled with nitrogen, and the $(\eta^5\text{-}C_5Me_5)Mn(CO)_3$ is collected from the cold finger in the air. Yield: 0.158 g (52%).

Properties

$(\eta^5$-Pentamethylcyclopentadienyl)tricarbonylmanganese(I) is a yellow, air-stable solid that sublimes at elevated temperatures. It dissolves in many nonpolar and moderately polar solvents (e.g., hexane, toluene, CH_2Cl_2, and acetone). The solution IR spectrum in dichloromethane displays $\nu(C\equiv O)$ bands at 2002 (m, sh) and 1913 (s, br) cm^{-1}. The ^{1}H NMR spectrum in CDCl$_3$ displays a singlet at δ 1.88.

Acknowledgment

P.N.K. thanks the West Virginia EPSCoR—Research Challenge Fund—for support in the form of a SURE Summer Fellowship.

References

1. E. O. Fischer and R. Jira, *Z. Naturforsch. B* **9**, 618 (1954).
2. (a) P. M. Treichel, in *Comprehensive Organometallic Chemistry*, G. Wilkinson, F. G. A. Stone, and E. W. Abel, eds., Pergamon Press, Oxford, 1982, Vol. 4, pp. 123–132; (b) P. M. Treichel, in *Comprehensive Organometallic Chemistry II*, E. W. Abel, F. G. A. Stone, and G. Wilkinson, eds., Pergamon Press, Oxford, 1995, Vol. 6, pp. 109–115.
3. G. E. Herberich, U. Englert, B. Ganter, and M. Pons, *Eur. J. Inorg. Chem.* 979 (2000).
4. W. L. Bell, C. J. Curtis, A. Miedaner, C. W. Eigenbrot Jr., R. C. Haltiwanger, C. G. Pierpont, and J. C. Smart, *Organometallics* **7**, 691 (1988).
5. M. Pons and G. E. Herberich, *Inorg. Synth.*, **36**, 148 (2014).
6. J. C. Smart and C. J. Curtis, *Inorg. Chem.* **16**, 1788 (1977).
7. D. J. Parker and M. H. B. Stiddard, *J. Chem. Soc. A* 2263 (1968).
8. J. M. Manriquez, E. E. Bunel, and B. Oelkers, *Inorg. Synth.*, **31**, 214 (1997).

14. 1,1'-DIAMINOFERROCENE

Submitted by ALEXANDR SHAFIR,[*] MAURICE P. POWER,[*]
GLENN D. WHITENER,[*] and JOHN ARNOLD[*]
Checked by PHILIP W. MILLER[†] and NICHOLAS J. LONG[†]

Diamines are widely used in coordination chemistry as chelating ligands and as precursors to a variety of other ligand systems.[1] The unique structural properties and presence of a redox-active iron(II) center in 1,1'-diaminoferrocene set it apart from common organic diamines, rendering it of interest as an electrochemically active ligand. Related ferrocene diphosphine ligands[2–9] are widely used in catalysis,[10] and substituted ferrocenyl amines have also been extensively studied.[11–16] As part of our own work on anionic N-donor ligands, we set out to prepare and study transition metal complexes supported by the dianionic ferrocenediamide ligands. At the outset of this work in 1999, we were surprised to find little mention of the parent 1,1'-diaminoferrocene.[17] At that time, one of the reasons its chemistry had not been examined in more detail was that simple routes to the compound were not available. Interestingly, 1,1'-diaminoferrocene had been prepared in situ as far back as 1961 by Knox and Pauson via reduction of 1,1'-diphenylazoferrocene and had been characterized as a urethane derivative.[18] Two years later, Nesmeyanov et al. reported the synthesis and isolation of small quantities of 1,1'-diaminoferrocene by reduction of 1,1'-diazidoferrocene.[19] Aiming to overcome these limitations, we described in 2000 an improved high-yield synthesis of diaminoferrocene from 1,1'-diazidoferrocene (a modified Nesmeyanov procedure) and chemical oxidation of 1,1'-diaminoferrocene to the isolable ferrocenium cation.[20] Since then, 1,1'-diaminoferrocene and its derivatives have found use in applications such as molecular recognition[21] and synthesis of ferrocene-containing peptides,[22] as well as catalysis in transfer hydrogenation,[23] olefin polymerization,[24] and cross-coupling,[25] to name a few.[26] In addition, diaminoferrocene-based chelating ligands have been used to prepare a range of main group ferrocenophanes (with Ga, Sn, Al, B, and P), and a variety of d- and f-metal complexes.[27] As part of our own work, we synthesized a series of group 4 metal complexes that are olefin polymerization catalysts; in addition, one of the activated cationic diamine-Ti species contains a short Fe–Ti interaction.[28]

Owing to the usefulness of 1,1'-diaminoferrocene as a building block, we report here its convenient multigram synthesis by a modified Nesmeyanov protocol. In addition, 1,1'-dibromoferrocene, the first intermediate in the process, is a valuable precursor in its own right, and we also provide a convenient route to multigram

[*]Department of Chemistry, University of California, Berkeley, CA 94720.
[†]Department of Chemistry, Imperial College of Science and Technology, London SW7 2AZ, UK.

quantities of this substance by an adaptation of an earlier method.[29] Finally, the preparation and isolation of two 1,1'-diaminoferrocenium salts are reported.

General Procedures

Standard Schlenk line and glove box techniques are used unless otherwise indicated. CH_3CN is distilled from CaH_2 under a nitrogen atmosphere. All other solvents are passed through a column of activated alumina and degassed with argon before use. $[Bu_4N][PF_6]$ is purchased from Aldrich, recrystallized from EtOH, and dried under vacuum for 18 h at room temperature. N,N,N',N'-Tetramethylethylenediamine (TMEDA) is purchased from Aldrich and distilled from Na metal. All other chemicals are purchased from Aldrich (ferrocene, NaN_3, AgOTf, $[Cp_2Fe][PF_6]$, tetracyanoethylene), Kodak ($Br_2HCCHBr_2$), Alfa (*n*-butyllithium), and Fisher (CuCl) and used as received. Tetracyanoethylene is sublimed before use. C_6D_6 is vacuum transferred from sodium/benzophenone. Melting points are determined in sealed capillary tubes under nitrogen. ^1H NMR spectra are recorded at ambient temperature on a Bruker AM-300 or DRX-500 spectrometer. ^1H NMR chemical shifts are given relative to C_6D_5H (δ 7.16) or $CHCl_3$ (δ 7.26). Magnetic susceptibilities are determined using the Evans NMR method with CD_3CN as solvent.[30] IR samples are prepared as Nujol mulls and taken between KBr plates.

A. 1,1'-DILITHIOFERROCENE *N,N,N',N'*-TETRAMETHYL-ETHYLENEDIAMINE

$$Fe(C_5H_5)_2 + 2Li(n\text{-Bu}) + \tfrac{2}{3} \text{TMEDA} \rightarrow Fe(C_5H_4Li)_2 \cdot \tfrac{2}{3}\text{TMEDA} + 2C_4H_{10}$$

Procedure[31]

An oven-dried 2 L three-necked round-bottomed flask is equipped with a mechanical stirrer, a N_2 inlet, and a 500 mL addition funnel. Ferrocene (85.6 g, 0.46 mol) is added and the flask is evacuated for 30 min, and then purged with N_2. Dry pentane (800 mL) is added to the flask. The addition funnel is charged with *n*-butyllithium (385 mL of a 2.6 M solution in hexane, 1.0 mol), TMEDA (155 mL, 1.0 mol), and pentane (50 mL). The Li(*n*-Bu)–TMEDA complex is allowed to form for 5 min and the resulting cloudy mixture is added dropwise to the stirring suspension of ferrocene. When the addition is complete, the mixture is stirred for an additional 14 h, over which time a bright yellow precipitate forms. This solid is collected on a Schlenk-type fritted glass filter and is washed with hexane (3×50 mL) to remove ferrocene. The resulting bright yellow powder is dried in vacuum for 5 h. Yield: 103 g (81% based on the formulation of the product as $Fe(C_5H_4Li)_2 \cdot \tfrac{2}{3}$TMEDA).

Properties

The dilithioferrocene–TMEDA complex is a bright yellow powder with very low solubility in toluene, Et_2O, and THF. The product can be stored for extended periods at room temperature under a nitrogen atmosphere. In our hands, no decomposition is observed after 12 months in a glove box. The complex is highly pyrophoric when exposed to air, however, and must be handled with care.

B. 1,1'-DIBROMOFERROCENE

$$Fe(C_5H_4Li)_2 \cdot \tfrac{2}{3}TMEDA + 2C_2H_2Br_4 \rightarrow Fe(C_5H_4Br)_2 + 2LiBr$$

$$+ 2C_2H_2Br_2 + \tfrac{2}{3}TMEDA$$

Procedure[29]

An oven-dried 2 L three-necked round-bottomed flask equipped with a mechanical stirrer, a 500 mL addition funnel, and a N_2 gas inlet is charged with Fe $(C_5H_4Li)_2 \cdot \tfrac{2}{3}$TMEDA (100 g, 0.36 mol). Et_2O (1 L) is added* and the resulting yellow suspension is cooled to $-78°C$ in a dry ice/acetone bath. The addition funnel is charged with 1,1,2,2-tetrabromoethane (93 mL, 0.8 mol) and Et_2O (350 mL), and the resulting solution is slowly added to the dilithioferrocene suspension with vigorous stirring over 6 h. Care should be taken to ensure that the flask is immersed in the cooling mixture and that copious amounts of dry ice are present throughout the addition. Failure to do so or the use of a higher-temperature dry ice/isopropanol mixture typically leads to a less efficient reaction.

The solution is allowed to warm slowly to ambient temperature, stirred for an additional 10 h, and the resulting dark red solution is hydrolyzed with H_2O (50 mL). From this point, the workup is done in air. The mixture is stirred for an additional 10 min, during which time a layer of black oil forms at the bottom of the flask. The red upper Et_2O layer is decanted off and the remaining black oil is extracted with additional Et_2O (2×300 mL). The Et_2O fractions are combined and the solvent is removed by rotary evaporation. The residue is dried under vacuum ($\sim 10^{-2}$ mmHg) at 50°C for 5 h to remove most of the TMEDA. This drying step is necessary to ensure complete removal of the LiBr in the following step. The resulting red sticky solid is extracted with Et_2O (300 mL) and the solution is filtered to remove the LiBr by-product. The solvent is once again removed by rotary evaporation to yield the crude product as red oil. The oil is redissolved in hot MeOH (150 mL). Cooling the solution slowly to room temperature and then storing it at $-30°C$ overnight results in large red crystals of 1,1'-dibromoferrocene. Yield: 89 g (71%).

---

*The checkers report that they dissolve the $Fe(C_5H_4Li)_2 \cdot \tfrac{2}{3}$TMEDA in freshly distilled 1,2-dimethoxy-ethane, rather than using it as a suspension in Et_2O. This change often increases the yields and enables easier workup.

Properties

1,1'-Dibromoferrocene is a red, crystalline, air-stable solid that is soluble in a wide range of polar and nonpolar organic solvents. ^{1}H NMR (C_6D_6): δ 4.16 (t, 4H), 3.71 (t, 4H). ^{1}H NMR ($CDCl_3$): δ 4.42 (t, 4H), 4.17 (t, 4H).

C. ONE-POT PREPARATION OF 1,1'-DIBROMOFERROCENE FROM FERROCENE

$$Fe(C_5H_5)_2 + 2Li(n\text{-}Bu) + \tfrac{2}{3}TMEDA \rightarrow Fe(C_5H_4Li)_2 \cdot \tfrac{2}{3}TMEDA + 2C_4H_{10}$$

$$Fe(C_5H_4Li)_2 \cdot \tfrac{2}{3}TMEDA + 2C_2H_2Br_4 \rightarrow Fe(C_5H_4Br)_2 + 2LiBr + 2C_2H_2Br_2 + \tfrac{2}{3}TMEDA$$

Procedure

An oven-dried 2 L three-necked round-bottomed flask is equipped with a mechanical stirrer, a N_2 inlet, and a 250 mL addition funnel. Ferrocene (44.8 g, 240 mmol) is added to the flask, which is then flushed with N_2 for 20 min. Dry pentane (1.0 L) is added to the flask, and the addition funnel is charged with *n*-butyllithium (227 mL of a 2.6 M solution in hexane, 590 mmol), TMEDA (89 mL, 590 mmol), and pentane (100 mL). The Li(*n*-Bu)–TMEDA complex is allowed to form for 5 min and the resulting cloudy mixture is added dropwise to the stirring suspension of ferrocene. After the addition is complete, the mixture is stirred for an additional 14 h, over which time a bright yellow precipitate of ferrocenyllithium forms. This solid is allowed to settle and the red supernatant solution is decanted with a cannula and discarded. Dry Et_2O (1.0 L) is then added to the remaining solid, and the flask is fitted with a 100 mL addition funnel. The funnel is charged with 1,1,2,2-tetrabromoethane (62 mL, 530 mmol) and the flask is cooled to $-78°C$ in a dry ice/acetone bath. The tetrabromoethane is added dropwise to the ferrocenyllithium suspension with vigorous stirring over a period of 6 h. Care should be taken to ensure that the flask is immersed in the cooling mixture and that copious amounts of dry ice are present throughout the addition. Failure to do so typically leads to a lower yield.

The solution is allowed to warm slowly to ambient temperature, and then is stirred for an additional 12 h. The resulting dark red solution is carefully hydrolyzed with H_2O (100 mL). From this point, the workup is done in air. The mixture is stirred for an additional 10 min, during which time a layer of black oil forms at the bottom of the flask. The red upper Et_2O layer is decanted off and the remaining black oil is extracted with additional Et_2O (2×300 mL). The Et_2O extracts are combined and the solvent is removed by rotary evaporation. The residue is dried in vacuum ($\sim 10^{-2}$ mmHg) at 50°C for 5 h to remove most of TMEDA. The resulting red sticky solid is extracted with Et_2O (200 mL), and the extract is filtered, leaving behind the LiBr by-product. The solvent is again removed by rotary evaporation to yield the crude product as red oil. The oil is crystallized from hot MeOH (200 mL) by first cooling the solution to room temperature and then to −30°C. Yield: 51 g (62%).

D. 1,1′-DIAMINOFERROCENE

$$Fe(C_5H_4Br)_2 + 2CuCl + 2NaN_3 \rightarrow Fe(C_5H_4N_3)_2 + 2CuBr + 2NaCl$$

$$Fe(C_5H_4N_3)_2 + 2H_2 \xrightarrow{Pd/C} Fe(C_5H_4NH_2)_2 + 2N_2$$

Procedure

■ **Caution.** *Although we have carried out this preparation without incident several times, copper azides, which are explosive, are intermediates in this reaction. This preparation should be conducted behind a safety shield. Pure 1,1′-diazidoferrocene has been reported to explode when heated rapidly above its melting point of 56°C.[20] We strongly recommend carrying out the hydrogenation of 1,1′-diazidoferrocene without isolating it from its ether solutions.*

A 2 L two-necked flask equipped with a mechanical stirrer is charged with FcBr$_2$ (45.0 g, 130 mmol), CuCl (29.2 g, 295 mmol), and 95% EtOH (1 L). A solution of NaN$_3$ (42.6 g, 655 mmol) in H$_2$O (200 mL) is added, resulting in the formation of a dark brown suspension. Nitrogen is blown over the mixture for 10 min, the reaction flask is wrapped in foil, and stirring is continued for 48 h under an N$_2$ atmosphere. The workup is performed in air. The reaction mixture is filtered to remove most of the solid, and the cloudy filtrate is partitioned between water (1 L) and Et$_2$O (700 mL) in a large separatory funnel. The organic layer is separated, and the aqueous layer is extracted with additional Et$_2$O (2 × 150 mL). The combined organic extracts are washed with water (2 × 100 mL), dried over MgSO$_4$, and concentrated to 150 mL. This solution of 1,1′-diazidoferrocene is used *immediately* in the next step.

The red solution of 1,1′-diazidoferrocene is transferred into a 2 L round-bottomed flask. Methanol (600 mL) is added, followed by 0.1 g of Pd/C (5% Pd). By means of a cannula, H$_2$ gas is bubbled through the solution at a moderate rate for 7–11 h.[*] The cannula is checked periodically to make sure that product deposition does not interfere with the gas flow. After the hydrogenation is complete (as gauged by TLC), the mixture is filtered and the filtrate is concentrated to ∼300 mL and cooled to −40°C. After 12 h, yellow crystalline solid weighing

---

[*]The long reaction times are especially necessary on large scales to ensure complete reduction of the azide.

14.0 g is isolated. Concentration of the mother liquor affords a second crop weighing 2.47 g. The solid is dried in vacuum overnight. Yield: 16.27 g (57%).

Anal. Calcd. for $C_{10}H_{12}N_2Fe$: C, 55.59; H, 5.60; N, 12.97. Found: C, 55.75; H, 5.76; N, 12.95.

Properties

1,1'-Diaminoferrocene is a yellow crystalline solid that is sparingly soluble in nonpolar hydrocarbons, and quite soluble in ethereal or aromatic solvents such as tetrahydrofuran or toluene. The compound is oxygen sensitive in solution, as evidenced by the appearance of a dark green color due to the Fe^{III} species $[Fe(C_5H_4NH_2)_2]^+$. The solid is stable in air for extended periods at room temperature, but is best stored in an inert atmosphere. 1,1'-Diaminoferrocene melts at 183–186°C and can be sublimed at 90°C/0.001 mmHg. ^1H NMR (C_6D_6): δ 3.74 (t, 4H), 3.65 (t, 4H), 1.87 (s, br, 4H). Melting point: 183–186°C. IR (Nujol, cm^{-1}): 3344 (s), 3266 (m), 3171 (m), 3098 (w), 3081 (w), 2954 (s), 2924 (s), 2854 (s), 1608 (m), 1485 (s), 1463 (s), 1377 (m), 1242 (m), 1039 (m), 1027 (m).

E. 1,1'-DIAMINOFERROCENIUM HEXAFLUOROPHOSPHATE

$$Fe(C_5H_4NH_2)_2 + [Cp_2Fe]PF_6 \rightarrow [Fe(C_5H_4NH_2)_2][PF_6] + Cp_2Fe$$

Procedure

A 50 mL Schlenk flask is charged with diaminoferrocene (150 mg, 0.690 mmol), $[Cp_2Fe]PF_6$ (229 mg, 0.690 mmol), and $CHCl_3$ (30 mL). The green suspension is stirred for 1 h. The solvent is removed under reduced pressure and the dark solid is washed with Et_2O (2 × 20 mL) to remove ferrocene. The remaining green solid is dried in vacuum to give the product as a dark green powder. Yield: 210 mg (84%). The product can be recrystallized from concentrated CH_2Cl_2 by addition of Et_2O.

Anal. Calcd. for $C_{10}H_{12}N_2F_6FeP$: C, 32.68; H, 3.44; N, 7.70. Found: C, 33.27; H, 3.35; N, 7.76.

Properties

$[Fe(C_5H_4NH_2)_2][PF_6]$ is a dark green crystalline solid that can be stored indefinitely in the air at room temperature. It is soluble in polar organic solvents such as CH_2Cl_2, from which it can be recrystallized by addition of Et_2O. The solid is paramagnetic with $\mu_{eff} = 2.1\mu_B$ in CD_3CN at 25°C.

F. 1,1′-DIAMINOFERROCENIUM TRIFLATE

$$Fe(C_5H_4NH_2)_2 + AgO_3SCF_3 \rightarrow [Fe(C_5H_4NH_2)_2][O_3SCF_3]$$

Procedure

A flask is charged with 1,1′-diaminoferrocene (350 mg, 1.62 mmol) and silver triflate (AgOTf) (385 mg, 1.50 mmol). CH_3CN (5 mL) is added and the resulting green solution is stirred for 1 h. The solvent is removed under reduced pressure and the green residue is washed with Et_2O (20 mL). The residue is extracted with CH_3CN (4 mL), and the extract is filtered, leaving behind silver nuggets. The solvent is removed under reduced pressure to afford the green product. Yield: 450 mg (82% based on AgOTf). The product can be recrystallized from concentrated CH_2Cl_2 by addition of Et_2O.

Anal. Calcd. for $C_{11}H_{12}N_2FeO_3S$: C, 36.18; H, 3.31; N, 7.67. Found: C, 36.32; H, 3.45; N, 7.55.

Properties

$[Fe(C_5H_4NH_2)_2][OTf]$ is a dark green crystalline solid that can be stored indefinitely in the air at room temperature. It is insoluble in nonpolar hydrocarbon solvents and is soluble in polar halogenated solvents such as CH_2Cl_2. The solid is paramagnetic with $\mu_{eff} = 2.1\mu_B$ in CD_3CN at 25°C. ^1H NMR (CD_3CN): δ 40.6 (s, br, 4H), 22.7 (s, br, 4H), −38.6 (s, br, 4H).

References

1. F. A. Cotton and G. Wilkinson, *Advanced Inorganic Chemistry*, 5th ed., Wiley, New York, 1988.
2. K. H. Shaughnessy, P. Kim, and J. F. Hartwig, *J. Am. Chem. Soc.* **121**, 2123 (1999).
3. O. Riant, O. Samuel, T. Flessner, S. Taudien, and H. B. Kagan, *J. Org. Chem.* **62**, 6733 (1997).
4. J. F. Hartwig, *Synlett* 329 (1997).
5. H. C. L. Abbenhuis, U. Burckhardt, V. Gramlich, C. Koellner, P. S. Pregosin, R. Salzmann, and A. Togni, *Organometallics* **14**, 759 (1995).
6. T. Hayashi, A. Ohno, S.-J. Lu, Y. Matsumoto, E. Fukuyo, and K. Yanagi, *J. Am. Chem. Soc.* **116**, 4221 (1994).
7. T. Katayama, and M. Umeno, *Chem. Lett.* 2073 (1991).
8. W. R. Cullen, T. J. Kim, F. W. B. Einstein, and T. Jones, *Organometallics* **4**, 346 (1985).
9. (a) J. D. Unruh and J. R. Christenson, *J. Mol. Catal.* **14**, 19 (1982); (b) N. J. Goodwin, W. Henderson, B. K. Nicholson, J. Fawcett, and D. R. Russell, *J. Chem. Soc., Dalton Trans.* 1785 (1999).
10. (a) X. L. Hou, S. L. You, T. Tu, W. P. Deng, X. W. Wu, M. Li, K. Yuan, T. Z. Zhang, and L. X. Dai, *Top. Catal.* **35**, 87 (2005); (b) P. Barbaro, C. Bianchini, G. Giambastiani, and S. L. Parisel, *Coord. Chem. Rev.* **248**, 2131 (2004); (c) T. J. Colacot, *Chem. Rev.* **103**, 3101 (2003);

(d) R. C. J. Atkinson, V. C. Gibson, and N. J. Long, *Chem. Soc. Rev.* **33**, 313 (2004); (e) P. Štěpnička, ed., *Ferrocenes: Ligands, Materials and Biomolecules*, Wiley, New York, 2008.

11. K.-P. Stahl, G. Boche, and W. Massa, *J. Organomet. Chem.* **277**, 113 (1984).

12. H. Plenio and D. Burth, *J. Organomet. Chem.* **519**, 269 (1996).

13. H. Plenio and D. Burth, *Organometallics* **15**, 4054 (1996).

14. H. Plenio and D. Burth, *Angew. Chem., Int. Ed. Engl.* **34**, 800 (1995).

15. H. Plenio, J. J. Yang, R. Diodone, and J. Heinze, *Inorg. Chem.* **33**, 4098 (1994).

16. H. Plenio, D. Burth, and P. Gockel, *Chem. Ber.* **126**, 2585 (1993).

17. A. Togni and R. L. Halterman, *Metallocenes*, Wiley-VCH, Weinheim, 1998, Vol. 2.

18. G. R. Knox and P. L. Pauson, *J. Chem. Soc.* 4615 (1961).

19. A. N. Nesmeyanov, V. N. Drozd, and V. A. Sazonova, *Dokl. Akad. Nauk SSSR* **150**, 321 (1963).

20. A. Shafir, M. P. Power, G. D. Whitener, and J. Arnold, *Organometallics* **19**, 3978 (2000).

21. P. D. Beer, J. J. Davis, D. A. Drillsma-Milgrom, and F. Szemes, *Chem. Commun.* 1716 (2002).

22. L. Barišić, M. Čakić, K. A. Mahmoud, Y.-N. Liu, H. B. Kraatz, H. Pritzkow, S. I. Kirin, N. Metzler-Nolte, and V. Rapić, *Chem. Eur. J.* **12**, 4965 (2006).

23. A. G. Elliott, A. G. Green, and P. L. Diaconescu, *Dalton Trans.* **41**, 7852 (2012).

24. V. C. Gibson, C. M. Halliwell, N. J. Long, P. J. Oxford, A. M. Smith, A. J. P. White, and D. J. Williams, *Dalton Trans.* 918 (2003).

25. Z. Weng, L. L. Koh, and T. S. A. Hor, *J. Organomet. Chem.* **689**, 18 (2004).

26. For a Highlight article, see: M. Herberhold, *Angew. Chem., Int. Ed.* **41**, 956 (2002).

27. For selected recent examples, see: (a) B. Wrackmeyer, E. V. Klimkina, and W. Milius, *Z. Anorg. Allg. Chem.* **635**, 2532 (2009); (b) J. S. Mugridge, D. Fiedler, and K. N. Raymond, *J. Coord. Chem.* **63**, 2779 (2010); (c) M. J. Monreal and P. L. Diaconescu, *J. Am. Chem. Soc.* **132**, 7676, (2010).

28. (a) A. Shafir and J. Arnold, *J. Am. Chem. Soc.* **123**, 9212 (2001); (b) A. Shafir and J. Arnold, *Organometallics* **22**, 567 (2003).

29. R. F. Kovar, M. D. Rausch, and H. Rosenberg, *Organomet. Chem. Synth.* **1**, 173 (1970/1971).

30. D. F. Evans, *J. Chem. Soc.* 2005 (1959).

31. J. J. Bishop, A. Davison, M. L. Katcher, D. W. Lichtenberg, R. E. Merrill, and J. C. Smart, *J. Organomet. Chem.* **27**, 241 (1971).

15. MONO(η^5-PENTAMETHYLCYCLOPENTADIENYL) COMPLEXES OF OSMIUM

Submitted by CHRISTOPHER L. GROSS,[*] JULIA L. BRUMAGHIM,[*] JESSE M. JEFFERIS,[*] PAUL W. DICKINSON,[*] and GREGORY S. GIROLAMI[*]
Checked by CHRISTOPHER W. GRIBBLE[†] and T. DON TILLEY[†]

The chemistry of transition metal mono(cyclopentadienyl) molecules of the type (η^5-C_5R_5)MX$_n$, where X is a halide ligand, has been widely explored.[1] Such complexes can serve as catalysts or catalyst precursors for a wide variety of organic reactions ranging from the Ziegler–Natta polymerization of olefins to the

[*]Department of Chemistry, University of Illinois at Urbana-Champaign, Urbana, IL 61801.
[†]Department of Chemistry, University of California at Berkeley, Berkeley, CA 94920.

hydrogenation of unsaturated hydrocarbons. These "half-sandwich" molecules are also useful starting materials for the synthesis of other cyclopentadienyl-containing species: they offer unparalleled opportunities to synthesize a wide variety of compounds that otherwise would be unobtainable. For example, the synthesis of $(\eta^5\text{-}C_5Me_5)_2Ru_2Cl_4$ from $RuCl_3 \cdot xH_2O$ and C_5Me_5H, which was discovered independently by two groups in 1984,[2,3] led to an explosion of interest in $(\eta^5\text{-}C_5Me_5)Ru$ chemistry.

By 1994, halide complexes of the general formula $(\eta^5\text{-}C_5R_5)MX_n$ were known for every transition element except osmium and the radioactive element technetium. In part for this reason, compounds containing the $(\eta^5\text{-}C_5Me_5)Os$ fragment were few relative to the number of $(\eta^5\text{-}C_5Me_5)M$ derivatives of other transition metals. Unfortunately, attempts to develop a simple preparation of the chloroosmium analog of $(\eta^5\text{-}C_5Me_5)_2Ru_2Cl_4$ have to date been unsuccessful. For example, the reaction of $OsCl_3$ with C_5Me_5H does not afford $(\eta^5\text{-}C_5Me_5)_2Os_2Cl_4$, but instead gives the metallocene derivative $[(\eta^5\text{-}C_5Me_5)_2OsCl]_2[OsCl_6]$.[4] A procedure to $(\eta^5\text{-}C_5R_5)OsX_n$ complexes starting from OsO_4 would be very convenient, but "one-pot" approaches, such as treatment of osmium tetroxide with hydrochloric acid followed by the addition of pentamethylcyclopentadiene or the tin reagent $(\eta^1\text{-}C_5Me_5)Sn(n\text{-}Bu)_3$, unfortunately do not afford $(\eta^5\text{-}C_5Me_5)OsCl_n$ products. Two-step approaches involving chloroosmium species prepared from OsO_4 were also explored without success. For example, the osmium complexes $[Os(cod)Cl_2]_n$, $OsCl_4$, and $(NH_4)_2OsCl_6$ can be obtained from OsO_4 in high yield, but no $(\eta^5\text{-}C_5Me_5)OsCl_n$ species could be isolated upon treatment of these materials with C_5Me_5H or $(\eta^1\text{-}C_5Me_5)Sn(n\text{-}Bu)_3$ in a variety of solvents. In contrast, for example, the reaction of Na_2OsCl_6 with C_5Me_5H in ethanol affords decamethylosmocene in high yield.[5]

The unsuccessful attempts to prepare a mono(C_5Me_5) osmium complex from chloroosmium starting materials prompted us to explore the use of bromoosmium reagents instead. Bromoosmic acid (H_2OsBr_6) is readily obtainable by heating OsO_4 in concentrated hydrobromic acid. Treatment of bromoosmic acid with C_5Me_5H in refluxing *tert*-butyl alcohol produces a brown microcrystalline precipitate of the osmium(III) dimer $(\eta^5\text{-}C_5Me_5)_2Os_2Br_4$.[6] Since its discovery, this compound has become a very useful starting material for the synthesis of other mono(C_5Me_5) osmium complexes. Its conversion to the 1,5-cyclooctadiene complex $(\eta^5\text{-}C_5Me_5)Os(\eta^2,\eta^2\text{-}C_8H_{12})Br$ is also described here.

General Procedures

Except where noted, all procedures were carried out under a dry argon atmosphere using Schlenk and cannula techniques.[*] Concentrated hydrobromic acid is

---

[*]The checkers used nitrogen instead of argon as an inert gas.

purchased from Aldrich and used as received. The *tert*-butyl alcohol (Aldrich) is dried over anhydrous $MgSO_4$, filtered, and distilled under nitrogen from magnesium turnings.[*] Ethanol is distilled from magnesium turnings under nitrogen. The pentamethylcyclopentadiene and 1,5-cyclooctadiene (Aldrich) are degassed by three successive freeze–pump–thaw cycles under nitrogen before use.

A. BROMOOSMIC ACID

$$OsO_4 + 8HBr \rightarrow \text{``}H_2OsBr_6\text{''} + 4H_2O + Br_2$$

Procedure

■ **Caution.** *Hydrobromic acid is corrosive. Osmium tetroxide OsO_4 is highly toxic if inhaled, ingested, or absorbed through the skin. Vapors can react with corneal tissues and cause blindness. There is a possible risk of irreversible effects. Wear appropriate gloves and safety goggles and always use in a chemical fume hood. Do not breathe the vapors.*

A solution of OsO_4 (2.0 g, 7.9 mmol) in concentrated HBr (80 mL) is heated to reflux for 2 h. The dark red solution is *promptly* taken to dryness on a rotary evaporator, keeping the bath temperature no higher than 50°C; allowing the solution to stand for more than 1 day before evaporation or overheating during the evaporation results in a product that is not as useful in the next step. The dried solid is further dried under vacuum at room temperature overnight. The very hygroscopic material should be kept under an atmosphere of nitrogen (or argon) from this point onward. In a dry box, the solid is ground into a powder with a mortar and pestle and then vacuum dried again at 35°C overnight. This material is used directly in the next step.

Properties

Bromoosmic acid is a dark brown substance whose composition in aqueous HBr is almost certainly a hydrate of H_2OsBr_6. Its solid-state composition is uncertain. Analytical data for a sample prepared and dried in a manner somewhat different[7] from that described above are consistent with the stoichiometry $(H_3O)Os_2Br_9 \cdot 5H_2O$.

---

[*]The checkers dried the *tert*-butyl alcohol over potassium hydride.

B. BIS(η^5-PENTAMETHYLCYCLOPENTADIENYL)-TETRABROMODIOSMIUM(III)

$$2\text{"}H_2OsBr_6\text{"} + 2C_5Me_5H \rightarrow (\eta^5\text{-}C_5Me_5)_2Os_2Br_4 + 6HBr + \text{"}Br_2\text{"}$$

Procedure

The bromoosmic acid from the previous step is dissolved under nitrogen in degassed *tert*-butyl alcohol (80 mL) and treated with pentamethylcyclopentadiene (1.80 mL, 11.8 mmol). The red solution is heated to reflux for 45 min, during which time a brown microcrystalline solid precipitates. If the reaction solution is heated longer than 45 min, undesired by-products, including the salt [(η^5-C_5Me_5)$_2$OsH]$_2$-[Os$_2$Br$_8$],[8] will coprecipitate with the desired product. The precipitate is collected by filtration and dried under vacuum. The product contains traces of (η^5-C_5Me_5)$_2$Os, which is volatile and may be removed by subliming it at 110°C in vacuum onto a water-cooled cold finger. Yield: 2.96 g (80%).

Anal. Calcd. for $C_{20}H_{30}Br_4Os_2$: C, 24.8; H, 3.12; Br, 32.9. Found: C, 24.6; H, 3.25; Br, 33.0.

Properties

The crystal structure of (η^5-C_5Me_5)$_2$Os$_2$Br$_4$ shows that two of the bromide ligands are bridging, with one bromide ligand and one C_5Me_5 group being terminal on each osmium atom.[6] The Os–Os distance is 2.970 Å. It is only slightly soluble in diethyl ether and alcohols, but is highly soluble in dichloromethane; it is air and moisture sensitive in solution, but can be handled briefly in air as a solid. The FI mass spectrum exhibits a peak at 890 (M^+ − Br). Its ^{1}H NMR spectrum in CD_2Cl_2 shows a peak at δ 2.56 (s, C_5Me_5),[*] and its ^{13}C{^{1}H} NMR spectrum has features at δ 13.0 (s, C_5Me_5) and 101.2 (s, C_5Me_5). Its IR spectrum shows bands at 1075 (w) and 1024 (m) cm^{-1}. Traces of the oxo compound (η^5-C_5Me_5)$_2$Os$_2$(μ-O)Br$_4$ are also present; more of this material is formed if the *tert*-butyl alcohol and the H_2OsBr_6 are not thoroughly dried and deoxygenated as described above. The presence of this oxo impurity sometimes has little effect on the usefulness of (η^5-C_5Me_5)$_2$Os$_2$Br$_4$ as a starting material; for example, (η^5-C_5Me_5)$_2$Os$_2$(μ-O)Br$_4$ is relatively inert toward Lewis bases. For comparison, the ^{1}H NMR spectra in CD_2Cl_2 of possible impurities are as follows: (η^5-C_5Me_5)$_2$Os (δ 1.70, s, C_5Me_5); [(η^5-C_5Me_5)$_2$OsH]$_2$[Os$_2$Br$_8$] (δ 1.98, s, 30H, C_5Me_5; −15.65, s, 1H, Os−H); (η^5-C_5Me_5)$_2$Os$_2$(μ-O)Br$_4$ (δ 2.15, s, C_5Me_5).

---

[*]The checkers found a chemical shift of δ 2.80 in C_6D_6.

The compound $(\eta^5\text{-}C_5Me_5)_2Os_2Br_4$ is a catalyst for the ring-opening metathesis polymerization of norbornadiene,[9] and a useful starting material for other organo-osmium complexes. It can be converted into such species as $(\eta^5\text{-}C_5Me_5)OsH_5$,[6,10] $(\eta^5\text{-}C_5Me_5)OsBr_2(L)$,[11] $(\eta^5\text{-}C_5Me_5)OsBr(L)_2$,[12] $(\eta^5\text{-}C_5Me_5)OsBr_2(\eta^3\text{-allyl})$,[13] and $(\eta^5\text{-}C_5Me_5)Os(\eta^5\text{-}C_5R_5)$,[14] several of which also serve as excellent synthetic entries into other osmium compounds.

C. (η^5-PENTAMETHYLCYCLOPENTADIENYL) (1,5-CYCLOOCTADIENE)BROMOOSMIUM(II)

$$\tfrac{1}{2}(\eta^5\text{-}C_5Me_5)_2Os_2Br_4 + C_8H_{12} \rightarrow (\eta^5\text{-}C_5Me_5)OsBr(\eta^2,\eta^2\text{-}C_8H_{12}) + \text{``Br''}$$

Procedure

To a slurry of $(\eta^5\text{-}C_5Me_5)_2Os_2Br_4$ (0.62 g, 0.64 mmol) in ethanol (40 mL) is added 1,5-cyclooctadiene (0.82 mL, 6.7 mmol). The solution is heated to reflux for 90 min, and the solution color changes to a clear orange and a light-colored precipitate forms. The solvent is removed under vacuum, the residue is extracted with diethyl ether (4×30 mL), and the extracts are filtered. The filtrates are combined, concentrated to $\sim$50 mL, and cooled to $-20°C$ to afford orange crystals. Additional crops of crystals can be obtained by further concentrating and cooling the supernatant. Yield: 0.48 g (74%).[*]

Anal. Calcd. for $C_{18}H_{27}BrOs$: C, 42.1; H, 5.30; Br, 15.6. Found: C, 42.2; H, 5.39; Br, 15.4.

Properties

The 1,5-cyclooctadiene compound $(\eta^5\text{-}C_5Me_5)OsBr(\eta^2,\eta^2\text{-}C_8H_{12})$ can be handled in air (although it is best stored under an inert atmosphere) and is soluble in most standard organic solvents. Its ^1H NMR spectrum in CD_2Cl_2 features two multiplets at δ 4.07 and 3.52 for the olefinic protons of the cod ligand: two environments are expected because these protons are either distal or proximal to the C_5Me_5 ring. The methylene protons appear as four multiplets at δ 2.59, 2.15, 1.91, and 1.68, and the C_5Me_5 resonance is at δ 1.65. The $^{13}C\{^1H\}$ NMR resonances for the sp^2 carbons of the cod ligand appear as two singlets at δ 68.8 and 67.5, which again reflect the presence of distal and proximal environments. The two resonances expected for the sp^3 carbon atoms of the cod ligand evidently overlap and appear as a single peak at δ 32.5. The C_5Me_5 resonances appear at δ 9.8 and 94.4. The field desorption mass

---

[*]The checkers obtained four successive crops by crystallization at $-38°C$, and obtained a total yield of 69%.

spectrum features a peak at 514 for the molecular ion. Its IR spectrum exhibits bands at 1514 (w), 1321 (m), 1295 (w), 1261 (w), 1239 (w), 1207 (w), 1152 (m), 1072 (w), 1026 (m), 1010 (m), 993 (m), 887 (w), 842 (m), 813 (w), 791 (w), 603 (w), 526 (w), and 487 (w) cm^{-1}.

The 1,5-cyclooctadiene ligand can be replaced readily with other Lewis bases to afford a variety of other organoosmium complexes.[12]

References

1. R. Poli, *Chem. Rev.* **91**, 509 (1991).
2. N. Oshima, H. Suzuki, and Y. Moro-oka, *Chem. Lett.* 1161 (1984).
3. T. D. Tilley, R. H. Grubbs, and J. E. Bercaw, *Organometallics* **3**, 274 (1984).
4. T. Sixt, W. Kaim, and W. Preetz, *Z. Naturforsch. B* **55**, 235 (2000).
5. M. O. Albers, D. C. Liles, D. J. Robinson, A. Shaver, E. Singleton, M. B. Wiege, J. C. A. Boeyens, and D. C. Levendis, *Organometallics* **5**, 2321 (1986).
6. (a) C. L. Gross, S. R. Wilson, and G. S. Girolami, *J. Am. Chem. Soc.* **116**, 10294 (1994); (b) C. L. Gross, J. L. Brumaghim, and G. S. Girolami, *Organometallics* **26**, 2258 (2007).
7. H. Moraht and C. Wischin, *Z. Anorg. Allg. Chem.* **3**, 153 (1893).
8. C. L. Gross, S. R. Wilson, and G. S. Girolami, *Inorg. Chem.* **34**, 2582 (1995).
9. J. L. Brumaghim and G. S. Girolami, *Organometallics* **18**, 1923 (1999).
10. (a) T. Shima and H. Suzuki, *Organometallics* **24**, 3939 (2005); (b) C. L. Gross and G. S. Girolami, *Organometallics* **26**, 160 (2007).
11. C. L. Gross and G. S. Girolami, *Organometallics* **25**, 4792 (2006).
12. C. L. Gross and G. S. Girolami, *Organometallics* **15**, 5359 (1996).
13. J. L. Brumaghim, J. G. Priepot, and G. S. Girolami, *Organometallics* **18**, 2139 (1999).
14. S. M. Arachchige, M. J. Heeg, and C. H. Winter, *J. Organomet. Chem.* **690**, 4356 (2005).

Chapter Three

COMPOUNDS WITH METAL–METAL BONDS

16. TETRA(ACETATO)DIMOLYBDENUM(II)

Submitted by RICHARD A. WALTON,[*] PHILLIP E. FANWICK,[*] and
GREGORY S. GIROLAMI[†]
Checked by CARLOS A. MURILLO[‡] and ERIK V. JOHNSTONE[§]

$$2Mo(CO)_6 + 4HO_2CCH_3 \rightarrow Mo_2(O_2CCH_3)_4 + 12CO + 2H_2$$

In 1960, Wilkinson discovered that several carboxylic acids, HO_2CR, react with molybdenum hexacarbonyl, $Mo(CO)_6$, at elevated temperatures to form complexes of the type "$Mo(O_2CR)_2$". The true nature of these bright yellow molybdenum(II) complexes was established in 1964 with the publication by Lawton and Mason of the X-ray crystal structure of tetra(acetato)dimolybdenum, which is depicted schematically in Fig. 1.[1] The Mo−Mo distance of 2.11 Å is considerably shorter than the Mo−Mo distance of 2.78 Å in molybdenum metal, and this fact is evidence that the Mo−Mo bond in $Mo_2(O_2CCH_3)_4$ has considerable multiple bond character.

In fact, because Mo(II) is isoelectronic with Re(III), Cotton proposed that $Mo_2(O_2CCH_3)_4$ contains a molybdenum–molybdenum quadruple bond. A vast

---

[*]Department of Chemistry, Purdue University, West Lafayette, IN 47907.
[†]Department of Chemistry, University of Illinois at Urbana-Champaign, Urbana, IL 61801.
[‡]Department of Chemistry, Texas A&M University, College Station, TX 77842.
[§]Department of Chemistry, University of Nevada-Las Vegas, Las Vegas, NV 89154.

Inorganic Syntheses, Volume 36, First Edition. Edited by Gregory S. Girolami and
Alfred P. Sattelberger.
© 2014 John Wiley & Sons, Inc. Published 2014 by John Wiley & Sons, Inc.

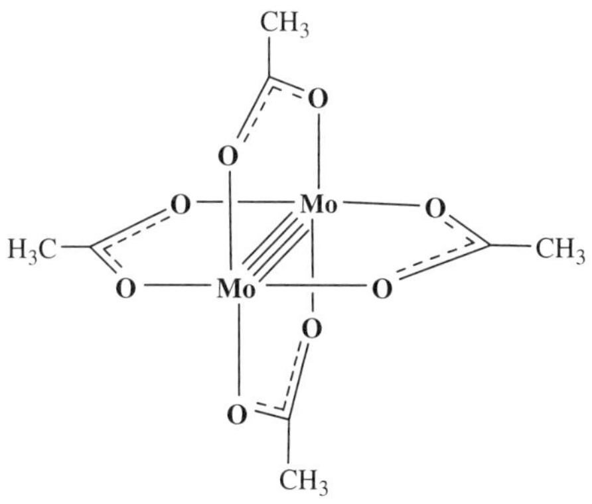

Figure 1. Structure of tetra(acetato)dimolybdenum(II).

body of experimental data has confirmed this proposal and $Mo_2(O_2CCH_3)_4$ has provided access to a large number of derivatives with a Mo_2^{4+} core. The reaction of $Mo_2(O_2CCH_3)_4$ with hydrochloric acid in the presence of inorganic cations was elaborated initially by Sheldon and later by Brencic and Cotton;[2] the products are red-colored salts containing the $[Mo_2Cl_8]^{4-}$ ion, which is isostructural and isoelectronic with $[Re_2Cl_8]^{2-}$. The preparation of molybdenum(II) acetate has been described in a previous volume of this series,[3] but the synthesis described here gives a much higher yield of the desired complex.

Procedure

■ **Caution.** *Molybdenum hexacarbonyl is highly toxic. When heated, it sublimes and partly decomposes to carbon monoxide and molybdenum. Never allow $Mo(CO)_6$ dust to form; never let any solution containing $Mo(CO)_6$ touch any part of the skin; keep the material away from any source of heat. Acetic acid and acetic anhydride are toxic and corrosive. 1,2-Dichlorobenzene is a suspected carcinogen. Do not let these chemicals come into contact with the skin; if they do, wash off the affected area with copious quantities of water. Follow local guidance/ protocols for the safe disposal of chlorinated waste.*

The following reaction should be performed in a well-ventilated fume hood. An oven-dried 1 L three-necked (24/40) round-bottomed flask equipped with a condenser (topped with a nitrogen gas inlet), stopper, nitrogen gas inlet, and a large Teflon®-coated stir bar is charged with 1,2-dichlorobenzene (300 mL). All joints are either greased with Apiezon-H or fitted with Teflon sleeves. The gas inlet atop the condenser is connected via a glass "tee" to a source of nitrogen gas and a mineral oil bubbler. Next, Mo(CO)$_6$ (14.0 g, 53 mmol), a 10:1 (v/v) mixture of acetic acid/acetic anhydride (35 mL), hexanes (35 mL),[*] and trimethylamine N-oxide[†] (about 0.3 g, 4 mmol) are added.

■   **Caution.**   *The added hexanes prevent molybdenum carbonyl from subliming into and clogging the condenser. A solid plug of Mo(CO)$_6$ is very dangerous because the gases (H$_2$ and CO) that are generated during the course of the reaction could build up pressure and lead to an explosion. It is equally important to ensure that the stopcock atop the condenser is open during the reflux. Even with the addition of hexanes, the authors recommend careful monitoring of the reaction over the first few hours. Should any Mo(CO)$_6$ sublime into the condenser during the early stages of the reaction, it should be carefully pushed back into the flask with a glass rod.*

The stopcock atop the condenser is closed and the flask is quickly evacuated and backfilled with nitrogen several times to remove dissolved air. The lower stopcock is closed, the stopcock atop the condenser is opened, and a slow flow of nitrogen is started to the bubbler. A heating mantle is placed around the flask and the mixture is refluxed overnight (12 h), during which period the solution darkens and a fine yellow crystalline solid precipitates. Stirring is stopped and the suspension is slowly cooled to room temperature (the power to the heating mantle is turned off, but it is kept wrapped around the round-bottomed flask to slow the cooling). The cooled mixture is filtered in air through a 60 mL medium- or coarse porosity

---

[*]An alternative to using hexanes to wash down the sublimed Mo(CO)$_6$ is to use a specialized piece of glassware designed by one of us (G.S.G.). It consists of a downflow condenser in which the condensate is returned to the flask by means of a siphon, as shown at right.

Any Mo(CO)$_6$ that condenses on the water-cooled condenser coils is automatically washed down and returned to the flask. A brace between the two vertical tubes makes the apparatus more robust, and the stopcock sidearm can be used to connect the apparatus by means of a hose to a source of N$_2$; this sidearm is slanted slightly upward and is higher than the top of the siphon so that the condensed liquid does not enter the hose. The hot solvent vapors pass by the drip tube and travel up the left side of the apparatus, which is wrapped with insulation such as glass wool; the vapors are condensed on the right side and return to the flask. This apparatus is not commercially available, but can easily be made by a good glassblower. It can be used for many reactions that involve heating a solution containing a volatile solid component such as I$_2$, phenol, and so on.

[†]Trimethylamine N-oxide is added to help initiate the reaction by oxidizing the carbonyl ligands to CO$_2$.

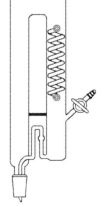

sintered glass frit inside a fume hood. The crystals are washed with ethanol ($3 \times 30\,\text{mL}$) followed by ether ($3 \times 30\,\text{mL}$) and dried under high vacuum for several hours. The yield is typically greater than $9.0\,\text{g}$ (80%).

Properties

Tetra(acetato)dimolybdenum(II) is a bright yellow crystalline solid that is sparingly soluble in organic solvents. The compound decomposes slowly (days to weeks) in air and quite rapidly in solution. It is best stored in an inert atmosphere or under vacuum in the absence of light. $Mo_2(O_2CCH_3)_4$ is a key starting material in dimolybdenum(II) chemistry and can be converted to many other quadruply bonded dimolybdenum(II) complexes, including $[Mo_2Cl_8]^{4-}$,[2] $Mo_2(allyl)_4$,[4] $[Mo_2(CH_3)_8]^{4-}$,[4] $Mo_2Cl_4(PR_3)_4$,[5] and $[Mo(\mu\text{-}t\text{-}Bu_2P)(t\text{-}Bu_2P)]_2$.[6]

References

1. D. Lawton and R. Mason, *J. Am. Chem. Soc.* **87**, 921 (1964).
2. (a) J. V. Brencic and F. A. Cotton, *Inorg. Chem.* **9**, 346 (1970); (b) J. V. Brencic and F. A. Cotton, *Inorg. Chem.* **9**, 351 (1970); (c) J. V. Brencic and F. A. Cotton, *Inorg. Chem.* **8**, 2898 (1969).
3. A. B. Brignole and F. A. Cotton, *Inorg. Synth.* **13**, 87 (1971).
4. F. A. Cotton, J. M. Troup, T. R. Webb, D. H. Williamson, and G. Wilkinson, *J. Am. Chem. Soc.* **96**, 3824 (1974).
5. F. A. Cotton, J. Czuchajowska, and R. L. Luck, *J. Chem. Soc., Dalton Trans.* 579 (1991).
6. R. A. Jones, J. G. Lasch, N. C. Norman, B. R. Whittlesey, and T. C. Wright, *J. Am. Chem. Soc.* **105**, 6184 (1983).

17. SUPRAMOLECULAR ARRAYS BASED ON DIMOLYBDENUM BUILDING BLOCKS

Submitted by F. ALBERT COTTON,[*†] **JAMES P. DONAHUE,**[*] **CHUN LIN,**[*] **and CARLOS A. MURILLO**[*]
Checked by R. THOMAS BAKER[‡] **and HONGBO LI**[‡]

The synthesis and characterization of large, discrete molecules having mononuclear coordinated metal ions have been growing at a rapid pace.[1] A few years

[*]Department of Chemistry, Texas A&M University, College Station, TX 77843.
[†]Deceased, February 20, 2007.
[‡]Department of Chemistry, D'Iorio Hall, 10 Marie Curie, Ottawa, Ontario, Canada K1N 6N5.

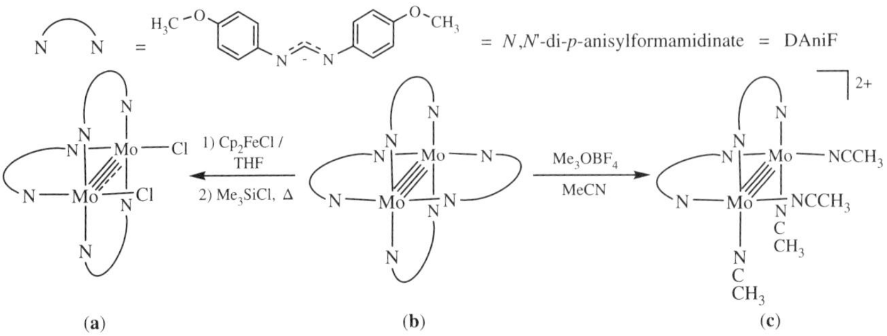

Figure 1. Dimolybdenum building blocks used in the preparation of supramolecular arrays.

ago, we pioneered the use of metal–metal bonded units in this same context, principally Mo_2^{4+} but also Rh_2^{4+}, Re_2^{5+}, and Ru_2^{5+}.[2] The dimetal units that we have employed as building blocks all have the general tetragonal paddlewheel motif illustrated in Fig. 1b for our initial starting material, dimolybdenum(II,II) tetrakis(N,N'-di-p-anisylformamidinate), $Mo_2(DAniF)_4$. One or two of the form-amidinate ligands may be removed to form $Mo_2(DAniF)_3Cl_2$ (Fig. 1a) or [cis-$Mo_2(DAniF)_2(MeCN)_4]^{2+}$ (Fig. 1c), respectively, which serve as end pieces or corner pieces for the construction of a range of supramolecules formed by joining these species with dianionic linkers, such as dicarboxylates. All of the molecules thus formed are neutral.

The simplest supramolecules built from Mo_2^{4+} units, referred to as pairs, are of two distinct types. Those with dicarboxylate or diamidate linkers generally have the Mo_2 units parallel or nearly so (Fig. 2a), whereas the molecules with tetrahedral oxoanions EO_4^{2-} as linkers have the Mo_2 units orthogonally disposed (Fig. 2b). The trigonal molecular propeller (Fig. 2c) is a molecule that forms with a tricarboxylate such as trimesate.

An array of closed structures, either polygons or polyhedra, may be formed with [cis-$Mo_2(DAniF)_2(MeCN)_4]^{2+}$. The simplest of these polygons, called loops (Fig. 3a), have two Mo_2^{4+} units joined by two dicarboxylate linkers. Molecular triangles and squares (Fig. 3b and c) are formed by procedures similar to that used to form loops. It should be noted that for loops, triangles, and squares, the ratio of the Mo_2^{4+} corner pieces to the dicarboxylate is the same, that is, 1:1. The formation of one polygon type versus another is determined principally by the shape of the linker, namely, the disposition of the carboxylate chelates relative to one another. Linear dicarboxylates favor the squares, while bent dicarboxylates tend to produce loops. Finally, closed polyhedra, or cages, may form when [cis-$Mo_2(DAniF)_2(MeCN)_4]^{2+}$ is used

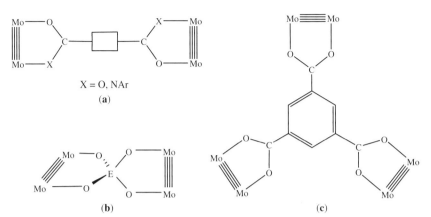

Figure 2. Structures of the molecular pairs and of the molecular propeller assembled from Mo_2^{4+} units. The coordination about each Mo_2^{4+} unit is completed by three DAniF ligands. (a) Pair with linear linkers; (b) pair with tetrahedral linkers; (c) molecular propeller with trimesate linker.

with a tricarboxylate such as trimesate. Figure 3d shows a simplified depiction of one such cage, in which the centers of the four trimesate ligands define the four vertices of a tetrahedron and the centers of the six Mo_2^{4+} units define an octahedron.

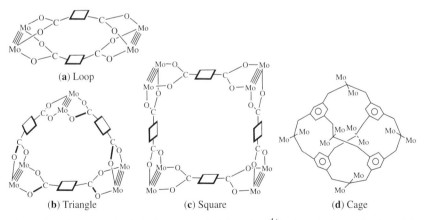

Figure 3. Polygons and polyhedra constructed from Mo_2^{4+} units and di- or tricarboxylate linkers. The coordination about each Mo_2^{4+} unit is completed by two DAniF ligands. (a) Loop; (b) triangle; (c) square; (d) cage.

We provide herein optimized syntheses for a selected array of supramolecular compounds containing quadruply bonded Mo_2^{4+} units, as well as the synthesis of the dimolybdenum compounds used as precursors. These syntheses are general and can be used for a variety of linkers, but only one example is provided for each structure type. However, an effort has been made to list other linkers that we used successfully.

General Procedures

$Mo(CO)_6$, *o*-dichlorobenzene, Me_3SiCl, Zn dust, Me_3OBF_4, 0.5 M NaOMe in MeOH, and 1 M $NaHBEt_3$ in toluene can be purchased from commercial sources and used without further purification. Acetonitrile is distilled twice under N_2, first from molecular sieves and then from CaH_2. Hexanes, Et_2O, THF, and toluene are distilled from Na/K-benzophenone, acetone from K_2CO_3, and CH_2Cl_2 and 1,2-dichloroethane from P_2O_5. All dicarboxylic or tricarboxylic acids are obtained from commercial sources, unless otherwise indicated, and converted into their corresponding bis- or tris(tetraalkylammonium) salts by neutralization with 2 or 3 equiv of Et_4NOH or $Bu^n_4NOH.$[*] Methanol solutions of Et_4NHSO_4, $(NH_4)_2MoO_4$, and $(NH_4)_2WO_4$ are treated with 1 or 2 equiv of Et_4NOH and reduced to dryness to afford the corresponding tetraalkylammonium salts of the tetrahedral oxoanion. *N,N'*-Di-*p*-anisylformamidine $(HDAniF)$[3] and *N,N'*-diphenylterephthaloyldiamide[4] are synthesized by the literature methods, and $Cp_2Fe[FeCl_4]$ is prepared at room temperature by bubbling Cl_2 through a hexanes solution of Cp_2Fe. Unless stated otherwise, all reactions and manipulations are conducted under an atmosphere of dry N_2.

A. TETRAKIS(*N,N'*-DI-*p*-ANISYLFORMAMIDINATO)-DIMOLYBDENUM(II)

$$2Mo(CO)_6 + 4HDAniF \xrightarrow[\text{hexanes}]{\textit{o}\text{-}C_6H_4Cl_2} Mo_2(DAniF)_4 + 12CO + 2H_2$$

---

[*]The checkers note that the tetraalkylammonium carboxylate salts tend to retain water and should be thoroughly dried before use in these syntheses.

■   **Caution.**   *The following procedure evolves appreciable amounts of odorless and toxic carbon monoxide and flammable H_2 and should be conducted in a well-ventilated fume hood. o-Dichlorobenzene is a carcinogenic substance and all contact either by skin or by inhalation of its vapors should be carefully avoided. The addition of hexanes is necessary as it prevents unreacted $Mo(CO)_6$ from subliming into the condenser. The plug of crystalline $Mo(CO)_6$ that could otherwise form in the condenser is very dangerous because the gases that are evolved during the course of the reaction will build up pressure and potentially explode.*

Procedure

An oven-dried 1 L Schlenk flask is charged with $Mo(CO)_6$ (17.17 g, 65 mmol), freshly prepared HDAniF (50.0 g, 195 mmol), hexanes (40 mL), o-dichloro-benzene (300 mL), and a large stir bar. This flask is fitted with a reflux condenser with an adaptor leading to a Schlenk line. The entire apparatus is degassed with several quick evacuations and backfills with N_2. The heterogeneous mixture is then heated to reflux (180–200°C) for 12 h, during which time it darkens and forms a bright yellow microcrystalline precipitate. Stirring is stopped, and the mixture is cooled slowly to ambient temperature. The crystalline product is collected on a glass frit opened to the air, washed with copious amounts of Et_2O to fully remove all traces of o-dichlorobenzene, and dried under vacuum. Yield: 37.7 g (96%).

Anal. Calcd. for $C_{60}H_{60}Mo_2N_8O_8$: C, 59.41; H, 4.99; N, 9.24. Found: C, 59.31; H, 5.06; N, 9.21.

Properties

$Mo_2(DAniF)_4$ is a moderately air-stable, bright yellow solid, but is best stored under N_2 for prolonged periods. It is insoluble in alkanes and Et_2O and polar solvents such as MeCN and MeOH. It is moderately soluble in chlorinated solvents such as CH_2Cl_2, but is air sensitive in solution and degrades to darkly colored species. $Mo_2(DAniF)_4$ undergoes a reversible one-electron oxidation at 0.142 V versus Ag/AgCl in CH_2Cl_2. Positive ion ESI: m/z 1214 $(M+H^+)$. ^1H NMR (CDCl$_3$): δ 8.41 (s, 4H, $-NCHN-$), 6.49 (d, 16H, aromatic $C-H$), 6.17 (d, 16H, aromatic $C-H$), 3.68 (s, 24H, $-OCH_3$). This procedure has been carried out successfully at one-tenth of this scale by undergraduate students (seniors) in an advanced inorganic laboratory course.

B. TRIS(*N*,*N*′-DI-*p*-ANISYLFORMAMIDINATO)DI(CHLORO)-DIMOLYBDENUM(II,III)

$$Mo_2(DAniF)_4 \xrightarrow[\text{2.Me}_3\text{SiCl},\Delta]{\text{1.Cp}_2\text{Fe[FeCl}_4\text{]/THF}} Mo_2(DAniF)_3Cl_2 + FeCp_2 + Me_3SiDAniF + FeCl_3(THF)_n$$

Procedure

In this procedure, all glassware is oven dried and assembled while hot. $Mo_2(DAniF)_4$ (14.5 g, 12.0 mmol) is suspended in distilled, degassed THF (2.25 L) in a 3 L two-necked flask equipped with a large stir bar and fitted with a reflux condenser leading to a Schlenk line. To this suspension is added a solution of $Cp_2Fe[FeCl_4]$ (2.70 g, 7.03 mmol) in THF (250 mL) over 2 h. The resulting dark solution is stirred for 2 h at ambient temperature, treated with 1 equiv of Me_3SiCl (1.55 mL, 12.2 mmol), and then vigorously refluxed for 4 h under N_2. After the solution is allowed to cool to room temperature, the solvent is evaporated under reduced pressure. The solid residue that remains is extracted with distilled acetone (300 mL) and the extract is filtered to remove unreacted $Mo_2(DAniF)_4$, which may be recovered for reuse. Upon standing for ~1 h, the filtrate produces a fine precipitate of additional $Mo_2(DAniF)_4$, the removal of which requires a filtration through Celite®. The dark brown filtrate is evaporated to dryness under vacuum and then washed with Et_2O (2 × 100 mL). The resulting dark brown microcrystalline $Mo_2(DAniF)_3Cl_2$ is dried under vacuum and is of sufficient purity for most applications. A further degree of purity may be obtained by redissolving $Mo_2(DAniF)_3Cl_2$ in a minimal amount of freshly distilled, degassed acetone (~250 mL) and slowly concentrating and cooling the solution, whereupon reprecipitation of microcrystalline material occurs. After the volume is reduced to 15–20% of the original quantity of acetone, the slow addition of several volume equivalents of Et_2O (~100 mL) completes the deposition of $Mo_2(DAniF)_3Cl_2$. Collection of the product on a glass frit, washing with Et_2O (2 × 50 mL), and drying under vacuum yields $Mo_2(DAniF)_3Cl_2$ as a dark coffee-colored microcrystalline solid. Yield (based on Mo): 5.40 g (44%).

Anal. Calcd. for $C_{45}H_{45}Cl_2Mo_2N_6O_6$: C, 52.54; H, 4.41; N, 8.17. Found: C, 52.36; H, 4.20; N, 8.02.

Properties

$Mo_2(DAniF)_3Cl_2$ is relatively air stable in the solid state, but is best stored under N_2 for long periods. It is insoluble in alkanes and Et_2O, but is very soluble in polar organic solvent such as CH_2Cl_2, acetone, THF, and MeCN. Absorption spectrum (CH_2Cl_2) λ (ϵ_M): 276 (34,400), 455 (5880), 570 (sh).[*]

[*]The checkers note that the compound is paramagnetic and gives no ^1H NMR signal.

C. *cis*-BIS(*N*,*N'*-DI-*p*-ANISYLFORMAMIDINATO)TETRAKIS (ACETONITRILE)DIMOLYBDENUM(II) BIS(TETRAFLUOROBOROATE)

$$Mo_2(DAniF)_4 + 4Me_3OBF_4 + 4H_2O \xrightarrow{MeCN/N_2/25°C}$$

$$cis-[Mo_2(DAniF)_2(MeCN)_4](BF_4)_2 + 2(H_2DAniF)BF_4 + 4Me_2O + 4MeOH$$

■ **Caution.** *Me₃OBF₄ is very corrosive. All contact with the skin should be carefully avoided.*

Procedure

A 100 mL three-necked round-bottomed flask is charged with $Mo_2(DAniF)_4$ (0.242 g, 0.200 mmol), Me_3OBF_4 (0.178 g, 1.20 mmol), and CH_3CN (60 mL). To the stirred suspension is added a small amount of deoxygenated H_2O (~2–3 drops). The resulting mixture is stirred at room temperature for ~5 h, after which time all of the yellow starting material has reacted to give a clear red solution. Further stirring of this solution will result in decomposition of the product. The solvent is completely removed under vacuum, and the red residue is washed with Et_2O (2 × 10 mL), and then extracted with CH_2Cl_2 (2 × 5 mL). To stabilize the product, CH_3CN (1 mL) is added to the extract. The resulting mixture is then dried and redissolved in CH_3CN (10 mL) in a 100 mL flask. Diethyl ether (75 mL) is then carefully added to the top of the extract, and the solution is stored for 24–48 h at room temperature until the diffusion is complete. Large orange-red block-shaped crystals form; they are collected by filtration, washed with Et_2O/CH_3CN (9:1), and dried for 2 h under vacuum. Yield: 0.166 g (80%).

Anal. Calcd. for $C_{38}H_{42}B_2F_8Mo_2N_8O_4$: C, 43.87; H, 4.07; N, 10.77. Found: C, 43.23; H, 4.07; N, 10.10.

Properties

$[cis-Mo_2(DAniF)_2(MeCN)_4](BF_4)_2$ can be dissolved in organic solvents such as CH_3CN, CH_2Cl_2, and acetone. It is not soluble in hexanes and ether. Prolonged standing in chlorinated solvents in the absence of CH_3CN results in slow decomposition. ^1H NMR (CD_2Cl_2): δ 8.73 (s, 2H, $-NCHN-$), 6.61 (m, 16H, aromatic), 3.73 (s, 12H, $-OCH_3$), 2.79 (s, 12H, MeCN). Single crystals suitable for X-ray analysis may be grown by diffusion of Et_2O into an acetonitrile solution.

D. (μ₂-SUCCINATO)BIS[TRIS(*N,N'*-DI-*p*-ANISYLFORMAMIDINATO)-DIMOLYBDENUM(II)]

$$2Mo_2(DAniF)_3Cl_2 + Zn + (Et_4N)_2O_2CCH_2CH_2CO_2 \xrightarrow[N_2]{MeCN}$$

$$[Mo_2(DAniF)_3](\mu\text{-}O_2CCH_2CH_2CO_2)[Mo_2(DAniF)_3] + (Et_4N)_2[ZnCl_4]$$

Procedure

A solution of $Mo_2(DAniF)_3Cl_2$ (0.500 g, 0.490 mmol) in MeCN (60 mL) is treated with Zn dust (7.5 g), and the resulting suspension is stirred vigorously for 1.5 h. The excess Zn dust is then separated from the yellow solution by filtration through Celite® into a flask containing $(Et_4N)_2O_2CCH_2CH_2CO_2$ (0.092 g, 0.243 mmol). The Celite® is washed with an additional portion of MeCN (5 mL). Within a few moments of stirring, the filtrate develops a voluminous yellow precipitate. Stirring is continued for 1 h, and the volume of the mixture is then reduced to 20 mL. The product is isolated by filtration and washed with MeCN (10 mL) followed by Et_2O (10 mL). After being briefly dried under vacuum, the product is extracted with CH_2Cl_2 (4 × 3 mL), the extracts being filtered and layered with hexanes (50 mL). The product crystallizes as bright yellow blocks within 24–48 h. Yield: 0.447 g (91%).*

Anal. Calcd. for $C_{92}H_{94}Mo_4N_{12}O_{16}$: C, 55.04; H, 4.72; N, 8.37. Found: C, 55.22; H, 4.72; N, 8.09.

Properties

$[Mo_2(DAniF)_3]_2(\mu\text{-succinate})$ is a bright yellow compound that is moderately air sensitive in the solid state and very much so in solution. It is insoluble in alkanes and Et_2O and such polar solvents as MeCN and H_2O. Solubility is good in chlorinated solvents such as CH_2Cl_2 and $CHCl_3$ and moderate in THF and acetone. ¹H NMR (CDCl₃): δ 8.45 (s, 4H, −NCHN−), 8.39 (s, 2H, −NCHN−), 6.64 (d, 16H, aromatic C−H), 6.53 (d, 16H, aromatic C−H), 6.43 (d, 8H, aromatic C−H), 6.21 (d, 8H, aromatic C−H), 3.72 (s, 24H, −OCH₃), 3.65 (s, 12H, −OCH₃), 3.50 (s, −CH₂CH₂−).

---

*The checkers report a yield of 85%.

The same procedure may be used to prepare virtually any dicarboxylate-linked pair of $Mo_2(DAniF)_3$ units.[5] The colors of these dimolybdenum pairs vary widely depending upon the chemical nature of the dicarboxylate linker, darker and more intense colors generally being produced by linkers with unsaturated bonds. Some variation in solubility is also observed, but in general those compounds with rigid dicarboxylate linkers are appreciably less soluble than those with flexible ones.

E. (μ_2-η^2,η^2-MOLYBDATO)BIS[TRIS(*N*,*N*′-DI-*p*-ANISYLFORMAMIDINATO)DIMOLYBDENUM(II)]

$$2Mo_2(DAniF)_3Cl_2 + Zn + (Et_4N)_2(MoO_4) \xrightarrow[\text{N}_2]{\text{MeCN}}$$

$$[Mo_2(DAniF)_3](\mu\text{-}MoO_4)[Mo_2(DAniF)_3] + (Et_4N)_2[ZnCl_4]$$

Procedure

A suspension of $Mo_2(DAniF)_3Cl_2$ (0.500 g, 0.490 mmol) and excess Zn dust (7.50 g) in MeCN (60 mL) is stirred under N_2 for 1.5 h. The Zn dust is removed from the heterogeneous mixture by filtering it through Celite® into a flask containing 0.5 equiv of solid $(Et_4N)_2MoO_4$ (0.102 g, 0.243 mmol). The Celite® is washed with MeCN (5 mL). The resulting orange solution is stirred for 1 h at ambient temperature and then concentrated under vacuum to a volume of 5 mL, whereupon an orange microcrystalline solid forms. The precipitate is then isolated by filtration and washed with MeCN (2×1 mL) followed by Et_2O (1×5 mL). The remaining solid is extracted with distilled, degassed 1,2-dichloroethane (4×2 mL), and the extracts are filtered into a Schlenk tube. The filtrate is layered with Et_2O (1 mL) followed by hexanes (50 mL). After several days, the crystalline product is isolated by filtration and dried under vacuum. Yield: 0.300 g (59%).

Anal. Calcd. for $C_{90}H_{90}Mo_5N_{12}O_{16}$: C, 52.08; H, 4.37; N, 8.10. Found: C, 51.93; H, 4.35; N, 8.12.

Properties

$[Mo_2(DAniF)_3]_2(\mu_2$-η^2,η^2-$MoO_4)$ crystallizes as dark brownish blocks from $ClCH_2CH_2Cl$ or CH_2Cl_2. Very dry CH_2Cl_2 solutions of analytically pure, recrystallized material have a dark pine-green color. $[Mo_2(DAniF)_3]_2(\mu_2$-η^2,η^2-$MoO_4)$ is very air sensitive both in solution and in the solid state and undergoes two reversible one-electron oxidations at 0.070 and 0.374 V versus Ag/AgCl in

CH_2Cl_2. ^1H NMR (CD_2Cl_2): δ 8.69 (s, 4H, $-NCHN-$), 8.11 (s, 2H, $-NCHN-$), 6.63 (d, 16H, aromatic $C-H$), 6.45 (d, 16H, aromatic $C-H$), 6.35 (d, 8H, aromatic $C-H$), 6.13 (d, 8H, aromatic $C-H$), 3.65 (s, 24H, $-OCH_3$), 3.59 (s, 12H, $-OCH_3$). Absorption spectrum (CH_2Cl_2) λ (ϵ_M): 278 (120,000), 295 (117,000), 592 (1640). The same procedure may be employed to prepare the analogous μ_2-η^2,η^2-SO_4 and μ_2-η^2,η^2-WO_4 complexes by using $(Bu^n_4N)_2SO_4$ or $(Et_4N)_2WO_4$ instead of $(Et_4N)_2MoO_4$.[6]

F. (μ_2-*N*,*N'*-DIPHENYLTEREPHTHALOYLDIAMIDATO)BIS[TRIS-(*N*,*N'*-DI-*p*-ANISYLFORMAMIDINATO)DIMOLYBDENUM(II)]

$$2Mo_2(DAniF)_3Cl_2 + Zn + PhN(H)COC_6H_4CON(H)Ph + 2NaOMe \xrightarrow[N_2]{MeCN/THF}$$

$$[Mo_2(DAniF)_3](\mu\text{-}(N(Ph)COC_6H_4CON(Ph))[Mo_2(DAniF)_3] + Na_2[ZnCl_4] + 2MeOH$$

Procedure

$Mo_2(DAniF)_3Cl_2$ (0.400 g, 0.389 mmol) is stirred with excess Zn dust (4 g) in MeCN (50 mL) under an N_2 atmosphere for 2 h. This heterogeneous mixture is then filtered through Celite® into a flask containing 0.5 equiv of solid *N*,*N'*-diphenylterephthaloyldiamide (0.038 g, 0.121 mmol). Into this solution is added by syringe an excess of $NaOCH_3$ (4.00 mL of a 0.50 M solution in MeOH, 2.0 mmol) followed by THF (50 mL) to render the mixture more homogeneous. After being stirred overnight at ambient temperature, the mixture is concentrated to dryness under vacuum, and extracted with CH_2Cl_2 (2×20 mL). These extracts are filtered to remove insoluble materials and then reduced to dryness. The resulting red-orange solid residue is extracted with CH_2Cl_2 (3×5 mL), and the extracts are filtered, combined, and layered with Et_2O (4 mL) followed by hexanes (50 mL). After several days, $[Mo_2(DAniF)_3]_2(\mu$-$NPhCOC_6H_4CONPh)$ crystallizes in the form of bright red needles. Yield (two crops): 0.176 g (41%).

Anal. Calcd. for $C_{110}H_{104}Mo_4N_{14}O_{14}$: C, 59.25; H, 4.70; N, 8.79. Found: C, 58.85; H, 4.56; N, 8.76.

Properties

$[Mo_2(DAniF)_3]_2(\mu$-$NPhCOC_6H_4CONPh)$ is a red solid that is stable in air for moderate durations (several days), but is best stored for prolonged periods under N_2. It is insoluble in alkanes and Et_2O and very polar solvents such as MeCN. Solubility is best in halogenated solvents such as CH_2Cl_2 and is relatively modest in other organic solvents such as THF and acetone. ^{1}H NMR (CD_2Cl_2): δ 8.57 (s, 4H, $-$NCHN$-$), 8.38 (s, 2H, $-$NCHN$-$), 7.29 (s, 4H, $-C_6H_4-$), 6.77 (q, 6H, aromatic C$-$H), 6.67 (d, 8H, aromatic C$-$H), 6.59 (d, 8H, aromatic C$-$H), 6.37–6.52 (m, 16H, aromatic C$-$H), 6.14–6.28 (m, 12H, aromatic C$-$H), 5.93 (d, 8H, aromatic C$-$H), 3.72 (s, 12H, $-OCH_3$), 3.68 (s, 12H, $-OCH_3$), 3.65 (s, 6H, $-OCH_3$), 3.63 (s, 6H, $-OCH_3$). Absorption spectrum (CH_2Cl_2) λ (ϵ_M): 241 (10,200), 279 (sh, 103,000), 293 (106,000), 462 (12,500). The same procedure has been applied successfully in the preparation of the corresponding μ-N(m-CF_3-Ph)COC$_6$H$_4$CON(m-CF_3-Ph) molecule.[4]

G. MOLECULAR PROPELLER: (μ_3-TRIMESATE)TRIS[TRIS(N,N'-DI-p-ANISYLFORMAMIDINATO)DIMOLYBDENUM(II)]

$$3Mo_2(DAniF)_3Cl_2 + 3NaHBEt_3 + (Bu^n{}_4N)_3(C_6H_3(CO_2)_3) \xrightarrow{CH_2Cl_2/N_2}$$

$$[Mo_2(DAniF)_3]_3(\mu_3\text{-}1,3,5\text{-}(O_2C)_3C_6H_3) + 3NaCl + 3Bu^n{}_4NCl + 3BEt_3 + \tfrac{3}{2}H_2$$

Procedure

To a mixture of $Mo_2(DAniF)_3Cl_2$ (0.154 g, 0.150 mmol) and $(Bu^n{}_4N)_3(1,3,5\text{-}C_6H_3(CO_2)_3)$ (0.140 g, 0.150 mmol) in CH_2Cl_2 (60 mL) is added NaHBEt$_3$ (1.0 mL of a 1 M solution in toluene, 1.00 mmol). The reaction mixture is stirred for 24 h at room temperature, and the volatile materials are removed under vacuum. The resulting residue is washed with Et$_2$O (10 mL), EtOH (2 × 20 mL), H$_2$O (2 × 10 mL), and EtOH (10 mL) and dried briefly under vacuum. It is then extracted with CH_2Cl_2 (3 × 5 mL). Hexanes (35 mL) are then carefully added to the extracts without stirring. The orange, block-shaped crystals that form after 1 week are collected by filtration and dried under vacuum. Yield: 0.082 g (53%).

Anal. Calcd. for $C_{144}H_{138}Mo_6N_{18}O_{24}$: C, 56.14; H, 4.52; N, 8.19. Found: C, 55.81; H, 4.49; N, 8.04.

Properties

The orange crystalline compound is soluble in various organic solvents such as CH_2Cl_2, benzene, toluene, and THF, and is sensitive to air. ^{1}H NMR (CD_2Cl_2): δ 8.52 (s, 3H, $-$NCHN$-$), 8.47 (s, 6H, $-$NCHN$-$), 6.56 (m, 48H, aromatic), 6.45 (d, 12H, aromatic), 6.23 (d, 12H, aromatic), 3.67 (s, 36H, $-OCH_3$), 3.64 (s, 18H, $-OCH_3$).

H. MOLECULAR LOOP: *closo*-BIS(μ_2-MALONATO)BIS[BIS(*N,N'*-DI-*p*-ANISYLFORMAMIDINATO)DIMOLYBDENUM(II)]

$$2[cis\text{-}Mo_2(DAniF)_2(MeCN)_4](BF_4)_2 + 2(Bu^n{}_4N)_2O_2CCH_2CO_2 \xrightarrow{MeCN/N_2}$$

$$[cis\text{-}Mo_2(DAniF)_2]_2(\mu\text{-}O_2CCH_2CO_2)_2 + 4Bu^n{}_4NBF_4$$

Procedure

To a stirred solution of $[cis\text{-}Mo_2(DAniF)_2(CH_3CN)_4](BF_4)_2$ (0.104 g, 0.100 mmol) in CH_3CN (20 mL) is added $(Bu^n{}_4N)_2(O_2CCH_2CO_2)$ (0.065 g, 0.110 mmol) in CH_3CN (10 mL). An immediate reaction takes place with formation of a bright yellow precipitate, which is collected by filtration, washed several times with CH_3CN, and dried under vacuum. The crude product is extracted with CH_2Cl_2 (3 × 4 mL). Hexanes are then carefully layered on the top of the solution to afford a yellow crystalline material after several days. Yield: 0.074 g (91%).

Anal. Calcd. for $C_{66}H_{64}Mo_4N_8O_{16}$: C, 49.26; H, 4.01; N, 6.97. Found: C, 48.82; H, 3.99; N, 6.94.

Properties

Single crystals of this compound can be grown by layering CH_2Cl_2 solution with hexanes. It is readily soluble in CH_2Cl_2, less soluble in benzene and acetone, and insoluble in hexanes or Et_2O. ^{1}H NMR (CD_2Cl_2): δ 8.57 (s, 4H, −NCHN−), 6.60 (dd, 32H, aromatic), 4.53 (s, 4H, −O_2CCH_2CO_2−), 3.69 (s, 24H, −OCH_3). A similar method can be used for the preparation of the loops in which the linkers are the dianions of 1,4-phenylenediacetic acid ($C_6H_4(CH_2CO_2H)_2$), homophthalic acid (α-carboxy-*o*-toluic acid, *o*-HO_2CCH_2C_6H_4CO_2H), or *trans*-cyclopentane-1,2-dicarboxylic acid (HO_2CC_5H_8CO_2H).[7]

I. MOLECULAR TRIANGLE: *closo*-TRIS(μ_2-*eq,eq*-1,4-CYCLOHEXANEDICARBOXYLATO)TRIS[BIS(*N,N'*-DI-*p*-ANISYLFORMAMIDINATO)DIMOLYBDENUM(II)]

$$3[cis\text{-}Mo_2(DAniF)_2(MeCN)_4](BF_4)_2 + 3(Bu^n{}_4N)_2(trans\text{-}1,4\text{-}O_2CC_6H_{10}CO_2) \xrightarrow{MeCN/N_2}$$

$$[cis\text{-}Mo_2(DAniF)_2]_3(\mu\text{-}eq,eq\text{-}1,4\text{-}O_2CC_6H_{10}CO_2)_3 + 6Bu^n{}_4NBF_4$$

Procedure

To a stirred solution of [*cis*-Mo$_2$(DAniF)$_2$(CH$_3$CN)$_4$](BF$_4$)$_2$ (0.104 g, 0.100 mmol) in CH$_3$CN (20 mL) is added (Bu^n$_4$N)$_2$(*trans*-1,4-O$_2$CC$_6$H$_{10}$CO$_2$) (0.0655 g, 0.100 mmol) in CH$_3$CN (10 mL). An immediate reaction takes place with formation of a bright yellow precipitate, which is collected by filtration, washed several times with CH$_3$CN, and dried under vacuum. The crude product is extracted with CH$_2$Cl$_2$ (2 × 5 mL). Hexanes are then carefully layered on the top of the solution to afford a yellow crystalline material after several days. The yield is essentially quantitative.

Anal. Calcd. for C$_{114}$H$_{120}$Mo$_6$N$_{12}$O$_{24}$: C, 52.30; H, 4.62; N, 6.42. Found: C, 51.95; H, 4.60; N, 6.35.

Properties

The yellow crystalline material is moderately soluble in most common organic solvents except for alkanes and Et$_2$O. ^{1}H NMR (CD$_2$Cl$_2$): δ 8.43 (s, 6H, −NCHN−), 6.63 (dd, 48H, aromatic), 3.70 (s, 36H, −OCH$_3$), 2.91 (br, 6H, cyclohexyl), 2.30 (m, 12H, cyclohexyl), 1.89 (m, 12H, cyclohexyl).

J. MOLECULAR SQUARE: *closo*-TETRAKIS(μ$_2$-OXALATO)-TETRAKIS[BIS(*N,N'*-DI-*p*-ANISYLFORMAMIDINATO)-DIMOLYBDENUM(II)]

$$4[cis\text{-Mo}_2(\text{DAniF})_2(\text{MeCN})_4](\text{BF}_4)_2 + 4(\text{Bu}^n{}_4\text{N})_2\text{O}_2\text{CCO}_2 \xrightarrow{\text{MeCN/N}_2}$$

$$[cis\text{-Mo}_2(\text{DAniF})_2]_4(\mu\text{-O}_2\text{CCO}_2)_4 + 8\text{Bu}^n{}_4\text{NBF}_4$$

Procedure

To a stirred solution of [Mo$_2$(*cis*-DAniF)$_2$(CH$_3$CN)$_4$](BF$_4$)$_2$ (0.312 g, 0.300 mmol) in CH$_3$CN (30 mL) is added (Bu^n$_4$N)$_2$C$_2$O$_4$ (0.180 g, 0.310 mmol) in CH$_3$CN (20 mL). An immediate reaction takes place with the formation of a bright red precipitate, which is collected by filtration, washed several times with CH$_3$CN, and dried under vacuum. The crude product is extracted with CH$_2$Cl$_2$ (3 × 5 mL). Hexanes are then carefully layered on the top of the solution to afford a bright red crystalline material after 7 days. Yield: 0.205 g (86%).

Anal. Calcd. for C$_{128}$H$_{120}$Mo$_8$N$_{16}$O$_{32}$: C, 48.62; H, 3.83; N, 7.09. Found: C, 48.27; H, 3.88; N, 7.10.

Properties

This compound is moderately soluble in organic solvents such as CH_2Cl_2, benzene, and toluene. Single crystals of $[(cis\text{-}Mo_2(DAniF)_2)(O_2CCO_2)]_4$ can be grown by layering a CH_2Cl_2 solution with hexanes. ^1H NMR (CD_2Cl_2): δ 8.57 (s, 8H, $-NCHN-$),[*] 6.65 (dd, 64H, aromatic), 3.70 (s, 48H, $-OCH_3$). A similar method can be used for the preparation of the squares in which the linkers are the dianions of fumarate, ferrocene dicarboxylate, 4,4'-biphenyldicarboxylate, acetylenedicarboxylate, tetrafluoroterephthalate, and carborane dicarboxylate.[8]

K. MOLECULAR CAGE: *closo*-TETRAKIS(μ_3-TRIMESATE)HEXAKIS-[BIS(*N,N'*-DI-*p*-ANISYLFORMAMIDINATO)DIMOLYBDENUM(II)]

$$6[cis\text{-}Mo_2(DAniF)_2(MeCN)_4](BF_4)_2 + 4(Bu^n{}_4N)_2(1,3,5\text{-}C_6H_3(CO_2)_3) \xrightarrow{\text{MeCN/N}_2}$$

$$[cis\text{-}Mo_2(DAniF)_2]_6(\mu\text{-}1,3,5\text{-}C_6H_3(CO_2)_3)_4 + 12Bu^n{}_4NBF_4$$

Procedure

To a stirred solution of $[Mo_2(cis\text{-}DAniF)_2(CH_3CN)_4](BF_4)_2$ (0.312 g, 0.300 mmol) in CH_3CN (30 mL) is added $(Bu^n{}_4N)_3[1,3,5\text{-}C_6H_3(CO_2)_3]$ (0.187 g, 0.200 mmol) in CH_3CN (20 mL). An immediate reaction takes place with the formation of a bright red precipitate, which is collected, washed several times with CH_3CN, and dried. The crude product is extracted with CH_2Cl_2 (3×5 mL). Then CH_3CN is carefully layered on top of the solution. Bright red crystals are collected after about 2 weeks. The yield is essentially quantitative.

Anal. Calcd. for $C_{216}H_{192}Mo_{12}N_{24}O_{48}$: C, 51.44; H, 3.84; N, 6.67. Found: C, 51.02; H, 4.19; N, 6.95.

Properties

The compound is very soluble in CH_2Cl_2, and moderately soluble in many common organic solvents. ^1H NMR (CD_2Cl_2): δ 9.39 (s, 12H, aromatic protons of the trimesate groups), 8.52 (s, 12H, $-NCHN-$), 6.62 (dd, 96H, aromatic protons of the formamidinate groups), 3.67 (s, 72H, $-OCH_3$).

---

[*] The checkers report that there is also a signal at δ 8.69, which integrates to two protons, whereas the peak at δ 8.57 integrates to six protons.

References

1. S. Leininger, B. Olenyuk, and P. J. Stang, *Chem. Rev.* **100**, 853 (2000).
2. (a) F. A. Cotton, C. Lin, and C. A. Murillo, *Acc. Chem. Res.* **34**, 759 (2001); (b) F. A. Cotton, C. Lin, and C. A. Murillo, *Proc. Natl. Acad. Sci. USA* **99**, 4810 (2002).
3. C. Lin, J. D. Protasiewicz, E. T. Smith, and T. Ren, *Inorg. Chem.* **35**, 6422 (1996).
4. F. A. Cotton, L. M. Daniels, J. P. Donahue, C. Y. Liu, and C. A. Murillo, *Inorg. Chem.* **41**, 1354 (2002).
5. (a) F. A. Cotton, J. P. Donahue, C. Lin, and C. A. Murillo, *Inorg. Chem.* **40**, 1234 (2001); (b) F. A. Cotton, J. P. Donahue, and C. A. Murillo, *Inorg. Chem. Commun.* **5**, 59 (2002); (c) F. A. Cotton, J. P. Donahue, and C. A. Murillo, *J. Am. Chem. Soc.* **125**, 5436 (2003).
6. F. A. Cotton, J. P. Donahue, and C. A. Murillo, *Inorg. Chem.* **40**, 2229 (2001).
7. F. A. Cotton, C. Lin, and C. A. Murillo, *Inorg. Chem.* **40**, 472 (2001).
8. F. A. Cotton, C. Lin, and C. A. Murillo, *Inorg. Chem.* **40**, 478 (2001).

18. DIMOLYBDENUM AND DITUNGSTEN HEXA(ALKOXIDES)

Submitted by ERIN M. BRODERICK,[*] SAMUEL C. BROWNE,[*] and
MARC J. A. JOHNSON[*]
Checked by TRACEY A. HITT[†] and GREGORY S. GIROLAMI[†]

The dimetal hexaalkoxides $M_2(OR)_6$ (R = alkyl) of molybdenum and especially tungsten display a rich chemistry.[1] For a given alkoxide ligand set, the tungsten complexes are significantly more prone to coordinate additional ligands or aggregate further than are the molybdenum complexes. Several ditungsten alkoxide complexes undergo simple and high-yielding $M \equiv M / C \equiv E$ (E = C, N) triple bond metathesis reactions with various alkynes and nitriles to afford alkylidyne and nitride complexes, $R'C \equiv W(OR)_3$ and $N \equiv W(OR)_3$,[1,2] that are efficient catalysts for alkyne metathesis[3] and related reactions such as nitrile–alkyne metathesis and degenerate N-atom exchange between nitriles. The dimolybdenum analogs are less prone to undergo this reaction with alkynes.

The original syntheses of $M_2(OR)_6$ reported by Chisholm entailed preparation of $M_2(NMe_2)_6$[4] from $MoCl_3$ or WCl_4, purification of these amido complexes by sublimation, and subsequent alcoholysis.[5,6] Alcoholysis of $M_2(NMe_2)_6$ is a very general synthetic route to $M_2(OR)_6$, although the use of insufficiently bulky alcohols or silanols can permit aggregation or further reaction of transiently generated $M_2(OR)_6$ species.[1] Simple, direct routes to certain $M_2(OR)_6$ complexes were enabled by the development of convenient syntheses of the well-defined

[*]Chemical Sciences and Engineering Division, Argonne National Laboratory, 9700 South Cass Avenue, Argonne, IL 60439.
[†]School of Chemical Sciences, University of Illinois at Urbana-Champaign, Urbana, IL 61801.

molecular halide complexes $MoCl_3(THF)_3$ [7] and $NaW_2Cl_7(THF)_5$.[8] $MoCl_3(THF)_3$ reacts readily with LiO-t-Bu and $LiOCMe_2CF_3$ to afford $Mo_2(O\text{-}t\text{-}Bu)_6$ [9] and $Mo_2(OCMe_2CF_3)_6$ as described below. These complexes can also be prepared in comparable yield from "$MoCl_3(DME)$".[10] Similarly, $NaW_2Cl_7(THF)_5$ reacts readily with NaO-t-Bu and $NaOCMe_2CF_3$ to produce $W_2(O\text{-}t\text{-}Bu)_6$ and $W_2(OCMe_2CF_3)_6$, respectively.

General Procedures

Experiments are carried out in a dry nitrogen atmosphere in a glove box except where noted. Standard Schlenk techniques also suffice, but the glove box is more convenient for most manipulations here. Solvents are first sparged with nitrogen and then dried[*] by passing through activated alumina columns.[11] Deuterated solvents are degassed by three freeze–pump–thaw cycles, and then stored over 4 Å molecular sieves in the glove box. $MoCl_3(THF)_3$ is prepared according to a literature procedure.[12] WCl_4 (Strem), $HOCMe_2CF_3$ (Alfa Aesar), LiO-t-Bu (Aldrich), and NaO-t-Bu (Aldrich) are used as received. $LiOCMe_2CF_3$ is prepared by deprotonation of $HOCMe_2CF_3$ with n-BuLi in pentane exactly analogously to a literature procedure for the deprotonation of $HOCMe(CF_3)_2$.[13] $NaOCMe_2CF_3$ is prepared by treatment of $HOCMe_2CF_3$ with a slight deficiency of NaH in THF, followed by filtration and removal of the volatile components from the filtrate under vacuum. Sodium amalgam (0.4% Na by weight) is prepared by a literature method for 0.5% amalgam,[14] followed by dilution to 0.4% with additional Hg. Alternatively, it can be prepared as needed by sequential addition of small pieces of solid Na to liquid Hg (249 times the total mass of Na) in a small flask inside the glove box, with swirling until each chunk of Na dissolves fully.

A. HEXA(*tert*-BUTOXY)DIMOLYBDENUM(III)

$$2MoCl_3(THF)_3 + 6LiO\text{-}t\text{-Bu} \rightarrow Mo_2(O\text{-}t\text{-Bu})_6 + 6LiCl + 6THF$$

Procedure

Under a N_2 atmosphere, a 100 mL round-bottomed flask that contains a Teflon®-coated magnetic stir bar is charged with $MoCl_3(THF)_3$ (1.00 g, 2.40 mmol) and toluene (30 mL), producing an orange slurry. The flask is capped with a rubber septum and cooled at $-35°C$ for 30 min. After this time, the septum is removed, and to the stirring mixture is added LiO-t-Bu (0.577 g, 7.21 mmol) all at once. The septum is replaced and the stirring mixture is allowed to warm to ambient temperature (28°C), resulting in a reddish purple reaction mixture after 1 h. After

---

[*]The checkers dried their solvents by distillation under nitrogen from sodium benzophenone.

a total of 4 h of stirring, the color of the mixture is brown. The reaction mixture is filtered through Celite® and washed with toluene (30 mL). The filtrate and washings are combined and the volatile components are removed under vacuum to afford a brown solid. The brown solid is extracted with pentane (100 mL), the extract is filtered again through Celite®, and the volatile materials are again removed from the filtrate under vacuum. The resulting solid is dissolved in diethyl ether (2 mL) and the resulting solution is stored at −35°C overnight, whereupon dark orange crystals form. The crystals are isolated by filtration. The filtered supernatant is concentrated to ∼1 mL total volume and again stored at −35°C overnight to obtain a second crop of crystals. A third crop is similarly obtained from <1 mL of diethyl ether. Yield (three crops): 0.429 g (56%).

Anal. Calcd. for $C_{24}H_{54}Mo_2O_6$: C, 45.71; H, 8.63. Found: C, 45.49; H, 8.37.

Properties

$Mo_2(O\text{-}t\text{-}Bu)_6$ is an orange-brown, diamagnetic, air-sensitive solid that can be sublimed. It is highly soluble in most common organic solvents, including pentane and diethyl ether. It is nearly insoluble in acetonitrile. ^1H NMR (C_6D_6): δ 1.57 (s, CH_3). $^{13}C\{^1H\}$ NMR (C_6D_6): δ 79.22 (s, OC), 33.17 (s, CH_3). Unlike its tungsten analog, $Mo_2(O\text{-}t\text{-}Bu)_6$ does not react quantitatively with 1 equiv of small internal alkynes to yield alkylidyne complexes formed by cleavage of the M≡M triple bond.

B. HEXAKIS(2-TRIFLUOROMETHYL-2-PROPOXY)-DIMOLYBDENUM(III)

$$2MoCl_3(THF)_3 + 6LiOCMe_2CF_3 \rightarrow Mo_2(OCMe_2CF_3)_6 + 6LiCl + 6THF$$

Procedure

Under a N_2 atmosphere, a 25 mL Schlenk flask containing a Teflon-coated stir bar is charged sequentially with $MoCl_3(THF)_3$ (1.0 g, 2.39 mmol) and toluene (8 mL). To the stirred reaction mixture is added $Li(OCMe_2CF_3)$ (0.961 g, 7.17 mmol), and the flask is heated to 45°C for 14 h, during which time the mixture acquires a red color. The reaction mixture is filtered through Celite®, and afterward the Celite® is washed with toluene (30 mL). The combined filtrate and washings are concentrated to 3 mL total volume and cooled to −35°C overnight, during which time a red-orange precipitate forms. The precipitate is isolated by vacuum filtration. The filtrate is concentrated slightly under vacuum and again cooled overnight to −35°C in order to obtain a second crop of red-orange powder. Yield (three crops): 0.800 g (70%).

Anal. Calcd. for $C_{24}H_{36}F_{18}Mo_2O_6$: C, 30.20; H, 3.80. Found: C, 29.97; H, 3.67.

Properties

This red-orange compound is very similar to $Mo_2(O\text{-}t\text{-}Bu)_6$ in its physical appearance and sensitivity to oxygen and water, but $Mo_2(OCMe_2CF_3)_6$ is noticeably less soluble in pentane at room temperature. $Mo_2(OCMe_2CF_3)_6$ is soluble in benzene and toluene and also dissolves readily in acetonitrile, with which it forms a 1:2 adduct. ^1H NMR (C_6D_6): δ 1.39 (s, CH_3). ^{19}F NMR (C_6D_6): δ −82.51 (s, CF_3). $^{13}C\{^1H\}$ NMR (C_6D_6): δ 125.69 (q, J_{CF} = 285 Hz, CF_3), 82.27 (q, J_{CF} = 29 Hz, OC), 24.81 (s, CH_3).

C. SODIUM HEPTACHLOROPENTAKIS(TETRAHYDROFURAN) DITUNGSTATE(III)

$$2WCl_4 + 2Na + 5THF \rightarrow NaW_2Cl_7(THF)_5 + NaCl$$

Procedure

Under a N_2 atmosphere, a 100 mL round-bottomed flask containing a Teflon-coated stir bar is charged sequentially with THF (50 mL) and WCl_4 (5.377 g, 16.5 mmol).[*] The resulting suspension is frozen solid in liquid nitrogen and allowed to thaw over a magnetic stir plate. As soon as the suspension has melted sufficiently that stirring begins, sodium amalgam (0.38 g Na in 7 mL of Hg; 16.5 mmol of Na) is added.[†] The resulting reaction mixture is stirred and gradually warmed to ambient temperature (~28°C in the glove box). Within 1 h, the mixture turns dark green. After 18 h, the reaction mixture is decanted away from the pool of Hg, and the green decantate is filtered through Celite®.[‡] The Celite® is further washed with THF (200 mL) until the filtrate is colorless. The combined filtrate and washings are concentrated to 25 mL total volume[§] and left to stand in a stoppered flask at ambient temperature overnight to crystallize. Green crystals and powder

---

[*]The checkers note that this preparation often affords lower yields if carried out on smaller scale.

[†]The checkers added the sodium amalgam in three portions over 30 min from a dropping funnel under argon.

[‡]The checkers note that, when carrying out this preparation on a Schlenk line, filtration through Celite® on a glass frit is preferred to filtration with a filter cannula. Owing to the presence of large amounts of solid, filtration through a filter cannula is very slow and usually leads to oxidative decomposition of the product.

[§]The checkers note that it is easier to obtain the product as a powder if the solution is first concentrated under vacuum to ~15 mL, rediluted to 25 mL with fresh THF to dissolve the oil that forms, and then cooled to −35°C. It is also important not to try to crystallize from an overconcentrated solution; the viscosity should be similar to that of pure THF.

form by the following morning. The flask is then cooled to −35°C for 3 h, after which time the mother liquor is decanted. The isolated yellow-green powder and crystals are dried under vacuum. The mother liquor is concentrated to ~10 mL volume and then treated similarly to obtain a second crop of yellow-green crystals and powder. Yield (two crops): 4.13 g (50%).[*]

Anal. Calcd. for $C_{20}H_{50}Cl_7NaO_5W_2$: C, 24.04; H, 4.03. Found: C, 23.53; H, 3.70.

Properties

The compound is air stable in the crystalline state for short periods and solidifies as yellowish green cubes or long yellowish green needles, depending on the rate of crystallization. Crystals of both morphologies are characterized by the same space group and unit cell dimensions.[8] Crystallization of $NaW_2Cl_7(THF)_5$ occurs more readily when the synthesis is performed on a larger scale. The compound is reasonably soluble in THF, sparingly soluble in CH_2Cl_2 and $CHCl_3$, and essentially insoluble in hydrocarbon solvents such as toluene, benzene, and hexane. ^{1}H NMR ($CDCl_3$): δ 3.75 (m, α-CH_2, 4H), 1.85 (m, β-CH_2, 4H). X-ray crystallography has shown that this complex adopts a confacial bioctahedral geometry in which three chloride ligands bridge two W centers separated by 2.403(1) Å; two additional chloride and one tetrahydrofuran ligand complete the coordination at each W center. A $Na(THF)_3^+$ unit bridges the $W_2(\mu\text{-}Cl)_3Cl_4(THF)_2^-$ core by coordination to one otherwise terminal chloride ligand at each W center. $NaW_2Cl_7(THF)_5$ reacts with 6 equiv of LiX (X = NMe$_2$, O-t-Bu, CH_2-t-Bu) in THF to afford complexes of the form W_2X_6,[8] although we prefer to use sodium alkoxides in order to avoid the formation of "ate" complexes through coordination of an additional equivalent of alkoxide ligand.

D. HEXA(*tert*-BUTOXY)DITUNGSTEN(III)

$$NaW_2Cl_7(THF)_5 + 6NaO\text{-}t\text{-}Bu \rightarrow W_2(O\text{-}t\text{-}Bu)_6 + 7NaCl + 5THF$$

Procedure

Under a N_2 atmosphere, a 100 mL round-bottomed flask containing a Teflon-coated stir bar is charged with $NaW_2Cl_7(THF)_5$ (1.00 g, 1.01 mmol) and THF (40 mL). The resulting green solution is cooled to −35°C, and then is treated with NaO-t-Bu (0.580 g, 6.03 mmol). The reaction mixture is stirred for 18 h at ambient

---

[*]On a slightly smaller scale, the checkers obtained a yield of 16%. Possibly, the yield depends on the temperature during the reduction step; "ambient" temperature in a glove box is usually about 10°C higher than when using a Schlenk line.

temperature (28°C), during which time it takes on a red color. The solvent is removed in vacuum, and the resulting red solid is slurried in pentane (60 mL). The slurry is filtered through Celite®, concentrated in vacuum to 3 mL, and cooled to −35°C overnight. The red crystals are isolated by decantation or vacuum filtration. The filtrate is concentrated slightly, and then again cooled to −35°C overnight to obtain a second crop. Yield (three crops): 0.505 g (62%).[*]

Anal. Calcd. for $C_{24}H_{54}O_6W_2$: C, 35.75; H, 6.71. Found: C, 35.68; H, 6.59.

Properties

$W_2(O\text{-}t\text{-}Bu)_6$ is a red crystalline compound that is highly soluble in pentane, diethyl ether, and many other common organic solvents. Like all tungsten(III) alkoxides, $W_2(O\text{-}t\text{-}Bu)_6$ is highly sensitive to both oxygen and water. ^1H NMR (C_6D_6): δ 1.63 (s, CH_3). $^{13}C\{^1H\}$ NMR (C_6D_6): δ 80.08 (s, OC), 33.40 (s, CH_3). Single-crystal X-ray diffraction reveals an unbridged dimer composed of two W centers with local pseudotetrahedral coordination. This compound reacts rapidly and cleanly with many internal alkynes to afford terminal alkylidyne complexes of the form $RC≡W(O\text{-}t\text{-}Bu)_3$.[1,2] $RC≡W(O\text{-}t\text{-}Bu)_3$ complexes are excellent catalysts for alkyne metathesis. Reaction of $W_2(O\text{-}t\text{-}Bu)_6$ with 1 equiv of a nitrile RCN typically leads to a 1:1 mixture of $RC≡W(O\text{-}t\text{-}Bu)_3$ and the nitrido complex $N≡W(O\text{-}t\text{-}Bu)_3$;[1,2] the latter is sparingly insoluble in alkane solvents at room temperature, forming a linear polymer with alternating short and long W−N bonds. Reaction of $W_2(O\text{-}t\text{-}Bu)_6$ with two or more equivalents of RCN typically leads to $N≡W(O\text{-}t\text{-}Bu)_3$ as the sole W-containing product, along with the alkyne RCCR. In a suitable solvent, $N≡W(O\text{-}t\text{-}Bu)_3$ catalyzes the degenerate exchange of N atoms between nitriles, which is a convenient method for transferring a ^{15}N label.

E. HEXAKIS(2-TRIFLUOROMETHYL-2-PROPOXY)DITUNGSTEN(III)

$$NaW_2Cl_7(THF)_5 + 6NaOCMe_2CF_3 \rightarrow W_2(OCMe_2CF_3)_6 + 7NaCl + 5THF$$

Procedure

Under a N_2 atmosphere, a 100 mL round-bottomed flask containing a Teflon-coated stir bar is charged with $NaW_2Cl_7(THF)_5$ (0.518 g, 0.522 mmol) and THF (40 mL). The resulting green solution is cooled to −35°C, and then is treated with $NaOCMe_2CF_3$ (0.469 g, 3.13 mmol). The stirred green reaction mixture is allowed to warm to ambient temperature (28°C). The reaction mixture is stirred for 15 h,

---

[*]On a slightly smaller scale, the checkers obtained a 38% yield in the first crop and 51% overall.

during which time it turns reddish brown in color. The volatile material is removed in vacuum to afford a brown-red solid. This solid is extracted with pentane (80 mL) and the resulting pentane suspension is filtered through Celite®. The filtrate is concentrated under vacuum to dryness, then redissolved in a minimal (~3 mL) amount of diethyl ether, and cooled to $-35°C$ overnight. The deep red crystals are isolated by vacuum filtration; the filtered supernatant is concentrated to ~1.5 mL volume and again cooled to $-35°C$ overnight to obtain a second crop of crystals. Yield (two crops): 0.443 g (75%).[*]

Anal. Calcd. for $C_{24}H_{36}F_{18}O_6W_2$: C, 25.50; H, 3.21. Found: C, 25.37; H, 3.24.

Properties

$W_2(OCMe_2CF_3)_6$ is dark red in the solid state and when dissolved in hydrocarbon solvents. It is highly soluble in benzene and toluene as well as ethereal solvents. ^1H NMR (C_6D_6): δ 1.51 (s, CH_3). ^{19}F NMR (C_6D_6): δ -82.69 (s, CF_3). $^{13}C\{^1H\}$ NMR (C_6D_6): δ 125.80 (q, $J_{CF} = 285$ Hz, CF_3), 82.99 (q, $J_{CF} = 29$ Hz, OC), 25.25 (s, CH_3). The W–W triple bond is rapidly cleaved by internal alkynes such as 3-hexyne to afford the corresponding alkylidyne complexes, which are highly active catalysts for alkyne metathesis.[1,2] Unlike $W_2(O-t-Bu)_6$, $W_2(OCMe_2CF_3)_6$ is not rapidly cleaved by acetonitrile, instead forms a 1:2 adduct, $W_2(OCMe_2CF_3)_6(NCMe)_2$, like its Mo analog.[1] However, unlike the molybdenum analog, $W_2(OCMe_2CF_3)_6(NCMe)_2$ slowly undergoes further reaction in the presence of ~10 equiv of MeCN over 2 weeks at room temperature to produce $N\equiv W(OCMe_2CF_3)_3$. This nitrido complex exists as a trimer in which the alkoxide ligands are terminal, and there are alternating long and short W–N bonds in the W_3N_3 core.[1] The reaction of $W_2(OCMe_2CF_3)_6$ with benzonitrile also reversibly affords a 1:2 adduct, but cleavage of the $C\equiv N$ bond is more rapid in this case. In contrast to reactions of benzonitrile with $W_2(O-t-Bu)_6$, the benzylidyne complex $PhC\equiv W(OCMe_2CF_3)_3$ is not observed by $^{13}C\{^1H\}$ NMR spectroscopy, even in the presence of 1 equiv of isotopically labeled $Ph^{13}CN$. Instead, $N\equiv W(OCMe_2CF_3)_3$ and unreacted $W_2(OCMe_2CF_3)_6$ are the only observable tungsten-containing compounds. This result, along with the observed formation of $Ph^{13}C^{13}CPh$, implies that $PhC\equiv W(OCMe_2CF_3)_3$ reacts much more rapidly with $Ph^{13}CN$ than does $W_2(OCMe_2CF_3)_6$ itself.[1] Catalysis of $C\equiv C/C\equiv N$ triple bond metathesis between nitriles and alkynes occurs with $N\equiv W(OCMe_2CF_3)_3$, $PhC\equiv W(OCMe_2CF_3)_3$, and other $RC\equiv W(OCMe_2CF_3)_3$ alkylidyne complexes.

---

[*]The checkers find that this compound is more soluble and more difficult to crystallize than the *tert*-butoxide compound. They obtained it as a reddish brown powder in 52% overall yield.

References

1. M. H. Chisholm and C. B. Hollandsworth, in *Multiple Bonds Between Metal Atoms*, 3rd ed., F. A. Cotton, R. A. Walton, and C. A. Murillo, eds., Springer, New York, 2005.
2. R. R. Schrock, *Angew. Chem., Int. Ed.* **45**, 3748 (2006).
3. W. Zhang and J. S. Moore, *Adv. Synth. Catal.* **349**, 93 (2007).
4. M. H. Chisholm, D. A. Haitko, and C. A. Murillo, *Inorg. Synth.* **21**, 51 (1982).
5. M. H. Chisholm, F. A. Cotton, B. A. Frenz, W. W. Reichert, L. W. Shive, and B. R. Stults, *J. Am. Chem. Soc.* **98**, 4469 (1976).
6. M. H. Chisholm, F. A. Cotton, M. Extine, and B. R. Stults, *J. Am. Chem. Soc.* **98**, 4477 (1976).
7. F. Stoffelbach, D. Saurenz, and R. Poli, *Eur. J. Inorg. Chem.* 2699 (2001).
8. M. H. Chisholm, B. W. Eichhorn, K. Folting, J. C. Huffman, C. D. Ontiveros, W. E. Streib, and W. G. Van der Sluys, *Inorg. Chem.* **26**, 3182 (1987).
9. C. E. Laplaza, A. R. Johnson, and C. C. Cummins, *J. Am. Chem. Soc.* **118**, 709 (1996).
10. T. M. Gilbert, A. M. Landes, and R. D. Rogers, *Inorg. Chem.* **31**, 3438 (1992).
11. A. B. Pangborn, M. A. Giardello, R. H. Grubbs, R. K. Rosen, and F. J. Timmers, *Organometallics* **15**, 1518 (1996).
12. S. Maria and R. Poli, *Inorg. Synth.* **36**, 15 (2014).
13. F. J. de la Mata and R. H. Grubbs, *Organometallics* **15**, 577 (1996).
14. D. J. Santure, A. P. Sattelberger, F. A. Cotton, and W. Wang, *Inorg. Synth.* **26**, 219 (1989).

19. LINEAR TRICHROMIUM, TRICOBALT, TRINICKEL, AND TRICOPPER COMPLEXES OF 2,2′-DIPYRIDYLAMIDE

Submitted by JOHN F. BERRY,* F. ALBERT COTTON,*† and CARLOS A. MURILLO*
Checked by ZHI-KAI CHAN,‡ CHUN-WEI YEH,‡ and JHY-DER CHEN‡

The structural and physical properties of compounds containing extended metal atom chains (EMACs) have been of interest for many years.[1] Although early studies focused on polymeric materials such as Krogmann salts ($[PtL_4^{x-}]_n$, L = halide, x = a noninteger),[2] studies of discrete oligomeric EMAC-containing molecules have recently been made possible by the use of polydentate ligands that can hold several metal atoms together in a straight line.

I   M = Cr, Co, Ni, Cu   2,2'-Dipyridylamide (dpa)

*Laboratory for Molecular Structure and Bonding, Department of Chemistry, Texas A&M University, College Station, TX 77842.
†Deceased, February 20, 2007.
‡Department of Chemistry, Chung-Yuan Christian University, Chung-Li, Taiwan, Republic of China.

Trinuclear complexes of the type $M_3(dpa)_4X_2$ (**I**, dpa $=$ 2,2'-dipyridylamide) are the best studied;[3] in contrast, the preparation of longer chains generally necessitates the use of harsh reaction conditions that are difficult to control.[4] Efficient high-yield syntheses of trichromium, tricobalt, trinickel, and tricopper species in crystalline form have been achieved, however, and are presented here. The physical properties of these complexes have been studied, and many derivatives are known, including one-electron oxidized species,[5] organometallic species,[6] and supramolecular polymeric chain-like species.[7] For the synthesis of such derivatives, the acetonitrile complexes $[M_3(dpa)_4(NCCH_3)_2](PF_6)_2$ are far more useful than the chloro precursors; thus, their syntheses are also given here.

General Procedures

All reactions and manipulations should be performed under an atmosphere of dry nitrogen using either a dry box or standard Schlenk techniques. Anhydrous $CrCl_2$, $CoCl_2$, and $CuCl_2$ (Strem Chemicals) are rigorously dried before use. Thus, $CrCl_2$ is heated under nitrogen in refluxing Me_3SiCl, and $CoCl_2$ and $CuCl_2$ are heated under nitrogen in refluxing $SOCl_2$. The ligand 2,2'-dipyridylamine (Hdpa, Pfaltz & Bauer) is recrystallized successively from hot hexanes until colorless needle-shaped crystals are obtained. The compound $Ni(Hdpa)_2Cl_2$ is prepared by the method of Hurley and Robinson from $NiCl_2 \cdot 6H_2O$ and 2 equiv of Hdpa in refluxing butanol; the solid is dried overnight under vacuum before use.[8] Anhydrous $TlPF_6$ (Strem Chemicals) is crushed with a mortar and pestle, and heated to $>100°C$ under vacuum overnight immediately before use. All solvents are dried by distillation over appropriate drying agents.

A. DICHLOROTETRAKIS(2,2'-DIPYRIDYLAMIDO)-
TRICHROMIUM(II)

$$3CrCl_2 + 4Li(dpa) \rightarrow Cr_3(dpa)_4Cl_2 + 4LiCl$$

Procedure

The ligand Hdpa (1.10 g, 6.40 mmol) is dissolved in THF (50 mL) and the colorless solution is cooled to $-78°C$ by means of a dry ice/acetone bath. Carefully, methyllithium (4.0 mL of a 1.6 M solution in Et_2O, 6.4 mmol) is added, whereupon a cloudy white precipitate is observed. The mixture is allowed to warm to room temperature for about 30 min, and is then transferred via cannula to a flask containing anhydrous $CrCl_2$ (630 mg, 5.10 mmol). A bright red mixture is obtained, which is stirred at room temperature for 1 h and then stirred at reflux for another 6–8 h. A deep green solid and a green filtrate result. The solid is collected by filtration, washed with Et_2O (15 mL), and extracted with CH_2Cl_2

$(4 \times 20 \, \text{mL})$. Diffusion of hexanes into the combined extracts yields a poly-crystalline sample of $\mathbf{1} \cdot CH_2Cl_2$. The crystals are collected, washed with hexanes, and dried under dynamic vacuum overnight. Yield: 1.30 g (87%).

Anal. Calcd. for $C_{41}H_{34}Cl_4Cr_3$ ($Cr_3(dpa)_4Cl_2 \cdot CH_2Cl_2$): C, 49.60; H, 3.43; N, 16.94. Found: C, 49.32; H, 3.55; N, 17.08.

Properties

The dark green solid is stable in air. The IR spectrum shows the characteristic bands of the triply bridging dpa ligand at 1606, 1595, 1468, and $1427 \, \text{cm}^{-1}$. The ^1H NMR spectrum in CD_2Cl_2 consists of three broad singlets of roughly equal intensity at δ 13.6, -1.4, and -15.2. Though not all of the resonances expected for the compound are observed (due to the paramagnetism of the compound), the ^1H NMR spectrum is a useful diagnostic in assessing the purity of the compound. The visible absorption spectrum consists of three distinct peaks at 673, 611, and 515 nm with molar absorptivities ranging from 1000 to $2000 \, M^{-1} \, cm^{-1}$. Also present is a high-energy shoulder at 437 nm. The solid-state molecular structure consists of an unsymmetrical trichromium chain having a short Cr–Cr distance of 2.24 Å and a long Cr $\cdots$ Cr nonbonded separation of 2.47 Å.[9]

B. DICHLOROTETRAKIS(2,2′-DIPYRIDYLAMIDO)TRICOBALT(II)

$$3CoCl_2 + 4Li(dpa) \rightarrow Co_3(dpa)_4Cl_2 + 4LiCl$$

Procedure

The ligand Hdpa (1.37 g, 8.00 mmol) is dissolved in THF (50 mL) and the resulting solution is cooled to $-78°C$ by means of a dry ice/acetone bath. Methyllithium (5 mL of a 1.6 M solution in Et_2O, 8.0 mmol) is carefully added, forming a cloudy white precipitate. This mixture is allowed to warm to room temperature with stirring for 30 min, whereupon it is transferred via cannula to a flask containing anhydrous $CoCl_2$ (0.780 g, 6.00 mmol). A dark brown mixture forms immediately, which is stirred at room temperature for 1 h and then heated to reflux for ~ 12 h. A dark black solid and a dark brown solution result. The solid is collected by filtration, washed with Et_2O (2×15 mL), and extracted with CH_2Cl_2 (3×20 mL). Slow diffusion of hexanes into the combined extracts yields large crystals of $\mathbf{2} \cdot CH_2Cl_2$ and $\mathbf{2} \cdot 2CH_2Cl_2$. The crystals are collected, washed with hexanes, and dried under vacuum overnight. Yield: 1.58 g (85%).

Anal. Calcd. for $C_{41}H_{34}N_{12}Cl_4Co_3$ ($Co_3(dpa)_4Cl_2 \cdot CH_2Cl_2$): C, 48.59; H, 3.38; N, 16.59. Found: C, 48.45; H, 3.35; N, 16.51. Calcd. for $C_{42}H_{36}N_{12}Cl_6Co_3$

$(Co_3(dpa)_4Cl_2 \cdot 2CH_2Cl_2)$: C, 45.93; H, 3.30; N, 15.31. Found: C, 45.38; H, 3.32; N, 15.07.

Properties

When the black solid is crystallized from CH_2Cl_2/hexanes, two different crystal morphologies are observed that have been shown to have significantly different physical properties.[10] Orthorhombic crystals of $Co_3(dpa)_4Cl_2$ are block shaped and contain one interstitial molecule of CH_2Cl_2 per formula unit; the Co−Co bond distances within each molecule are equal. Tetragonal crystals of $Co_3(dpa)_4Cl_2$ are needle-like in appearance and contain *two* interstitial molecules of CH_2Cl_2 per formula unit; the Co−Co distances within each molecule are unequal (2.29 and 2.47 Å). It is not necessary to separate these crystals before further reactions. The IR spectra of these crystalline forms are essentially the same and contain strong absorptions at 1607, 1593, 1469, and 1423 cm^{-1} due to the dpa ligand, though some minor differences in the spectra in the 600–1200 cm^{-1} region are observed. Brown solutions of either the tetragonal crystals or the orthorhombic crystals show the same properties. The ^1H NMR spectrum in CD_2Cl_2 consists of four sharp signals at δ 48.0 (s), 33.2 (d, $J = 7.9\,Hz$), 16.8 (d, $J = 6.7\,Hz$), and 16.0 (t) due to the pyridyl protons. There are two main features in the visible–NIR spectrum: a weak band at 1315 nm ($\varepsilon = 70\,M^{-1}\,cm^{-1}$) and an intense band at 567 nm ($\varepsilon = 2000\,M^{-1}\,cm^{-1}$). It is likely that the former absorption is due to a spin-forbidden $^4A \leftarrow {}^2A$ transition, whereas the latter absorption band is probably the HOMO–LUMO transition.

C. DICHLOROTETRAKIS(2,2′-DIPYRIDYLAMIDO)TRINICKEL(II)

$$3Ni(Hdpa)_2Cl_2 + 4LiCH_3 \rightarrow Ni_3(dpa)_4Cl_2 + 4CH_4 + 4LiCl + 2Hdpa$$

Procedure

A pale bright blue suspension of $Ni(Hdpa)_2Cl_2$ (2.83 g, 6.00 mmol) in THF (60 mL) is cooled to −78°C by means of a dry ice/acetone bath. To this suspension is slowly added methyllithium (5.0 mL of a 1.6 M solution in diethyl ether, 4.0 mmol), which causes the suspension to become lime green in color. The suspension is then allowed to warm to room temperature, at which point it becomes dark brown. After the brown mixture is stirred at room temperature for 30 min, it is heated to reflux for ~12 h, which causes the color to change to dark purple. The solvent is removed under reduced pressure leaving a dark purple residue. Extraction of this solid with dichloromethane affords a purple solution, from which the dark purple-red product is crystallized by diffusion of hexanes. Crystals were collected by filtration and washed with hexanes. Yield: 1.57 g (85%).

Anal. Calcd. for $C_{42}N_{12}H_{36}Cl_6Ni_3$ ($Ni_3(dpa)_4Cl_2 \cdot 2CH_2Cl_2$): C, 45.96; H, 3.31; N, 15.31. Found: C, 45.58; H, 3.21; N, 14.74.

Properties

The dark purple-red compound is stable in air both in the solid state and in solution. The IR spectrum shows strong peaks at 1602, 1592, 1467, and 1423 cm^{-1} characteristic of the triply bridging dpa ligand. The ^1H NMR spectrum in CD_2Cl_2 consists of four resonances at δ 48.8, 32.3, 15.7, and 9.1 due to the pyridyl protons. The UV–vis spectrum shows a single absorption at 520 nm ($\epsilon = 3000\ M^{-1}\ cm^{-1}$).

D. DICHLOROTETRAKIS(2,2′-DIPYRIDYLAMIDO)TRICOPPER(II)

$$3CuCl_2 + 4Li(dpa) \rightarrow Cu_3(dpa)_4Cl_2 + 4LiCl$$

Procedure

A flask is charged with Hdpa (0.548 g, 3.20 mmol) dissolved in THF (35 mL). The solution is cooled to −78°C, and methyllithium (2 mL of 1.6 M solution in THF, 3.2 mmol) is added dropwise. The flask is removed from the cold bath, and the solution is stirred until it reaches room temperature, when a white turbidity is observed. The mixture is added via cannula to a flask containing anhydrous $CuCl_2$ (0.323 g, 2.40 mmol). A dark blue color appears immediately. The reaction mixture is stirred at room temperature for 30 min and then heated to reflux for 6 h, resulting in a dark blue solution with a blue precipitate. The mixture is filtered, and the precipitate is washed with THF (2 × 15 mL) and ether (2 × 15 mL). Finally, the blue precipitate is extracted with dichloromethane (2 × 20 mL), and the dark blue solution is layered with hexanes. After a week, large, block-shaped crystals are collected by filtration and washed with hexanes. Yield: 0.755 g (80%).

Anal. Calcd. for $C_{41}H_{34}N_{12}Cl_4Cu_3$ ($Cu_3(dpa)_4Cl_2 \cdot CH_2Cl_2$): C, 47.94; H, 3.34; N, 16.36. Found: C, 47.55; H, 3.33; N, 16.45.

Properties

The dark blue microcrystalline solid is stable in air indefinitely, though solutions slowly decompose. The IR spectrum shows peaks at 1605, 1593, 1473, and 1423 cm^{-1}, which are characteristic of the triply bridging dpa ligand. The ^1H NMR spectrum in CD_2Cl_2 consists of four resonances at δ 47.8, 27.4, 13.8, and 1.6, which are significantly shifted from their normal positions due to the paramagnetism of the compound. The peaks that are shifted farthest downfield are extremely broad, whereas the resonance at δ 1.6 is relatively sharp. There are two main absorptions in the visible region at 556 nm ($\epsilon = 900\,M^{-1}\,cm^{-1}$) and 666 nm ($\epsilon = 1200\,M^{-1}\,cm^{-1}$), which have been assigned as arising from the two different Cu^{2+} chromophores (one square planar Cu^{2+} and two square pyramidal Cu^{2+}).

E. BIS(ACETONITRILE)TETRAKIS(2,2′-DIPYRIDYLAMIDO)- TRICHROMIUM(II) BIS(HEXAFLUOROPHOSPHATE)

$$Cr_3(dpa)_4Cl_2 + 2TlPF_6 + 2CH_3CN \rightarrow [Cr_3(dpa)_4(NCCH_3)_2](PF_6)_2 + 2TlCl$$

Procedure

■ **Caution.** *Thallium hexafluorophosphate is toxic and should be handled in a hood.*

To a flask containing polycrystalline $Cr_3(dpa)_4Cl_2$ (814 mg, 0.897 mmol) and $TlPF_6$ (627 mg, 1.79 mmol) is added acetonitrile (25 mL). The resulting green mixture is stirred for 4 h, and a fine white precipitate is observed. The mixture is filtered through Celite®, taking care to avoid exposing the solution to air. The resulting clear dark olive green solution is layered with ether (100 mL). After the ether diffuses into the acetonitrile solution, large green block-shaped crystals of $[Cr_3(dpa)_4(NCCH_3)_2](PF_6)_2 \cdot 2CH_3CN$ form. These are collected, washed with hexanes, and dried under vacuum. Some of the axially coordinated CH_3CN ligands are lost upon placing the crystals under vacuum. Yield: 0.803 g (74%).

Anal. Calcd. for $C_{41}H_{33.5}N_{12.5}Cr_3F_{12}P_2$ ($[Cr_3(dpa)_4(NCCH_3)_{0.5}](PF_6)_2$): C, 42.93; H, 2.94; N, 15.26. Found: C, 42.44; H, 3.29; N, 15.54.

Properties

The large green crystals are stable when stored in a nitrogen atmosphere but decompose in air. Solutions are very air sensitive, and turn from green to orange upon exposure to the atmosphere. Thus, the filtration step above must be done very carefully to ensure good yields. The IR spectrum consists of strong bands at 1608, 1468, and 1429 cm^{-1} due to the dpa ligand. The C≡N stretch appears at 2268 cm^{-1} and a large broad peak at 840 cm^{-1} is due to the presence of the PF_6 anions. The ^1H NMR spectrum in CD_3CN consists of three broadened peaks at δ 7.3, 0.9, and −13.4. Although these cannot be specifically assigned, the spectrum does allow the purity of the compound to be assessed. The visible spectrum taken in CH_3CN consists of two main bands at 679 nm ($\epsilon = 2000\,M^{-1}\,cm^{-1}$) and 561 nm ($\epsilon = 1000\,M^{-1}\,cm^{-1}$) as well as two shoulders at higher energies: 515 nm ($\epsilon = 2000\,M^{-1}\,cm^{-1}$) and 435 nm ($\epsilon = 4000\,M^{-1}\,cm^{-1}$). When the compound is exposed to air, the lowest-energy band shifts to ~730 nm.

F. BIS(ACETONITRILE)TETRAKIS(2,2′-DIPYRIDYLAMIDO)-
TRICOBALT(II) BIS(HEXAFLUOROPHOSPHATE)

$$Co_3(dpa)_4Cl_2 + 2TlPF_6 + 2CH_3CN \rightarrow [Co_3(dpa)_4(NCCH_3)_2](PF_6)_2 + 2TlCl$$

Procedure

■ **Caution.** *Thallium hexafluorophosphate is toxic and should be handled in a hood.*

To a flask containing a polycrystalline sample of $Co_3(dpa)_4Cl_2$ (0.825 g, 0.889 mmol) and $TlPF_6$ (0.621 g, 1.777 mmol) is added acetonitrile (25 mL). The brown mixture is stirred for ~3 h, after which a white precipitate settles at the bottom of the flask. The mixture is filtered through Celite® and layered with Et_2O (100 mL). The crystals are collected by filtration, washed with hexanes, and dried under vacuum. Yield: 0.929 g (85%).

Anal. Calcd. for $C_{44}N_{14}H_{38}Co_3F_{12}P_2$ ([$Co_3(dpa)_4(NCCH_3)_2](PF_6)_2$): C, 42.98; H, 3.11; N, 15.95. Found: C, 43.12; H, 3.24; N, 16.34.

Properties

The black microcrystalline solid is stable in air and is best stored either in a desiccator or under nitrogen. Solutions are also somewhat stable, but decompose over time. The IR spectrum of the solid shows the characteristic stretches of the dpa ligand at 1603, 1473, and 1431 cm^{-1} as well as a $\tilde{\nu}(C{\equiv}N)$ band at 2270 cm^{-1} and a broad PF_6 band at 839 cm^{-1}. The ^1H NMR spectrum (CD_3CN) consists of the four resonances at δ 10.55, 3.55, 2.00, and −2.87 due to the pyridyl protons. Often a broad peak at δ ~3 is observed, which appears to be due to water exchanging in some of the axial sites. The vis–NIR spectrum consists of two bands at 562 nm ($\epsilon = 2000\,M^{-1}\,cm^{-1}$) and 1277 nm ($\epsilon = 300\,M^{-1}\,cm^{-1}$).

G. BIS(ACETONITRILE)TETRAKIS(2,2′-DIPYRIDYLAMIDO)-
TRINICKEL(II) BIS(HEXAFLUOROPHOSPHATE)

$$Ni_3(dpa)_4Cl_2 + 2TlPF_6 + 2CH_3CN \rightarrow [Ni_3(dpa)_4(NCCH_3)_2](PF_6)_2 + 2TlCl$$

Procedure

■ **Caution.** *Thallium hexafluorophosphate is toxic and should be handled in a hood.*

To a flask containing a crystalline sample of $Ni_3(dpa)_4Cl_2$ (0.319 g, 0.342 mmol) and $TlPF_6$ (0.240 g, 0.687 mmol) is added acetonitrile (25 mL). The purple mixture is stirred for ~3 h, whereupon a white precipitate is observed. The mixture is filtered through Celite® and layered with Et_2O (100 mL). Large purple crystals form overnight, which are collected, washed with hexanes, and dried under vacuum. Yield: 0.342 g (81%).

Anal. Calcd. for $C_{44}H_{38}F_{12}Ni_3P_2$ ($[Ni_3(dpa)_4(NCCH_3)_4](PF_6)_2$): C, 42.99; H, 3.09; N, 15.96. Found: C, 43.20; H, 3.32; N, 15.57.

Properties

The large purple crystals are stable in air as are solutions of the compound, though the compound is best stored in a desiccator or under nitrogen to avoid hydration. The IR spectrum shows the prominent peaks of the dpa ligand at 1604, 1595, 1469, 1460, and 1425 cm^{-1} as well as a $\tilde{\nu}(C\equiv N)$ band at 2291 cm^{-1} and a broad PF_6 band at 839 cm^{-1}. The ^1H NMR spectrum in CD_3CN consists of four resonances due to the pyridyl protons at δ 8.8, 15.2, 27.6, and 42.5. The resonances for the CH_3CN ligands are not seen, most likely because they are in rapid exchange with the CD_3CN solvent. The visible spectrum consists of two main absorptions at 515 nm ($\epsilon = 3000$ M^{-1} cm^{-1}) and 616 nm ($\epsilon = 2000$ M^{-1} cm^{-1}).

The axially coordinated acetonitrile ligands are lost under prolonged exposure to vacuum, and the resulting solid easily acquires water from the atmosphere, as attested by the following analysis taken on a sample that had been vacuum dried for over 24 h.

Anal. Calcd. for $C_{40}H_{34}N_{12}F_{12}Ni_3OP_2$ ($[Ni_3(dpa)_4(H_2O)](PF_6)_2$): C, 41.25; H, 2.94; N, 14.43. Found: C, 41.54; H, 3.41; N, 14.88.

References

1. J. S. Miller, ed., *Extended Linear Chain Compounds*, Plenum Press, New York, 1982–1983, 3 vols.
2. (a) K. Krogmann and P. Dodel, *Chem. Ber.* **99**, 3402 (1966); (b) K. Krogmann and P. Dodel, *Chem. Ber.* **99**, 3408, (1966); (c) K. Krogmann, *Z. Anorg. Allg. Chem.* **358**, 97 (1968); (d) K. Krogmann, *Angew. Chem., Int. Ed. Engl.* **8**, 35 (1969).
3. J. F. Berry, in *Multiple Bonds Between Metal Atoms*, F. A. Cotton, C. A. Murillo, and R. A. Walton, eds., Springer, New York, 2005, Chapter 15.
4. (a) S. Aduldecha and B. Hathaway, *J. Chem. Soc., Dalton Trans.* 993 (1991); (b) E.-C. Yang, M.-C. Cheng, M.-S. Tsai, and S.-M. Peng, *J. Chem. Soc., Chem. Commun.* **20**, 2377 (1994); (c) J.-T. Sheu, C.-C. Lin, I. Chao, C.-C. Wang, and S.-M. Peng, *Chem. Commun.* **3**, 315 (1996); (d) F. A. Cotton, L. M. Daniels, C. A. Murillo, and I. Pascual, *J. Am. Chem. Soc.* **119**, 10223 (1997); (e) S.-J. Shieh, C.-C. Chao, G.-H. Lee, C.-C. Wang, and S.-M. Peng, *Angew. Chem., Int. Ed. Engl.* **36**, 56

(1997). (f) S.-Y. Lai, T.-W. Lin, Y.-H. Chen, C.-C. Wang, G.-H. Lee, M.-H. Yang, M.-K. Leung, and S.-M. Peng, *J. Am. Chem. Soc.* **121**, 250 (1999); (g) S.-M. Peng, C.-C. Wang, Y.-L. Jang, Y.-H. Chen, F.-Y. Li, C.-Y. Mou, and M.-K. Leung, *J. Magn. Magn. Mater.* **209**, 80 (2000).

5. J. F. Berry, F. A. Cotton, L. M. Daniels, C. A. Murillo, and X. Wang, *Inorg. Chem.* **42**, 2418 (2003), and references therein.
6. (a) J. F. Berry, F. A. Cotton, C. A. Murillo, and B. K. Roberts, *Inorg. Chem.* **43**, 2277 (2004); (b) J. F. Berry, F. A. Cotton, and C. A. Murillo, *Organometallics* **23**, 2503 (2004); (c) T. Sheng, R. Appelt, V. Comte, and H. Vahrenkamp, *Eur. J. Inorg. Chem.* 3731 (2003).
7. (a) H. Li, G.-H. Lee, and S.-M. Peng, *Inorg. Chem. Commun.* **6**, 1 (2003); (b) C.-H. Peng, C.-C. Wang, H.-C. Lee, W.-C. Lo, G.-H. Lee, and S.-M. Peng, *J. Chin. Chem. Soc. (Taipei)* **48**, 987 (2001); (c) T.-B. Tsao, G.-H. Lee, C.-Y. Yeh, and S.-M. Peng, *J. Chem. Soc., Dalton Trans.* 1465 (2003).
8. T. J. Hurley and M. A. Robinson, *Inorg. Chem.* **7**, 33 (1968).
9. J. F. Berry, F. A. Cotton, T. Lu, C. A. Murillo, B. K. Roberts, and X. Wang, *J. Am. Chem. Soc.* **126**, 7082 (2004).
10. R. Clérac, F. A. Cotton, L. M. Daniels, K. R. Dunbar, K. Kirschbaum, C. A. Murillo, A. A. Pinkerton, A. J. Schultz, and X. Wang, *J. Am. Chem. Soc.* **122**, 6226 (2000).

20. BIS(TETRABUTYLAMMONIUM) OCTACHLORODITECHNETATE(III)

Submitted by FREDERIC POINEAU,[*] ERIK V. JOHNSTONE,[*] and
KENNETH R. CZERWINSKI[*]
Checked by ALFRED P. SATTELBERGER[†]

The coordination chemistry of quadruply metal–metal bonded technetium(III) dimers is not well developed and, at present, only a small number of dinuclear technetium(III) compounds are well characterized.[1] Among these compounds, $(n\text{-}Bu_4N)_2Tc_2Cl_8$ (**I**) is the most studied and is a potential precursor to many other multiply bonded technetium dimers. This compound has been claimed by several different laboratories using a variety of synthetic methods, but its preparation in pure form has proven difficult and in some instances has been controversial.[2–5] We tested the two procedures reported by Preetz et al.[4] The first procedure, which involves a reduction of $(n\text{-}Bu_4N)_2TcCl_6$ with 2 equiv of $(n\text{-}Bu_4N)BH_4$ in dichloromethane, gives very poor yields of the desired compound, which was only identified in solution by its characteristic UV–vis bands. The second procedure, presented here, involves the two-electron reduction of $(n\text{-}Bu_4N)TcOCl_4$ with $(n\text{-}Bu_4N)BH_4$ in THF, followed by treatment of the resultant brown intermediate

[*]Department of Chemistry, University of Nevada-Las Vegas, Las Vegas, NV 89154.
[†]Energy Engineering and Systems Analysis Directorate, Argonne National Laboratory, Lemont, IL 60439.

with gaseous hydrogen chloride. $(n\text{-}Bu_4N)_2Tc_2Cl_8$ is obtained reproducibly in $\sim$30% overall yield, based on NH_4TcO_4.

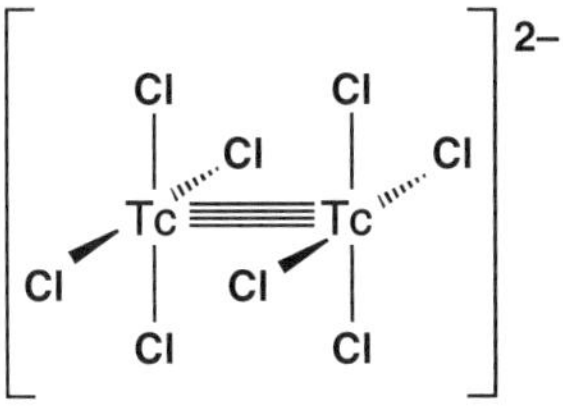

I

General Procedures

NH_4TcO_4 is purchased from Oak Ridge National Laboratory. The compound, which should be pure white, usually contains a black TcO_2 impurity. The solvents and reagents (THF, acetone, ether, $(n\text{-}Bu_4N)BH_4$, $(n\text{-}Bu_4N)HSO_4$, CH_2Cl_2, hexane, and 12 M HCl) are all purchased from Sigma-Aldrich and used as received. A lecture bottle of HCl(g) is purchased from Sigma-Aldrich.

■ **Caution.** *Technetium-99 is a weak beta-emitter ($E_{max} = 292\,keV$). All manipulations must be performed in a laboratory designed for low-level radioactivity using efficient HEPA-filtered fume hoods, Schlenk techniques, and following locally approved nuclear chemistry handling and monitoring procedures. Lab coats, disposable gloves, and protective eyewear must be worn at all times.*

A. TETRABUTYLAMMONIUM PERTECHNETATE(VII)

$$NH_4TcO_4 + (n\text{-}Bu_4N)HSO_4 \rightarrow (n\text{-}Bu_4N)TcO_4 + (NH_4)HSO_4$$

Procedure

A sample of impure NH_4TcO_4 (0.617 g, 3.409 mmol) is dissolved/suspended in deionized water (12 mL) in a 100 mL round-bottomed flask, and 30% aqueous hydrogen peroxide (30 μL) and concentrated NH_4OH (300 μL) are added to the black suspension. The flask is placed on a rotary evaporator and heated for 30 min while rotating in a 100°C oil bath. After 30 min, the solution is *carefully* evaporated to dryness in vacuum while heating. After being dried, the NH_4TcO_4 in the 100 mL flask is redissolved in water (12 mL) and the solution is carefully transferred by disposable pipette into four centrifuge glass tubes ($\sim$4 mL in each, including washings). Then $(n\text{-}Bu_4N)HSO_4$ (1.72 g, 5.05 mmol) dissolved in deionized water (16 mL) is added dropwise (4 mL to each tube), and voluminous white $(n\text{-}Bu_4N)$ TcO_4 precipitates. The suspensions are centrifuged, the supernatant is removed with a disposable glass pipette, and the solid in each tube is washed two times with ice-cold water (5 mL); the washings are discarded. The tubes are placed in a desiccator under vacuum and the product is dried over Drierite® for 24 h. The white $(n\text{-}Bu_4N)TcO_4$ product is used directly in the next step.

B. TETRABUTYLAMMONIUM OXOTETRACHLOROTECHNETATE(V)

$$(n\text{-}Bu_4N)TcO_4 + 6HCl \rightarrow (n\text{-}Bu_4N)TcOCl_4 + 3H_2O + Cl_2$$

Procedure

To each tube containing $(n\text{-}Bu_4N)TcO_4$ from the previous step is added 12 M HCl (4 mL) and a small magnetic stir bar. The suspensions are shaken briefly and then stirred for 30 min at room temperature. Voluminous, green $(n\text{-}Bu_4N)TcOCl_4$ precipitates. The tubes are centrifuged and the supernatant is removed and discarded. Each tube is treated a second time with fresh, cold (5°C) concentrated HCl (4 mL) for 30 min. After the tubes are centrifuged, the clear, pale yellow supernatant is decanted and discarded, and the green $(n\text{-}Bu_4N)TcOCl_4$ is thoroughly dried in a vacuum desiccator overnight. Then, CH_2Cl_2 (2 mL) is added to each tube, the solutions are combined in two *preweighed* 20 mL glass vials, and hexane (4 mL) is added to induce crystallization. The gray-green $(n\text{-}Bu_4N)TcOCl_4$ is dried in a stream of Ar. Yield: 1.449 g (85% based on NH_4TcO_4).

Properties

The compound $(n\text{-}Bu_4N)TcOCl_4$ has been characterized by extended X-ray absorption fine structure (EXAFS) spectroscopy.[6]

C. BIS(TETRABUTYLAMMONIUM) OCTACHLORODITECHNETATE(III)

$$2(n\text{-}Bu_4N)TcOCl_4 + 4(n\text{-}Bu_4N)BH_4 + 16HCl \rightarrow$$

$$(n\text{-}Bu_4N)_2Tc_2Cl_8 + 2H_2O + 14H_2 + \text{``}4(n\text{-}Bu_4N)BCl_4\text{''}$$

Procedure

The reaction is performed in a 250 mL round-bottomed flask equipped with a three-hole rubber stopper. A portion of $(n\text{-}Bu_4N)TcOCl_4$ (933 mg, 1.866 mmol) is

TABLE 1. Optical Absorption Spectra of $M_2Cl_8^{2-}$ (M = Tc, Re) in Degassed Dichloromethane.[6]

Compound	ν, cm^{-1} (ϵ, M^{-1} cm^{-1})
$Tc_2Cl_8^{2-}$	14,713 (1391), 25,623 (8674), 33,016 (10,564)
$Re_2Cl_8^{2-}$	14,589 (2556), 32,003 (11,430), 39,447 (10,742)

dissolved in THF (20 mL). A solution of (n-Bu$_4$N)BH$_4$ (967 mg, 3.764 mmol) in THF (20 mL) is added under a countercurrent of Ar. Some effervescence is observed during the addition. Five minutes after the addition, the brown supernatant is removed with a pipette, leaving a brown solid at the bottom of the flask. The solid is washed with THF (2 × 10 mL) and ether (2 × 10 mL). After the washing, the solid (whose color has changed to yellow-green) is dried (5 min) in the flask under an Ar stream. Then, CH$_2$Cl$_2$ (20 mL) is added to the flask, the solution is stirred, and an HCl(g) stream is passed slowly through the solution for 7 min. A color change from yellow-green to emerald green occurs immediately. Hexane (20 mL) is added to the flask and a green solid precipitates. The supernatant is removed by pipette and the crude (n-Bu$_4$N)$_2$Tc$_2$Cl$_8$ is washed with hexane (20 mL) and dried with an Ar stream. Acetone (20 mL) is added to the flask, the green solution is transferred into four centrifuged tubes, and the (n-Bu$_4$N)$_2$Tc$_2$Cl$_8$ product is induced to crystallize by adding a minimal amount of ether to each tube. After the tubes are centrifuged, the supernatant is removed, and the remaining solid is dried with an Ar stream. Yield: 333 mg (37%).

Properties

Green (n-Bu$_4$N)$_2$Tc$_2$Cl$_8$ is soluble in dichloromethane and acetone, but insoluble in THF, hexane, and ether. In the absence of air, the compound is stable in organic solvents for a couple of hours, but rapidly oxidizes to TcCl$_6^{2-}$ in air. Analysis of the solid by X-ray diffraction, EXAFS spectroscopy, and UV–vis spectroscopy confirms the presence of (n-Bu$_4$N)$_2$Tc$_2$Cl$_8$ as a single phase without impurities.[6,7] The Tc$_2$Cl$_8^{2-}$ ion possesses a quadruple bond (Tc–Tc = 2.147(4) Å) and is isostructural with Re$_2$Cl$_8^{2-}$. The UV–vis spectra of Re$_2$Cl$_8^{2-}$ and Tc$_2$Cl$_8^{2-}$ exhibit similar features with the presence of three major absorption bands. The $\delta \rightarrow \delta^*$ transitions (characteristic of the quadruple bond) are nearly superimposable, whereas the other bands are shifted to higher energy for the Re compound (Table 1).

References

1. A. P. Sattelberger, in *Multiple Bonds Between Metal Atoms*, 3rd ed., F. A. Cotton, C. A. Murillo, and R. A. Walton, eds., Springer, New York, 2005.

2. F. A. Cotton, L. Daniels, A. Davison, and C. Orvig, *Inorg. Chem.* **20**, 3051 (1981).
3. K. Schwochau, K. Hedwig, H. J. Schenk, and O. Greis, *Inorg. Nucl. Chem. Lett.* **13**, 77 (1977).
4. W. Preetz and G. Peters, *Z. Naturforsch. B* **35**, 797 (1980).
5. W. Preetz, G. Peters, and D. Bublitz, *J. Cluster Sci.* **5**, 83 (1994).
6. F. Poineau, A. P. Sattelberger, S. D. Conradson, and K. R. Czerwinski, *Inorg. Chem.* **47**, 1991 (2008).
7. F. Poineau, E. V. Johnstone, P. M. Forster, L. Ma, A. P. Sattelberger, and K. R. Czerwinski, *Inorg. Chem.* **51**, 9563, (2012).

21. DIRUTHENIUM FORMAMIDINATO COMPLEXES

Submitted by PANAGIOTIS A. ANGARIDIS,[*] F. ALBERT COTTON,[*†] and CARLOS A. MURILLO[*]
Checked by ALEXANDER S. FILATO[‡] and MARINA A. PETRUKHINA[‡]

Among the plethora of dinuclear compounds of the paddlewheel framework, those of the diruthenium core are of special interest due to their rich redox activity and their paramagnetic behavior. Ru_2 compounds have been isolated in three formal oxidation states, Ru_2^{4+}, Ru_2^{5+}, and Ru_2^{6+}, with a variety of O,O'-, N,O-, and N,N'-donor bridging ligands.[1] The type of the bridging ligands (as well as the type of the axial ligands as has been recently shown[2]) is of great importance because they affect the structural characteristics and the electronic structures of these compounds, which in turn have a large influence on their magnetic properties.[3]

One of the less studied compounds of the Ru_2^{5+} core with N,N'-donor bridging ligands are those with formamidinates, despite the fact that such ligands have been extensively used in complexes of other metals, such as Mo and Rh. The first Ru_2^{5+} formamidinate, $Ru_2(DToIF)_4Cl$ (DToIF = N,N'-di-p-tolylformamidinate), was reported by Cotton and Ren in 1995.[4] Since then, a few other similar compounds have been synthesized and studied.[5] The majority of them are of the type $Ru_2(formamidinate)_4Cl$,[6] and only a few examples of Ru_2^{5+} compounds with a mixed set of bridging ligands have been reported.[7]

Here, we present the syntheses of a series of Ru_2^{5+} complexes of the type $Ru_2(acetate)_{4-n}(formamidinate)_nCl$ ($n = 1$, 2, 3, 4), in which the number of formamidinates can be varied systematically;[8] see the scheme below. Compounds of this type exhibit interesting magnetic properties, such as spin crossover[6] or spin admixture,[9] and also have the potential to be used as precursors for the syntheses of more complex architectures, such as molecular pairs,[10] molecular squares,[11] or

[*]Laboratory for Molecular Structure and Bonding, Department of Chemistry, Texas A&M University, PO Box 30012, College Station, TX 77842.
[†]Deceased, February 20, 2007.
[‡]Department of Chemistry, University at Albany, State University of New York, Albany, NY 12222.

other polymeric assemblies[12] (exploiting the fact that bridging acetate ligands are labile and they can be substituted by other polyfunctional ligands).

General Procedures

$Ru_2(O_2CMe)_4Cl$ is prepared by a modification of the procedure[13] in which $RuCl_3 \cdot nH_2O$ and LiCl are heated to reflux in a mixture of glacial acetic acid and acetic anhydride for 15 h in a flask open in air. $RuCl_3 \cdot nH_2O$ (Strem) is used as received. LiCl (Aldrich) is rigorously dried before use. The formamidines HDAniF (*N,N'*-di-*p*-anisylformamidine) and $DXyl^{2,6}F$ (*N,N'*-di-2,6-xylylformamidine) are prepared according to a published procedure[14] from the reactions of triethylorthoformate with the corresponding amines in a 1 : 2 ratio.

All reactions and manipulations are performed under a nitrogen atmosphere, using standard Schlenk line techniques. Commercial grade solvents are dried and deoxygenated by reflux under an N_2 atmosphere for at least 24 h over appropriate drying agents and are freshly distilled before use. Infrared spectra are recorded as KBr pellets. Positive ion electrospray ionization mass spectra (+ESI-MS) are recorded on CH_2Cl_2 solutions. Cyclic voltammetry experiments are recorded in THF solutions that contained 0.1 M $^nBu_4NPF_6$ as the supporting electrolyte. All potentials are referenced to the Ag/AgCl electrode; the $E_{1/2}$ for the Fc/Fc$^+$ couple under our experimental conditions occurred at +0.600 V.

A. CHLOROTRIS(ACETATO)(*N,N'*-DI-2,6-XYLYLFORMAMIDINATO)-DIRUTHENIUM(II,III)

$$Ru_2(O_2CMe)_4Cl + HDXyl^{2,6}F \rightarrow Ru_2(O_2CMe)_3(DXyl^{2,6}F)Cl + HO_2CMe$$

Procedure

To a Schlenk flask containing $Ru_2(O_2CMe)_4Cl$ (0.105 g, 0.220 mmol) and $HDXyl^{2,6}F$ (0.551 g, 2.19 mmol) is added toluene (20 mL). The reaction mixture is stirred and refluxed gently (the temperature should not be higher than 110°C) for 36 h resulting in a red mixture. That mixture is filtered in order to remove a small amount of unreacted $Ru_2(O_2CMe)_4Cl$. To the filtrate a mixture of isomeric hexanes (60 mL) is added under stirring, resulting in the precipitation of a red solid.[*] The supernatant liquid is carefully removed by cannula filtration and the residue is washed with a 1:1 mixture of isomeric hexanes and Et_2O (2 × 40 mL). The resulting solid is dried under vacuum and extracted with THF (20 mL). Slow diffusion of isomeric hexanes (40 mL) into the resulting red THF solution yields a dark orange-brown crystalline sample of $Ru_2(O_2CMe)_3(DXyl^{2,6}F)Cl(THF)\cdot$ 0.5THF over a period of 1 week.[†] Drying the crystals results in loss of the THF. Yield: 0.04 g (28%).[‡]

Anal. Calcd. for $C_{23}H_{28}N_2ClO_6Ru_2$: C, 41.47; H, 4.21; N, 4.21. Found: C, 41.78; H, 4.29; N, 4.09.

Properties

The dark orange-brown crystalline compound is stable in air and soluble in CH_2Cl_2, THF, and EtOH. Its positive ion ESI-MS shows only one peak at $m/z = 631$, with an isotopic distribution consistent with the fragment $[M-Cl]^+$. Positive ion electrospray ionization mass spectrometry is very useful in characterizing these compounds, but the axially coordinated Cl^- anion is easily lost during the fragmentation process, and only fragments that correspond to the species $[M-Cl]^+$ are observed.

   The IR spectrum of the polycrystalline compound shows the following bands (cm^{-1}): 3449 (w), 3198 (w), 3126 (w), 2959 (w), 1627 (w), 1502 (s), 1441 (s), 1351 (sh), 1294 (m), 1200 (m), 1097 (m), 1034 (m), 771 (m), 689 (s), 625 (w). Its cyclic voltammogram displays an irreversible metal-centered oxidation at $E_{ap}(1) = +1.295$ V and a quasireversible metal-centered reduction with $E_{ap}(2) = -0.142$ V and $E_{cp}(2) = -0.228$ V. An additional cathodic peak is observed at $E_{cp}(3) = -0.627$ V. Its room-temperature magnetic moment is 3.74 BM per Ru_2^{5+} unit. The solid-state molecular structure of $Ru_2(O_2CMe)_3(DXyl^{2,6}F)Cl(THF)\cdot$ 0.5THF shows a metal–metal bond distance of 2.305(1) Å.[12]

---

[*]The checkers induced precipitation by placing the mixture in an ice bath for 10–15 min.
[†]The checkers note that no crystals were observed even after 1 week. However, crystallization occurred after placing the mixture in a freezer for 2 h.
[‡]The checkers obtained a yield of 37%.

If the solid isolated from the reaction (after it is washed with a 1:1 mixture of isomeric hexanes and Et$_2$O) is dissolved in a noncoordinating solvent, such as toluene, and the resulting solution is layered with hexanes, crystals of Ru$_2$(O$_2$C-Me)$_3$(DXyl^{2,6}F)Cl(HDXyl^{2,6}F) · toluene may grow, apparently, due to the incomplete purification of the residue from the excess of unreacted neutral HDXyl^{2,6}F.

B. *trans*-CHLOROBIS(ACETATO)BIS(*N*,*N*′-DI-2,6-XYLYLFORMAMIDINATO) DIRUTHENIUM(II,III)

Ru$_2$(O$_2$CMe)$_4$Cl + 2HDXyl^{2,6}F → *trans*-Ru$_2$(O$_2$CMe)$_2$(DXyl^{2,6}F)$_2$Cl + 2HO$_2$CMe

Procedure

A Schlenk tube containing a mixture of Ru$_2$(O$_2$CMe)$_4$Cl (0.100 g, 0.210 mmol) and HDXyl^{2,6}F (0.517 g, 2.05 mmol) is heated at ∼150°C for 8 h. During that time, the molten reaction mixture turns dark red. After the mixture is cooled to room temperature, the solid is transferred* into a sublimation apparatus and the excess of HDXyl^{2,6}F is removed by vacuum sublimation onto a water-cooled cold finger by heating the mixture at 135°C under vacuum for 8 h. The resulting dark solid is extracted with toluene (3 × 10 mL) and to the combined red extracts is added hexanes (60 mL). The mixture is stirred until a red microcrystalline solid precipitates. The supernatant liquid is removed by cannula filtration and the residue is washed with isomeric hexanes (40 mL). The resulting solid is dried under vacuum and extracted with THF (20 mL). Slow diffusion of hexanes (40 mL) into the red THF solution yields large dark red crystals of *trans*-Ru$_2$(O$_2$CMe)$_2$(DXyl^{2,6}F)$_2$Cl (THF) over a period of a few days. Drying of the crystals results in loss of the THF. Yield: 0.13 g (71%).

Anal. Calcd. for C$_{38}$H$_{44}$N$_4$ClO$_4$Ru$_2$: C, 53.18; H, 5.13; N, 6.53. Found: C, 53.56; H, 5.21; N, 6.41.

Properties

The dark red crystalline material is stable in air and soluble in CH$_2$Cl$_2$, THF, EtOH, and toluene. The positive ion ESI-MS of this compound shows only one peak at $m/z = 824$ with an isotopic distribution consistent with the fragment [M − Cl]$^+$. The IR spectrum of the polycrystalline compound exhibits the

---

*The checkers did not transfer the solid but instead sublimed out the excess HDXyl^{2,6}F by fitting the reaction flask with a cold finger. The yield of product was 77%.

following bands (cm^{-1}): 2955 (w), 1528 (s), 1465 (m), 1439 (s), 1318 (m), 1254 (w), 1194 (s), 1095 (m), 1040 (m), 887 (m), 774 (m), 686 (m), 455 (m). Its cyclic voltammogram displays a quasireversible metal-centered oxidation with $E_{ap}(1) = +0.978$ V and $E_{cp}(1) = +0.888$ V and a quasireversible metal-centered reduction with $E_{ap}(2) = -0.303$ V and $E_{cp}(2) = -0.412$ V. An additional cathodic wave is observed at $E_{cp}(3) = -0.830$ V. Its room-temperature magnetic moment is 3.74 BM per Ru_2^{5+} unit. The solid-state molecular structure of *trans*-$Ru_2(O_2CMe)_2(DXyl^{2,6}F)_2Cl(THF)$ reveals a metal–metal bond distance of 2.326(1) Å.[12]

If the solid isolated from the reaction mixture (after it is washed with isomeric hexanes) is dissolved in a noncoordinating solvent, such as toluene, and the resulting solution is layered with hexanes, crystals with composition *trans*-$Ru_2(O_2CMe)_2(DXyl^{2,6}F)_2Cl \cdot 2toluene$ may grow.

C. *cis*-CHLOROBIS(ACETATO)BIS(*N*,*N*′-DI-*p*-ANISYLFORMAMIDINATO)DIRUTHENIUM(II,III)

$$Ru_2(O_2CMe)_4Cl + 2HDAniF \rightarrow cis\text{-}Ru_2(O_2CMe)_2(DAniF)_2Cl + 2MeCO_2H$$

Procedure

Method 1. To a mixture of $Ru_2(O_2CMe)_4Cl$ (0.237 g, 0.500 mmol), HDAniF (0.256 g, 1.00 mmol), NEt$_3$ (0.15 g, 1.5 mmol), and LiCl (0.500 g) is added THF (35 mL). The red mixture is heated to reflux for 18 h resulting in a very dark solution. The volatile materials are removed under vacuum. Benzene (2 × 15 mL) is added to the solid and the mixture is filtered through Celite®. The volume of the benzene solution is reduced to 15 mL and then a mixture of isomeric hexanes (50 mL) is added with stirring, which results in the precipitation of a dark solid. This solid is collected by filtration, dried under vacuum, and dissolved in CH$_2$Cl$_2$ (25 mL). Slow diffusion of isomeric hexanes (60 mL) into the resulting CH$_2$Cl$_2$ solution yields large dark green-purple crystals of *cis*-$Ru_2(O_2CMe)_2(-DAniF)_2Cl \cdot H_2O$ over a period of 1 week. Yield: 0.35 g (81%). Column chromatography (silica gel; eluent (v/v): CH$_2$Cl$_2$/acetone = 5/2) may also be used for the purification of the solid in those cases where a pure solid cannot be isolated.

Method 2. To a mixture of $Ru_2(O_2CMe)_4Cl$ (0.157 g, 0.330 mmol) and HDAniF (0.211 g, 0.820 mmol) is added THF (15 mL). The brown suspension is stirred and refluxed gently (the temperature should not be higher than 75°C) for 18 h, resulting in a color change to dark purple. Some unreacted $Ru_2(O_2C-Me)_4Cl$ is filtered off and the solvent is removed under vacuum leaving a dark residue. The residue is dissolved in MeCN (5 mL) and precipitated by addition of a mixture of Et$_2$O (20 mL) and isomeric hexanes (70 mL). The supernatant liquid is removed by cannula filtration, and the solid is dried under vacuum. The procedure of dissolution in MeCN and cannula filtration is then repeated. The product is purified by column chromatography (green band; silica gel; eluent (v/v): CH_2Cl_2/acetone = 5/2). A dark green solid is isolated. Yield: 0.11 g (38%).

Anal. Calcd. for $C_{34}H_{36}N_4ClO_8Ru_2$: C, 47.14; H, 4.16; N, 6.47. Found: C, 46.73; H, 4.08; N, 6.32.

Properties

The compound *cis*-$Ru_2(O_2CMe)_2(DAniF)_2Cl$ is stable in air and soluble in various organic solvents, such as CH_2Cl_2, THF, EtOH, and benzene, but insoluble in hexanes. Its positive ion ESI-MS shows only one peak at $m/z = 831$, with an isotopic distribution consistent with the fragment $[M - Cl]^+$. The IR spectrum of the crystalline compound exhibits the following bands (cm^{-1}): 3447 (m), 2938 (w), 2835 (w), 1605 (m), 1529 (s), 1500 (s), 1438 (s), 1316 (m), 1297 (m), 1247 (s), 1216 (s), 1177 (m), 1109 (w), 1034 (m), 835 (m), 770 (w), 693 (m), 594 (w), 545 (w), 460 (w). Its cyclic voltammogram displays a reversible metal-centered oxidation with $E_{1/2}(1) = +0.817$ V and an irreversible metal-centered reduction with $E_{cp}(3) = -0.668$ V. An additional anodic wave is observed at $E_{ap}(2) = -0.297$ V. Its room-temperature magnetic moment is 3.75 BM per Ru_2^{5+} unit. The solid-state molecular structure of *cis*-$Ru_2(O_2CMe)_2(DAniF)_2Cl \cdot H_2O$ shows a Ru–Ru distance of 2.319(1) Å.[11]

D. CHLORO(ACETATO)TRIS(*N,N'*-DI-*p*-ANISYLFORMAMIDINATO)-DIRUTHENIUM(II,III)

$$Ru_2(O_2CMe)_4Cl + 3HDAniF \rightarrow Ru_2(O_2CMe)(DAniF)_3Cl + 3MeCO_2H$$

Procedure

To a mixture of $Ru_2(O_2CMe)_4Cl$ (0.237 g, 0.500 mmol), HDAniF (0.384 g, 1.50 mmol), NEt_3 (0.202 g, 2.00 mmol), and LiCl (0.500 g) is added THF (30 mL). The mixture is heated to reflux for 36 h. The color of the solution changes gradually to dark purple. After the volatile materials are removed under vacuum, benzene (30 mL) is added to the solid and the mixture is filtered through Celite®. The volume of the solution is reduced to ~15 mL and a mixture of isomeric hexanes (50 mL) is added while stirring, which results in the precipitation of a dark purple solid. This solid is collected by filtration and dissolved in benzene (20 mL). Diffusion of hexanes (60 mL) into the resulting solution results in the formation of dark purple polycrystalline $Ru_2(O_2CMe)(DAniF)_3Cl$ within 3–4 days. Yield: 0.45 g (85%).[*]

Anal. Calcd. for $C_{47}H_{48}N_6ClO_8Ru_2$: C, 53.13; H, 4.52, N, 7.91. Found: C 52.79, H, 4.55, N, 7.77.

Properties

The dark purple polycrystalline material is soluble in CH_2Cl_2, THF, EtOH, and benzene. Its positive ion ESI-MS shows only one signal at $m/z = 1028$, with an isotopic distribution consistent with the fragment $[M - Cl]^+$. The IR spectrum of the polycrystalline compound exhibits the following bands (cm^{-1}): 2917 (vs), 2849 (s), 1607 (w), 1542 (m), 1503 (vs), 1467 (m), 1440 (w), 1378 (w), 1316 (m), 1291 (m), 1246 (s), 1215 (s), 1172 (m), 1106 (w), 1033 (m), 941 (w), 831 (m), 790 (w), 766 (w), 721 (w), 691 (w), 652 (w), 621 (w), 591 (w), 547 (w), 515 (w). Its cyclic voltammogram displays a reversible metal-centered oxidation with $E_{1/2}(1) = +0.624$ V and an irreversible metal-centered reduction with $E_{cp}(3) = -0.709$ V. An additional anodic wave is observed at $E_{ap}(2) = -0.253$ V. Its room-temperature magnetic moment is 3.79 BM per Ru_2^{5+} unit.

E. CHLOROTETRAKIS(*N*,*N'*-DI-*p*-ANISYLFORMAMIDINATO)-DIRUTHENIUM(II,III)

$$Ru_2(O_2CMe)_4Cl + 4HDAniF \rightarrow Ru_2(DAniF)_4Cl + 4MeCO_2H$$

Procedure

To a mixture of $Ru_2(O_2CMe)_4Cl$ (0.097 g, 0.200 mmol) and HDAniF (0.551 g, 2.15 mmol) is added toluene (15 mL). The mixture is stirred and refluxed vigorously for 3 days resulting in a dark-colored mixture. After the solid is allowed to settle, the supernatant liquid is removed by cannula filtration. The microcrystalline solid left in the flask is washed with toluene (2×20 mL), EtOH (2×20 mL), and subsequently dried under vacuum. The solid is dissolved in CH_2Cl_2 (20 mL) and the resulting green solution is layered with isomeric hexanes (30 mL). Dark green-purplish crystals of $Ru_2(DAniF)_4Cl \cdot 0.5CH_2Cl_2$ grow over a period of 3–4 days. Drying the solid results in loss of the dichloromethane. Yield: 0.20 g (79%).

Anal. Calcd. for $C_{60}H_{60}N_8ClO_8Ru_2$: C, 57.24; H, 4.77; N, 8.90. Found: C, 57.04; H, 4.69; N, 8.95.

---

[*]The checkers obtained a yield of 91%.

Properties

The dark purple-greenish crystals of $Ru_2(DAniF)_4Cl$ are stable in air and soluble in CH_2Cl_2 and THF, but are insoluble in EtOH and toluene. The positive ion ESI-MS of this compound shows only one peak at $m/z = 1224$, with an isotopic distribution consistent with the fragment $[M - Cl]^+$. The IR spectrum of the crystalline compound exhibits the following bands (cm^{-1}): 3447 (w), 2957 (w), 2828 (w), 2368 (w), 1603 (m), 1547 (m), 1503 (vs), 1457 (m), 1337 (m), 1291 (m), 1243 (s), 1214 (s), 1172 (m), 1106 (m), 1031 (s), 930 (w), 827 (m) 830 (sh), 588 (w). Its cyclic voltammogram displays a reversible metal-centered oxidation with $E_{1/2}(1) = +0.431$ V and an irreversible metal-centered reduction with $E_{cp}(3) = -0.754$ V. Its room-temperature magnetic moment is 3.58 BM per Ru_2^{5+} unit. The solid-state molecular structure of $Ru_2(DAniF)_4Cl \cdot 0.5CH_2Cl_2$ reveals a metal–metal bond distance of 2.453(1) Å.[12]

References

1. P. Angaridis, in *Multiple Bonds Between Metal Atoms*, 3rd ed., F. A. Cotton, C. A. Murillo, and R. A. Walton, eds., Springer, New York, 2005, Chapter 9.
2. F. A. Cotton, C. A. Murillo, J. H. Reibenspies, D. Villagrán, X. Wang, and C. C. Wilkinson, *Inorg. Chem.* **43**, 8373 (2004).
3. (a) A. J. Lindsay, G. Wilkinson, M. Motevalli, and M. B. Hursthouse, *J. Chem. Soc., Dalton Trans.* 2321 (1985); (b) F. A. Cotton and T. Ren, *Inorg. Chem.* **30**, 3675 (1991).
4. F. A. Cotton and T. Ren, *Inorg. Chem.* **34**, 3190 (1995).
5. T. Ren, *Coord. Chem. Rev.* **175**, 43 (1998).
6. (a) C. Lin, T. Ren, E. J. Valente, J. D. Zubkowski, and E. T. Smith, *Chem. Lett.* 753 (1997); (b) J. L. Bear, B. Han, S. Huang, and K. M. Kadish, *Inorg. Chem.* **35**, 3012 (1996).
7. (a) T. Ren, V. DeSilva, G. Zou, C. Lin, L. M. Daniels, C. F. Campana, J. C. Alvarez, *Inorg. Chem. Commun.* **2**, 301 (1999); (b) M. C. Barral, S. Herrero, R. Jiménez-Aparicio, M. R. Torres, and F. A. Urbanos, *Inorg. Chem. Commun.* **7**, 42 (2004).
8. P. Angaridis, F. A. Cotton, C. A. Murillo, D. Villagrán, and X. Wang, *J. Am. Chem. Soc.* **127**, 5008 (2005).
9. M. C. Barral, S. Herrero, R. Jiménez-Aparicio, M. R. Torres, and F. A. Urbanos, *Angew. Chem.* **44**, 305 (2005).
10. P. Angaridis, J. F. Berry, F. A. Cotton, P. Lei, C. Lin, C. A. Murillo, and D. Villagrán, *Inorg. Chem. Commun.* **7**, 9 (2004).
11. P. Angaridis, J. F. Berry, F. A. Cotton, C. A. Murillo, and X. Wang, *J. Am. Chem. Soc.* **125**, 10327 (2003).
12. P. Angaridis, J. F. Berry, F. A. Cotton, C. A. Murillo, D. Villagrán, and X. Wang, *Inorg. Chem.* **43**, 8290 (2004).
13. T. A. Stephenson, and G. Wilkinson, *J. Inorg. Nucl. Chem.* **28**, 2285 (1966).
14. (a) C. Lin, J. D. Protasiewicz, E. T. Smith, and T. Ren, *Inorg. Chem.* **35**, 6422 (1996); (b) F. A. Cotton, L. M. Daniels, C. A. Murillo, and P. Schooler, *J. Chem. Soc., Dalton Trans.* 2001 (2000).

22. HEPTACARBONYL(DISULFIDO)DIMANGANESE(I)

Submitted by RICHARD D. ADAMS,[*] O-SUNG KWON,[*] and SHAOBIN MIAO[*]
Checked by CAMERON W. SPAHN[†] and THOMAS B. RAUCHFUSS[†]

$$+ 2CO + MeCN + 2CH_2=CH_2$$

Sulfur-containing transition metal carbonyl complexes have attracted much attention because of their versatile bonding mode and reactivities.[1] Compounds containing disulfido ligands are also of great interest due to their potential to attach to other metal centers as inorganic mimics of organic disulfides.[2] Oxidative addition of low-valent metal complexes to complexes containing disulfido ligands is a practical method for the preparation of higher nuclearity complexes containing sulfido ligands.[3]

Mixed-metal sulfido cluster complexes have attracted attention because of their application in catalysis.[4] Although disulfido ligands exhibit great versatility as bridging ligands, dinuclear metal disulfides are still relatively rare. We have synthesized the new dimanganese disulfide complex, $Mn_2(CO)_7(\mu\text{-}S_2)$.[5]

In this synthesis, a convenient procedure for the preparation of heptacarbonyl-disulfidodimanganese, $Mn_2(CO)_7(\mu\text{-}S_2)$, from the acetonitrile derivative of $Mn_2(CO)_{10}$ and the strained cyclic thioether known as thiirane (SCH_2CH_2), which serves as a source of sulfur, is described.

General Procedures

Thiirane (ethylene sulfide) can be purchased from Aldrich Chemical Company and should be vacuum distilled before use. The $Mn_2(CO)_9(NCMe)$ is readily prepared by treatment of $Mn_2(CO)_{10}$ with trimethylamine N-oxide in acetonitrile.[6]

■   **Caution.**   *Thiirane is strongly malodorous. All manipulations should be performed in a well-ventilated fume hood.*

[*]Department of Chemistry and Biochemistry, University of South Carolina, Columbia, SC 29208.
[†]Department of Chemistry, University of Illinois at Urbana-Champaign, Urbana, IL 61801.

Procedure

A 1 L three-necked round-bottomed flask is equipped with a Teflon®-coated magnetic stir bar, a nitrogen inlet, a condenser, and a nitrogen outlet adapter on the condenser. Distilled hexane (720 mL) is added and the flask is purged with N_2 for 30 min by using a gas dispersion tube. Crystalline $Mn_2(CO)_9(NCMe)$ (600 mg, 1.489 mmol) is added under a N_2 counterflow. To the yellow solution is added thiirane (240 μL, 4 mmol) via a syringe. The reaction apparatus is wrapped completely in aluminum foil to minimize possible effects of light on the reaction. The solution is allowed to stir at 25°C for 20 h. The solution is reduced in volume to 100 mL and then filtered through Celite®. The volatile materials are removed in vacuum, and the products are separated by using column chromatography (SiO_2, 3 cm × 20 cm) with hexane solvent as eluant to yield $Mn_2(CO)_7(\mu\text{-}S_2)$. Yield: 250 mg (45%). *Note*: Thin layer chromatography causes partial decomposition of $Mn_2(CO)_7(\mu\text{-}S_2)$ resulting in a lower yield. This product is readily recrystallized by dissolving in a minimum amount of hexane/methylene chloride (9/1, v/v) solvent mixture and then cooling to −78°C for 5 h.

Anal. Calcd. for $Mn_2S_2C_7O_7$: C, 22.71. Found: C, 23.29.

Properties

$Mn_2(CO)_7(\mu\text{-}S_2)$ is an orange-red crystalline solid at room temperature and is stable in air for long periods without significant decomposition. It is readily soluble in common organic solvents (dichloromethane, hexane, and benzene) to give red solutions that are slightly air sensitive. ^{13}C NMR (toluene-d_8): δ 212.5, 213.5, 236.5. IR [ν_{CO} (cm^{-1} in hexane)]: 2079 (w), 2044 (vs), 2007 (m), 1982 (vs), 1894 (w). Like its iron analog $Fe_2(CO)_6(\mu\text{-}S_2)$,[7] $Mn_2(CO)_7(\mu\text{-}S_2)$ can be used for the preparation of mixed metal complexes by insertion of low-valent metal complexes into the S−S bond.[8]

References

1. (a) M. Hidai, S. Kuwata, and Y. Mizobe, *Acc. Chem. Res.* **33**, 46 (2000); (b) M. Okazaki, M. Yuki, K. Kuge, and H. Ogino, *Coord. Chem. Rev.* **198**, 67 (2000); (c) P. Mathur, *Adv. Organomet. Chem.* **41**, 243 (1997); (d) L. C. Roof and J. W. Kolis, *Chem. Rev.* **93**, 1037 (1993).
2. (a) J. Wachter, *Angew. Chem., Int. Ed. Engl.* **28**, 1613 (1989); (b) R. B. King and T. B. Bitterwolf, *Coord. Chem. Rev.* **206–207**, 563 (2000); (c) M. Okazaki, M. Yuki, K. Kuge, and H. Ogino, *Coord. Chem. Rev.* **198**, 367 (2000); (d) M. Yuki, M. Okazaki, S. Inomata, and H. Ogino, *Organometallics* **18**, 3728 (1999); (e) K. H. Whitmire, Iron compounds without hydrocarbon ligands, in *Comprehensive Organometallic Chemistry II*, G. Wilkinson, F. G. A. Stone, and E. Abel, eds., Pergamon Press, New York, 1995, Vol. 7, Chapter 1, Section 1.11.2.2, p. 62, and references therein.
3. (a) D. Seyferth, R. S. Henderson, and L.-C. Song, *Organometallics* **1**, 125 (1982); (b) M. J. Don and M. G. Richmond, *Inorg. Chim. Acta* **210**, 129 (1993); (c) M. Cowie, R. L. Dekock, T. R.

Wagenmaker, D. Seyferth, R. S. Henderson, and M. K. Gallagher, *Organometallics* **8**, 119, (1989); (d) V. W. Day, D. A. Lesch, and T. B. Rauchfuss, *J. Am. Chem. Soc.* **104**, 1290 (1982); (e) M. D. Curtis, P. D. Williams, and W. M. Butler, *Inorg. Chem.* **27**, 2853 (1988).
4. M. Hidai, S. Kuwata, and Y. Mizobe, *Acc. Chem. Res.* **33**, 46 (2000).
5. R. D. Adams, O.-S. Kwon, and M. D. Smith, *Inorg. Chem.* **41**, 6281 (2002).
6. V. Koelle, *J. Organomet. Chem.* **155**, 53 (1978).
7. W. Hieber and J. Gruber, *Z. Anorg. Allg. Chem.* **296**, 91 (1958).
8. (a) R. D. Adams, O.-S. Kwon, and M. D. Smith, *Inorg. Chem.* **41**, 1658 (2002); (b) R.D. Adams, O.-S. Kwon, and M. D. Smith, *Organometallics* **21**, 1960 (2002).

23. DI(CARBIDO)TETRACOSA(CARBONYL)-DECARUTHENATE(2–) SALTS

Submitted by MICHAEL L. BRUCE,[*] AURORA CASTRO,[†] PAUL A. HUMPHREY,[*] PETER M. MAITLIS,[†] and NATASHA N. ZAITSEVA[*]
Checked by LYDIE VIAU[‡] and MARK G. HUMPHREY[‡]

The chemistry of metal cluster carbonyls containing carbido ligands dates back to the discovery and characterization of the first example, $Fe_5C(CO)_{15}$, in 1962.[1] For the metal ruthenium, many examples of clusters with different nuclearity are known.[2] In 1982, the decanuclear anion $[Ru_{10}C_2(CO)_{24}]^{2-}$ was obtained by pyrolysis of $[Ru_6C(CO)_{16}]^{2-}$ in tetraglyme at 210–230°C for 80 h.[3] The $Ru_{10}C_2$ cluster is formed by edge sharing of two Ru_6C clusters, the apical ruthenium atoms being 3.1020 and 3.2279(5) Å apart. Several subsequent reports described aspects of its CO substitution chemistry with ligands such as diaryl-acetylenes, bicyclo[2.2.0]hepta-2,5-diene, allene, and diazomethane.[4] In some of these compounds, the apical ruthenium atoms become close enough to bond to one another. However, extensive development of the chemistry of the $Ru_{10}C_2$ cluster has not occurred, perhaps because of its relative inaccessibility. Herein, we describe a synthesis of this anion from $Ru_3(CO)_{12}$ and calcium carbide in up to 50% yield, which is based on the method described earlier.[5]

A. CALCIUM DI(CARBIDO)TETRACOSA(CARBONYL)-DECARUTHENATE(2 –)

$$10Ru_3(CO)_{12} + 3CaC_2 \rightarrow 3Ca[Ru_{10}C_2(CO)_{24}] + 48CO$$

[*]School of Chemistry and Physics, University of Adelaide, Adelaide, South Australia 5005, Australia.
[†]Department of Chemistry, University of Sheffield, Sheffield S3 7HF, UK.
[‡]Department of Chemistry, Australian National University, Canberra, A.C.T. 0100, Australia.

Procedure

■   **Caution.**   *Carbon monoxide is a highly poisonous, colorless, and odorless gas. Calcium carbide reacts with water to produce flammable, poisonous ethyne gas. This reaction should be carried out in a well-ventilated hood and in the absence of water.*

A 100 mL round-bottomed Schlenk flask is charged with a mixture of $Ru_3(CO)_{12}$ (200 mg, 0.313 mmol)[6] and freshly ground CaC_2* (Aldrich, 80%; 200 mg, 2.50 mmol) and THF (40 mL). The stirred mixture is heated at reflux for 18 h during which time a dark brown solution over a dark-colored precipitate forms. If some $Ru_3(CO)_{12}$ remains, as determined via spot TLC, a further amount of CaC_2 (200 mg) can be added and heating continued until the $Ru_3(CO)_{12}$ is consumed.[†] The solvent is removed under vacuum, and the solid residue is extracted with acetone (2×20 mL, 6×10 mL) until the extracts, initially dark brown, are colorless. The combined purple-brown extracts are filtered by gravity and evaporated to an oil in a rotary evaporator. The components of the oil are separated by column chromatography (silica gel 60 (Merck), 5 cm $\times 2.5$ cm column[‡]). Elution with acetone/hexane affords several cluster complexes, previously identified, in small yields.[5] Elution with acetone gives a purple fraction, which is evaporated using a rotary evaporator to give a purple solid. Yield: 114 mg (70%).[§]

Properties

The product may contain THF or acetone coordinated to the calcium ion. IR (THF, cm^{-1}): 2049 (vw), 2004 (vs), 1960 (w), 1930 (w), 1796 (m).

B. BIS[BIS(TRIPHENYLPHOSPHORANYLIDENE)AMMONIUM] DI(CARBIDO)TETRACOSA(CARBONYL)DECARUTHENATE(2 –)

$$Ca[Ru_{10}C_2(CO)_{24}] + 2[PPN]Cl \rightarrow [PPN]_2[Ru_{10}C_2(CO)_{24}] + CaCl_2$$

---

*The yield of this procedure seems to depend on the quality of the calcium carbide used. The purity of the calcium carbide may be estimated from the amount of ethyne produced upon hydrolysis of a sample with dilute HCl. For example, pure CaC_2 (64 mg, 1 mmol) should afford 22.4 mL of C_2H_2. The checker manipulated the CaC_2 inside a glove box.
[†]The checker reports that, in two separate reactions, some unreacted $Ru_3(CO)_{12}$ remains even after adding additional CaC_2 and heating.
[‡]Preparative thin layer chromatography (silica gel plate 0.2 mm $\times 20$ cm $\times 20$ cm, acetone/hexane 3/7) enables separation of several minor products, including $Ru_4(\mu_4\text{-}HC_2H)(\mu\text{-}CO)_2(CO)_9$ and $Ru_6C(\mu_3\text{-}HC_2Me)(CO)_{15}$.[5] These by-products are probably formed from reactions involving small amounts of ethyne, formed from water and CaC_2.
[§]The checkers obtained a yield of 38–47%.

Procedure

A 100 mL round-bottomed Schlenk flask is charged with $Ca[Ru_{10}C_2(CO)_{24}]$ (114 mg, 0.066 mmol) and acetone (10 mL). To this mixture is added bis(triphenylphosphoranylidene)ammonium chloride, [PPN]Cl (85 mg, 0.15 mmol). After the mixture is stirred at room temperature for 30 min, it is filtered by gravity filtration, evaporated to dryness using a rotary evaporator, and crystallized as dark purple crystals from a 1:1 mixture of CH_2Cl_2 and MeOH. Yield: 127 mg (70%).

Anal. Calcd. for $C_{98}H_{60}N_2O_{24}P_4Ru_{10}$: C, 42.28; H, 2.17; N, 1.01. Found: C, 42.30; H, 2.04; N, 0.97.

Properties

The compound $[PPN]_2[Ru_{10}C_2(CO)_{24}]$ is a dark purple crystalline solid that is insoluble in hydrocarbons but is soluble in most other organic solvents. IR (CH_2Cl_2, cm^{-1}): 2050 (vw), 2004 (vs), 1961 (w), 1929 (w), 1783 (m, br).

References

1. E. H. Braye, L. F. Dahl, W. Hübel, and D. L. Wampler, *J. Am. Chem. Soc.* **84**, 4633 (1962).
2. (a) M. Tachikawa and E. L. Muetterties, *Prog. Inorg. Chem.* **28**, 203 (1981); (b) J. S. Bradley, *Adv. Organomet. Chem.* **22**, 1 (1983); (c) P. J. Dyson, *Adv. Organomet. Chem.* **43**, 43 (1998).
3. (a) C.-M. T. Hayward, J. R. Shapley, M. R. Churchill, C. Bueno, and A. L. Rheingold, *J. Am. Chem. Soc.* **104**, 7347 (1982); (b) M. R. Churchill, C. Bueno, and A. L. Rheingold, *J. Organomet. Chem.* **395**, 85 (1990).
4. (a) L. Ma, D. P. S. Rodgers, S. R. Wilson, and J. R. Shapley, *Inorg. Chem.* **30**, 3591 (1991); (b) J. W. Benson, T. Ishida, K. Lee, S. R. Wilson, and J. R. Shapley, *Organometallics* **16**, 4929 (1997); (c) K. Lee and J. R. Shapley, *Organometallics* **17**, 4030 (1998); (d) K. Lee and J. R. Shapley, *Organometallics* **17**, 4368 (1998); (e) K. Lee and J. R. Shapley, *Organometallics* **17**, 4113 (1998).
5. M. I. Bruce, N. N. Zaitseva, B. W. Skelton, and A. H. White, *J. Chem. Soc., Dalton Trans.* 3879 (2002).
6. M. I. Bruce, C. M. Jensen, and N. L. Jones, *Inorg. Synth.* **28**, 216 (1990).

Chapter Four

GENERAL TRANSITION METAL COMPOUNDS

24. BIS(1,2-BIS(DIMETHYLPHOSPHANO)ETHANE)-TRICARBONYLTITANIUM(0) AND HEXACARBONYLTITANATE(2–)

Submitted by ROBERT E. JILEK,[*] PAUL J. FISCHER,[*] and JOHN E. ELLIS[*]
Checked by ALEXANDER F. R. KILPATRICK[†] and F. GEOFFREY N. CLOKE[†]

Although titanocene dicarbonyl, $(\eta^5\text{-}C_5H_5)_2Ti(CO)_2$, has been known since 1959[1] and a wealth of chemistry has been discovered for this readily accessible divalent titanium carbonyl and its relatives,[2] lower-valent titanium carbonyl chemistry remains rather poorly investigated due in part to the limited availability of suitable starting materials. To help remedy this situation, syntheses of $Ti(CO)_3(DMPE)_2$, DMPE = 1,2-bis(dimethylphosphano)ethane, and $[Ti(CO)_6]^{2-}$ are presented. The former compound is of interest because it was the first isolable and well-characterized derivative of the unknown $Ti(CO)_7$ and is a key precursor to other zero-valent titanium carbonyls.[3–5] Wreford and coworkers originally prepared this Ti(0) carbonyl by rather difficult high-pressure (~ 70 bar) carbonylations of $Ti(\eta^4\text{-}butadiene)_2(DMPE)$ or $TiCl_4(THF)_2$, THF = tetrahydrofuran, with Na–Hg in the presence of DMPE at $0°C$,[3] and later established its formulation by a single-crystal X-ray study.[4] Subsequently, an atmospheric pressure and higher-yield synthesis for this compound was developed,[5] which provides the basis for the present method. Hexacarbonyltitanate(2–) has been obtained by reductive carbonylations of

[*]Department of Chemistry, University of Minnesota, Minneapolis, MN 55455.
[†]Department of Chemistry, University of Sussex, Brighton BN1 9RH, UK.

Inorganic Syntheses, Volume 36, First Edition. Edited by Gregory S. Girolami and Alfred P. Sattelberger.

$Ti(CO)_3(DMPE)_2$,[6] $TiCl_4(THF)_2$,[6] or $Ti(CO)_4(TRMPE)$,[7] TRMPE = 1,1,1-tris (dimethylphosphanomethyl)ethane, and by the carbonylation of $[Ti(biphenyl)_2]^-$.[8] Unlike corresponding syntheses of the isoelectronic $[V(CO)_6]^-$ from $VCl_3(THF)_3$,[9] more effective alkali metal complexants than THF, DME (1,2-dimethoxyethane), or diglyme (diethylene glycol dimethyl ether) must be present during the carbonylation step to afford $[Ti(CO)_6]^{2-}$.[10] Originally, the dianion was isolated as $[KL_n]^+$ salts, where L = 15-crown-5, $n = 2$, or L = 2.2.2-cryptand, $n = 1$.[6] However, due to the substantial cost of these potassium ion complexants, several less expensive ones were examined and 18-crown-6 was discovered to give a stable potassium salt, $[K(18$-crown-6$)(CH_3CN)]_2[Ti(CO)_6]$,[11] the synthesis of which is described herein. Surprisingly, attempts to obtain a similar $[Na(18$-crown-6$)]^+$ salt failed.[12] Also, no acyclic polyether, polyamine, or mixed amine ether complexants, including the promising tris(2-(2-methoxyethoxy)ethyl)amine, "an acyclic cryptand,"[13] have provided isolable alkali metal salts of $[Ti(CO)_6]^{2-}$ to date.[14]

General Procedures

All operations are conducted under an atmosphere of 99.9% argon or 99.5% carbon monoxide, further purified by passage through columns of activated BASF catalyst, 13X molecular sieves, and Ascarite®,* which is a trade name for a self-indicating sodium hydroxide-coated nonfibrous silicate formulation for the quantitative absorption of carbon dioxide. Noteworthy is that a major impurity in commercial sources of CO is carbon dioxide, which functions as an effective oxidant in these reactions. These syntheses do not work as effectively when conducted initially under a dinitrogen atmosphere. Standard Schlenk techniques are employed with a double-manifold vacuum line.[15] Unless otherwise stated, all starting materials and solvents are purchased from Aldrich or Strem Chemicals. Tetrachlorobis(tetrahydrofuran)titanium[16] and DMPE[17] are purchased or prepared by literature methods. Commercial samples of DMPE should be stirred with a glass-enclosed magnetic stir bar for at least 12 h at room temperature over NaK. (■ **Caution.** *Sodium–potassium alloy reacts violently with water and may ignite upon exposure to air.*) The DMPE is then distilled in vacuum to minimize contamination by moisture and other impurities. The 18-crown-6 is either used as received (if it is relatively free of water, alcohol, and/or hydroperoxides, as shown by a qualitative test under anaerobic conditions with dilute solutions of sodium benzophenone in THF) or purified as follows. It is dissolved in a minimum volume of THF under argon, cooled to 0°C, and treated dropwise with NaK until the solution color turns to a homogeneous deep blue (due to the formation of $[K(18$-crown-6$)][Na]$).

---

*Similar carbon dioxide absorbants should also provide satisfactory results. The checkers directly used carbon dioxide of very high purity (99.999%), which was supplied by Union Carbide and dispensed by means of a high-purity regulator.

Additional NaK must be avoided because this will decompose the 18-crown-6. Subsequently, THF and 18-crown-6 are removed sequentially by vacuum distillation (the latter at $\sim100°C$ and $0.1\,mmHg$). Although 18-crown-6 has a melting point of $\sim42°C$, it forms a relatively stable supercooled liquid at room temperature, which permits use of a standard vacuum distillation apparatus. Lower yields of $[Ti(CO)_6]^{2-}$ are often obtained when 18-crown-6 is purified by other methods.[18*] Naphthalene is sublimed in vacuum. THF is distilled from purple disodium benzophenone and other solvents are purged with nitrogen gas and purified by standard procedures. Unless otherwise stated, solid reagents are transferred in an inert gas-filled dry box. Glassware and filter aid (i.e., diatomaceous earth or Kieselguhr) are dried at about $160°C$ for $\sim12\,h$ before use. Medium-porosity Schlenk frits have a nominal maximum pore size of $10–15\,\mu m$ and are equivalent to analogous porosity 3 units.

A. BIS(1,2-BIS(DIMETHYLPHOSPHANO)ETHANE)-TRICARBONYLTITANIUM(0)

$$TiCl_4(DMPE)_2 + 4NaC_{10}H_8 \xrightarrow[-60\,to+20°C]{\overset{THF}{CO\,(1\,atm)}} Ti(CO)_3(DMPE)_2 + 4NaCl + 4C_{10}H_8$$

Procedure

■   **Caution.**   *Tetrahydrofuran is flammable and forms explosive peroxides on standing in air; only dry and anhydrous peroxide-free solvent should be used in this procedure. Sodium metal reacts violently with water and may ignite upon exposure to air. 1,2-Bis(dimethylphosphano)ethane is toxic and can spontaneously ignite in air. Tetrachlorobis(tetrahydrofuran)titanium undergoes slow hydrolysis in air to generate toxic fumes of hydrogen chloride. Carbon monoxide is very toxic and flammable. All reagents should be handled under an inert atmosphere in a well-ventilated fume hood. Also, small amounts of solid product may cause spontaneous ignition of flammable solvents. For this reason, final workup of the product is most safely performed in an argon-filled glove box. Product residues on frits must be destroyed quickly and carefully in a well-ventilated fume hood, upon exposure to air.*

A 1 L three-necked Morton flask,[†] equipped with a mechanical stirrer and glass paddle, is charged with sodium metal (1.20 g, 52.3 mmol), cut into about ~30

---

[*]The checkers precipitated the 18-crown-6 from acetonitrile, dried it in vacuum, and crystallized it from heptane.

[†]The reaction can also be conducted in a round-bottomed flask with stirring by a glass-enclosed magnetic stir bar, but lower yields of product (45–65%) are generally obtained.

small pieces, and naphthalene (13.2 g, 103 mmol). Tetrahydrofuran (250 mL) is added, and dark green $NaC_{10}H_8$ should begin to form within seconds.[*] The mixture is stirred for 5 h[†] at room temperature under an argon atmosphere. An orange-red solution of $TiCl_4(DMPE)_2$[19] is obtained by the addition of DMPE (5.0 mL, 4.5 g, 30 mmol) to a stirred solution of $TiCl_4(THF)_2$ (4.31 g, 12.9 mmol) in THF (250 mL) at room temperature.[‡] This solution is cooled to −70°C with stirring (often deep red microcrystals form) and transferred via cannula to the efficiently stirred cold (−70°C) solution/slurry of $NaC_{10}H_8$. After about 30 min, the reaction mixture changes from a green-black to a reddish brown color and the stirring is stopped. Argon gas is removed in vacuum and replaced by carbon monoxide (1 atm). The flask should be connected to a constant supply of carbon monoxide, that is, left open to the CO source throughout the reaction sequence, with the pressure maintained by a mercury bubbler. Almost immediately after stirring is resumed, the color of the solution changes to a much brighter red. The reaction mixture is then slowly warmed (while immersed in a low-form 2500 mL Dewar) with stirring for about 12 h to 0 (±15)°C under a CO atmosphere.[§] Stirring is stopped, the CO is replaced by an atmosphere of argon (or dinitrogen) gas, and after about 1 h (to permit settling of finely divided solids) the chilled deep red supernatant is filtered through a 2.5 cm pad of filter aid in a large (70–90 mm diameter) medium- or 3 porosity filtration apparatus into a receiver, chilled to 0°C. The filter cake is thoroughly washed with 30 mL portions of THF until the washings are very pale red. Without delay, all solvent is removed in vacuum and the resulting red solid, heavily contaminated by naphthalene, is thoroughly washed with pentane (3 × 100 mL), and transferred to a coarse or 2 porosity filtration unit. Then it is vigorously washed with pentane (3 × 20 mL) and dried in vacuum at room temperature to afford a highly air-sensitive, and often pyrophoric, crystalline deep red to burgundy, free-flowing solid. Yield: 4.48 g (80%).[¶]

■ **Caution.** *This product is likely to be quite toxic and requires the utmost care in handling and storing. Small amounts of solid product may cause spontaneous ignition of flammable solvents.*

---

[*]If the green color does not appear immediately, the naphthalene, solvent, and/or atmosphere is contaminated and a lower yield of product will result.

[†]Longer periods of stirring should be avoided to minimize decomposition of $NaC_{10}H_8$, which slowly reacts with THF at room temperature.

[‡]Formation of an orange flocculent solid during this step may indicate that the DMPE is impure.

[§]The warming should not be abrupt and the product should not be kept in solution longer than necessary at room temperature. The yield is reduced somewhat if the solution is warmed to room temperature overnight instead of being warmed to 0°C.

[¶]The checkers carried out the reaction on half scale in a 1 L round-bottomed flask equipped with a large glass-enclosed stir bar. They obtained a yield of 69%.

Anal. Calcd. for $C_{15}H_{32}O_3P_4Ti$: C, 41.68, H, 7.46; P, 28.67. Found: C, 41.60; H, 6.84; P, 28.85.

Properties

Crystalline bis(1,2-bis(dimethylphosphano)ethane)tricarbonyltitanium(0) is thermally stable indefinitely at room temperature under an inert atmosphere in the dark. The product may be recrystallized from THF–pentane, but solid from the initial isolation is of sufficient purity for subsequent use. Melting point: 122–124°C (dec.). It has the following spectroscopic properties. IR ν(CO) in THF: 1848 (s), 1757 (sh), 1747 (s) cm^{-1}. ^{1}H NMR (THF-d_8, 300 MHz): δ 1.60 (apparent d, $J = 15.2$ Hz, P-CH$_2$), 1.29 (d, $J = 4.6$ Hz, P-CH$_3$). ^{13}C{^{1}H} NMR (THF-d_8, 75 MHz): δ 282 (very br, CO), 30.5 (apparent dd, $J = 2$, 14 Hz, P-CH$_2$), 18.3 (d, $J = 10$ Hz, P-CH$_3$). The product is extremely air sensitive in solution. It is very slightly soluble in saturated hydrocarbons or diethyl ether and quite soluble in toluene, THF, DME, and pyridine to give deep red solutions that persist for hours at room temperature. It is also soluble in acetonitrile and dichloromethane, but these solutions deteriorate markedly within 1 h and are entirely decomposed within 12 h at room temperature. Chemical properties of Ti(CO)$_3$(DMPE)$_2$ remain poorly explored; however, it is quite resistant toward nucleophilic attack. For example, it remains unchanged after several hours in hot (65°C), neat DMPE. In contrast, under a CO atmosphere at 20°C, in THF, it is in equilibrium with Ti(CO)$_5$(DMPE), which is quite reactive toward nucleophilic attack,[5] and thereby functions in many reactions as an effective synthon for neutral Ti(CO)$_6$, an exceedingly unstable cryomolecule (dec. −200°C), characterized only by IR spectroscopy in a CO/Kr matrix.[20]

B. BIS[18-CROWN-6)(ACETONITRILE)POTASSIUM] HEXACARBONYLTITANATE(2−)

$$\text{TiCl}_4(\text{THF})_2 + 6\text{KC}_{10}\text{H}_8 + 2 \text{ 18-crown-6} \xrightarrow[-60°\text{C}]{\text{THF}} \xrightarrow[-60 \text{ to}+20°\text{C}]{\text{CO (1 atm)}}$$

$$[\text{K(18-crown-6)(THF)}_x]_2[\text{Ti(CO)}_6] + 6\text{C}_{10}\text{H}_8 + 4\text{KCl}$$

$$[\text{K(18-crown-6)(THF)}_x]_2[\text{Ti(CO)}_6] + 2\text{CH}_3\text{CN} \xrightarrow[0°\text{C}]{\text{CH}_3\text{CN/Et}_2\text{O}}$$

$$[\text{K(18-crown-6)(CH}_3\text{CN)}]_2[\text{Ti(CO)}_6] + 2x\text{THF}$$

Procedure

■   **Caution.**   *Carbon monoxide is a very toxic gas, and potassium metal easily ignites in a humid atmosphere. This reaction must be carried out in a well-*

ventilated hood and in an inert atmosphere. Excess potassium should be carefully destroyed with isopropyl alcohol under a protective blanket of nitrogen gas to prevent ignition of potassium and alcohol.

A 1 L round-bottomed flask, equipped with a large glass-enclosed magnetic stir bar, is charged with potassium metal (3.54 g, 90.6 mmol), cut into ~20 small pieces, and naphthalene (15.4 g, 120 mmol). Tetrahydrofuran (250 mL) is added and the resulting dark green mixture is stirred for 4 h at room temperature under an argon atmosphere, and then cooled to −60°C with stirring. A cold solution of $TiCl_4(THF)_2$ (5.00 g, 15.0 mmol) in THF (225 mL, −60°C) is transferred to the cold solution of $KC_{10}H_8$ with efficient stirring. Within minutes, a cold solution of 18-crown-6 (7.92 g, 30 mmol) in THF (150 mL, −60°C) is added. The resulting red-brown solution is stirred for 1 h at −60°C and then the cold bath is replaced by an ice bath (0°C). Stirring is resumed for 2 h to help ensure full complexation of soluble K^+ by the 18-crown-6. After the reaction mixture is recooled to −60°C and stirred for 15 min, the argon is removed in vacuum followed by addition of carbon monoxide (1 atm). The flask should be connected to a constant supply of carbon monoxide, that is, left open to the CO source throughout the reaction sequence, with the pressure maintained by a mercury bubbler. Red swirls of product begin to form almost immediately and the mixture is kept under CO at near atmospheric pressure for 16 h while it slowly warms from −60 to about 10°C. The carbon monoxide is exchanged for an atmosphere of argon gas and the reaction mixture is transferred to a large (70–90 mm diameter) medium- or 3 porosity filtration unit. Red solid, contaminated by KCl, is separated from the filtrate, which is generally pale orange to nearly colorless. The solid is washed with THF (2 × 50 mL) and dried in vacuum overnight. Acetonitrile is introduced in portions (3 × 100 mL, 3 × 50 mL) to dissolve the product on the frit. A Teflon® stir bar is added to facilitate the dissolution process. Agitation of the slurry with the stir bar is effected by a magnetic stirrer (oriented parallel to the filtration apparatus) or by hand with a permanent magnet. (■ **Caution.** *Agitation must not be so vigorous that it causes the glass frit to break!*) The deep red filtrate is collected in a 1 L round-bottomed flask kept at 0°C. The final wash with acetonitrile is very pale red. Acetonitrile is removed in vacuum without delay until about 20 mL of solution remains. Beautiful red microcrystals form during this process. Next, diethyl ether (300 mL) is added to precipitate the product, which is separated by filtration, washed with ether (2 × 50 mL), and dried in vacuum to give moderately air-sensitive, deep red microcrystalline product. Yield: 11.2 g (83% based on $TiCl_4(THF)_2$).[*]

Anal. Calcd. for $C_{34}H_{54}O_{18}N_2K_2Ti$: C, 45.13; H, 6.01; N, 3.10. Found: C, 44.76; H, 6.03; N, 3.07.

---

[*]The checkers carried out the procedure on half scale and obtained a yield of 60%.

Properties

Crystalline [K(18-crown-6)(CH$_3$CN)]$_2$[Ti(CO)$_6$] is thermally stable indefinitely at room temperature when stored in the dark under an inert atmosphere and has been structurally characterized by X-ray crystallography.[11] Melting point: 194–195°C (dec.). It has the following spectroscopic properties. IR ν(CO) in pyridine and acetonitrile: 1745 cm^{-1}. ^{1}H NMR (pyridine-d_5, 300 MHz): δ 3.42 (s, 48H, 18C6), 1.87 (s, 6H, CH$_3$CN). ^{13}C{^{1}H} NMR (pyridine-d_5, 75 MHz): δ 248 (s, CO), 118 (s, CH$_3$CN), 70 (s, 18C6), 1.0 (s, CH$_3$CN). Although the crystalline product usually requires several minutes to oxidize completely in air, solutions of this substance are extremely air sensitive and moderately light sensitive. The compound is insoluble in hydrocarbons, ethers, and liquid ammonia. Solutions of [Ti(CO)$_6$]$^{2-}$ in aceto-nitrile are significantly less stable than those in pyridine and decompose entirely within 48 h at 23°C. In contrast to the isoelectronic [V(CO)$_6$]$^{-}$, dichloromethane rapidly decomposes [Ti(CO)$_6$]$^{2-}$ to unknown products. Also, water, methanol, ethanol, and phenol cause rapid oxidation of [Ti(CO)$_6$]$^{2-}$ to afford the dimeric species [Ti(CO)$_4$(μ-OR)]$_2^{2-}$, where R = H,[21] Me, Et, or Ph.[22] Other titanium carbonyls obtained from [Ti(CO)$_6$]$^{2-}$ include [Ti(CO)$_6$(SnR$_3$)]$^{-}$,[23] [Ti(CO)$_6$(Au-PR$_3$)]$^{-}$,[24] [Ti(CO)$_4$(η^5-trityl)]$^{-}$,[25] and [Ti(CO)$_4$(η^3-BH$_4$)]$^{-}$.[26]

Acknowledgments

We thank the National Science Foundation and the Petroleum Research Fund for their support of research that led to these preparations. Ms. Christine Lundby is thanked for her expert assistance in the preparation of this manuscript.

References

1. J. G. Murray, *J. Am. Chem. Soc.* **81**, 752 (1959).
2. (a) D. J. Sikora, D. W. Macomber, and M. D. Rausch, *Adv. Organomet. Chem.* **25**, 318 (1986); (b) D. J. Sikora, K. J. Moriarty, and M. D. Rausch, *Inorg. Synth.* **28**, 248 (1990).
3. S. S. Wreford, M. B. Fischer, J.-S. Lee, and S. C. Nyburg, *J. Chem. Soc, Chem. Commun.* 458 (1981).
4. P. J. Domailee, R. L. Harlow, and S. S. Wreford, *Organometallics* **1**, 935 (1982).
5. K.-M. Chi, S. R. Frerichs, B. K. Stein, D. W. Blackburn, and J. E. Ellis, *J. Am. Chem. Soc.* **110**, 163 (1988).
6. K.-M. Chi, S. R. Frerichs, S. B. Philson, and J. E. Ellis, *J. Am. Chem. Soc.* **110**, 303 (1988).
7. J. E. Ellis and K.-M. Chi, *J. Am. Chem. Soc.* **112**, 6022 (1990).
8. D. W. Blackburn, J. D. Britton, and J. E. Ellis, *Angew Chem., Int. Ed. Engl.* **31**, 1495 (1992).
9. X. Liu and J. E. Ellis, *Inorg. Synth.* **34**, 96 (2004).
10. J. E. Ellis, *Organometallics* **22**, 3322 (2003).
11. P. J. Fischer, Ph.D. thesis, University of Minnesota, 1998.
12. R. E. Jilek, and J. E. Ellis, unpublished research.
13. G. Soula, *J. Org. Chem.* **50**, 3717 (1985).
14. J. M. Allen, P. J. Fischer, J. A. Seaburg, and J. E. Ellis, unpublished research.

15. (a) D. F. Shriver and M. A. Drezdon, *The Manipulation of Air-Sensitive Compounds*, 2nd ed., Wiley, New York, 1986; (b) A. L. Wayda and M. Y. Darensbourg, eds., *Experimental Organometallic Chemistry*, ACS Symposium Series 357, American Chemical Society, Washington, DC, 1982.
16. L. E. Manzer, *Inorg. Synth.* **21**, 135 (1982).
17. R. G. Burt, J. Chatt, W. Hussian, and G. J. Leigh, *J. Organomet. Chem.* **1982**, 203 (1979).
18. D. D. Perrin and W. L. F. Armarego, *Purification of Laboratory Chemicals*, 3rd ed., Pergamon Press, Oxford, 1988.
19. F. A. Cotton, J. H. Matonic, C. A. Murillo, and M. A. Petrukhina, *Inorg. Chim. Acta* **267**, 173 (1998); also see footnote 37 of Ref. 5.
20. R. Busby, W. Klotzbücher, and G. A. Ozin, *Inorg. Chem.* **16**, 822 (1977).
21. G. Tripepi, V. G. Young, Jr., and J. E. Ellis, *J. Organomet. Chem.* **593**, 354 (2000).
22. P. J. Fischer, P. Yuen, V. G. Young, Jr., and J. E. Ellis, *J. Am. Chem. Soc.* **119**, 5980 (1997).
23. J. E. Ellis and P Yuen, *Inorg. Chem.* **32**, 4998 (1993).
24. P. J. Fischer, V. G. Young, Jr., and J. E. Ellis, *Chem. Commun.* 1249 (1997).
25. P. J. Fischer, K. A. Ahrendt, V. G. Young, Jr., and J. E. Ellis, *Organometallics* **17**, 13 (1998).
26. P. J. Fischer, V. G. Young, Jr., and J. E. Ellis, *Angew. Chem., Int. Ed. Engl.* **39**, 189 (2000).

25. TUNGSTEN BENZYLIDYNE COMPLEXES

Submitted by JIBIN SUN,[*] CHESLAN K. SIMPSON,[*] and
MICHAEL D. HOPKINS[*]
Checked by ADAM S. HOCK[†] and RICHARD R. SCHROCK[†]

Triple-bond metathesis reactions between $M_2(OR)_6$ (M = Mo, W) complexes and alkynes or nitriles are among the most convenient and high-yield synthetic routes to metal–alkylidyne complexes, providing the d^0 Schrock-type compounds of the form $M(CR)(OR)_3$.[1] In view of the ease with which these compounds can be prepared, it would be desirable to use them as precursors for other metal–alkylidyne complexes. We have discovered conditions under which W(CR) $(OBu^t)_3$ compounds can be transformed, in one pot and high yield, into d^2 tungsten–alkylidyne complexes of the type $W(CR)L_4X$ (L = phosphine; X = Cl, Br). These latter complexes are valuable building blocks for functional conjugated materials by virtue of their rich and controllable redox and photophysical properties.[2] Isolable intermediates in the reaction sequence are d^0 complexes of the type $W(CR)X_3(DME)$ (DME = 1,2-dimethoxyethane),[3,4] which are versatile precursors for a variety of high- and low-oxidation-state tungsten–alkylidyne complexes and have a rich reaction chemistry with unsaturated organic molecules.[1] Here we describe representative synthetic procedures for

[*]Department of Chemistry, The University of Chicago, 929 East 57th Street, Chicago, IL 60637.
[†]Department of Chemistry, Massachusetts Institute of Technology, Cambridge, MA 02139.

W(CPh)Cl$_3$(DME) and W(CPh)(dppe)$_2$Cl (dppe = 1,2-bis(diphenylphosphino)-ethane), which are more convenient and higher yielding than previously reported procedures for these types of complexes.

General Procedures

All experiments are performed under nitrogen atmosphere using standard Schlenk and glove box techniques. HPLC-grade solvents, stored under nitrogen in stainless steel cylinders, are purified by passing them under nitrogen pressure through a stainless steel system consisting of either two 4.5 in. × 24 in. (1 gal) columns of activated A2 alumina (THF) or one column of activated A2 alumina and one column of activated BASF R3-11 catalyst (pentane, toluene).[5] 1,2-Dimethoxyethane (DME) is purified by stirring over Na/K (1:2) alloy, from which it is transferred under vacuum. CD$_2$Cl$_2$ is refluxed over CaH$_2$ for 18 h, from which it is then transferred under vacuum. Boron trichloride (1 M solution in heptane, Aldrich), mercury (99.9995%, Aldrich), 1,2-bis(diphenylphosphino)ethane (98%, Strem), and sodium (99.8%, Strem) are used as received. The compound W(CPh)(OBu^t)$_3$ [6] is prepared from the reaction between W$_2$(OBu^t)$_6$ [7] (1 equiv), diphenylacetylene (1 equiv), and 3-hexyne (2–4 mol %) in pentane, by analogy to the procedure for W(CMe)(OBu^t)$_3$;[6] it is recrystallized from pentane at −33°C. Na/Hg 0.4% (w/w) amalgam is prepared by slowly adding freshly cut sodium slivers to mercury at room temperature in a glove box.

A. TRICHLORO(1,2-DIMETHOXYETHANE)-BENZYLIDYNETUNGSTEN(VI)

$$W(CPh)(OBu^t)_3 + 3BCl_3 + DME \rightarrow W(CPh)Cl_3(DME) + 3BCl_{3-x}(OBu^t)_x$$

Procedure

■ **Caution.** *Boron trichloride fumes in air and should be handled under inert atmosphere in a well-ventilated fume hood.*

In a glove box, a 250 mL round-bottomed Schlenk flask (possessing a sidearm with a vacuum stopcock and a 14 mm Solv-Seal® connector) is charged with a PTFE-coated magnetic stir bar, W(CPh)(OBu^t)$_3$ (1.50 g, 3.05 mmol), pentane (80 mL), and DME (4.1 g, 45.5 mmol). The flask is sealed with a septum and removed to a fume hood, where it is attached to a nitrogen/vacuum Schlenk manifold. The yellow solution is cooled to −78°C with stirring using a dry ice/acetone bath. Boron trichloride (9.1 mL of a 1 M solution in heptane, 9.1 mmol) is added to the flask quickly through the septum via syringe. A dark red precipitate forms immediately and the color of the solution reddens slightly. The dry ice/acetone bath is removed and the reaction mixture is allowed to warm to room temperature with stirring. The reaction mixture changes color dramatically as it warmed; it exhibits a series of precipitates that are brown, olive green, blue-green, and purple in color, which are accompanied by solution-phase colorations of red-orange, orange, brown, and colorless. The last of these color changes occurs within 2 h, at which time it is inferred that the reaction is complete. In a glove box, the purple precipitate is isolated by filtration on a fine-porosity sintered glass filter; the filtrate is discarded. The receiver flask is changed and the product extracted by slowly washing the material on the frit with toluene* until only a white powder remains; this powder is discarded. The toluene is removed from the purple solution under vacuum at room temperature until ~10 mL remains. (*Note*: The solution should not be warmed to speed solvent removal because this results in decomposition of the product, as indicated by a change in its color from purple to brown.) Pentane (100 mL) is added to the purple solution, which induces precipitation of a fine purple powder; this is collected by filtration. Crude yield: 1.10 g (2.34 mmol, 77%). The product is recrystallized by dissolving it in DME (8 mL) and cooling the resulting solution to −55°C for 3 h. Yield: 0.98 g (69%).†

Anal. Calcd. for $C_{11}H_{15}Cl_3O_2W$: C, 28.14; H, 3.22. Found: C, 28.21; H, 3.24.

Properties

The compound is a purple powder that is stable under inert atmosphere at room temperature; it is air sensitive and decomposes upon heating in solution. It is soluble in toluene, DME, THF (presumably with replacement of DME by THF), and dichloromethane and is insoluble in pentane. ^{1}H NMR (CD_2Cl_2, 400.13 MHz, 25°C): δ 7.57 (t, 2H, C_6H_5), 6.83 (d, 2H, C_6H_5), 6.74 (t, 1H, C_6H_5), 4.44 (s, 3H, OCH_3), 4.15 (m, 2H, OCH_2), 3.90 (m, 2H, OCH_2), 3.76 (s, 3H, OCH_3); also, 3.48 (s, OCH_2, free DME), 3.33 (s, OCH_3, free DME). $^{13}C\{^1H\}$ NMR (CD_2Cl_2, 100.63 MHz, 25°C): δ 320.1 (W≡C), 138.7, 137.0, 131.9, 126.1, 79.3, 76.4, 70.3, 60.1.

B. CHLORO BIS[1,2-BIS(DIPHENYLPHOSPHINO)ETHANE]-BENZYLIDYNETUNGSTEN(IV)

$$W(CPh)(OBu^t)_3 + 3BCl_3 + DME \rightarrow W(CPh)Cl_3(DME) + 3BCl_{3-x}(OBu^t)_x$$

$$W(CPh)Cl_3(DME) + 2dppe + 2Na \rightarrow W(CPh)(dppe)_2Cl + 2NaCl$$

---

*The checkers noted that the volume of toluene needed for the extraction is 100–150 mL. They suggest that decantation of the colorless pentane phase from the product, extraction of the latter into toluene, and filtration of the resulting solution through Celite® supported on the frit may avoid potential clogging of the frit.

†The checkers reported a yield of 0.9 g (63%).

Procedure

■ **Caution.** *Mercury is toxic and should be handled with appropriate precautions.*

Following the procedure for the synthesis of W(CPh)Cl$_3$(DME) set out in part A, W(CPh)(OBu^t)$_3$ (1.50 g, 3.05 mmol) is treated with BCl$_3$ (9.1 mL of a 1 M solution in heptane, 9.1 mmol) in a mixture of pentane (80 mL) and 1,2-dimethoxyethane (4.1 g, 45.5 mmol). After 2 h, the reaction vessel is returned to the glove box and the colorless solution is decanted from the purple precipitate, which contains W(CPh)Cl$_3$(DME). To the reaction vessel is added THF (150 mL) with stirring. The solid rapidly dissolves to give a blue-green solution. A solution of dppe (2.18 g, 5.47 mmol; 2 equiv, assuming the yield of W(CPh)Cl$_3$(DME) is 90%) in THF (10 mL) is added with stirring. Within 10 min, a yellow precipitate is formed. To this suspension is added Na/Hg amalgam (0.4% (w/w), 2.5 mL, 5.89 mmol Na; 2.1 equiv, assuming the yield of W(CPh)Cl$_3$(DME) is 90%) and the reaction mixture is stirred vigorously for 4 h at room temperature.[*] The precipitate slowly dissolves and the color of the reaction mixture changes to orange. The solution phase is decanted from the mercury and the volatile components are removed under vacuum, leaving an orange-gray powder. This powder is extracted into toluene (150 mL), leaving a white solid (presumably NaCl and boron-containing reaction by-products); the solid is discarded. The solution is filtered through Celite® on a sintered glass filter and the toluene is removed under vacuum, yielding W(CPh)(dppe)$_2$Cl as an orange powder. The powder is washed with pentane (3 × 20 mL) and dried under vacuum. Yield: 2.79 g (83%). The identity of the product is established by comparison to published NMR spectroscopic data.[8] If, instead of this one-pot procedure, W(CPh)Cl$_3$(DME) is prepared and isolated as in part A and then treated with dppe and Na/Hg as above, the yield of W(CPh)(dppe)$_2$Cl after workup is 62% based on W(CPh)(OBu^t)$_3$.

Properties

The compound is an orange powder that is stable at room temperature. It is soluble in toluene, THF, and dichloromethane and very slightly soluble in pentane. Solutions of the compound that are exposed to air for 1 h showed no evidence of decomposition by ^{31}P NMR spectroscopy. ^{1}H NMR (CD$_2$Cl$_2$, 400.13 MHz, 25°C): δ 7.42 (br m, 8H, *o*-PPh$_2$), 7.17 (m, 16H, *o*-PPh$_2$, *m*-PPh$_2$), 7.01 (m, 8H, *o*-PPh$_2$), 6.93 (m, 8H, *p*-PPh$_2$), 6.82 (t, 1H, C$_6$H$_5$), 6.52 (t, 2H, C$_6$H$_5$), 5.67 (d, 2H, C$_6$H$_5$), 2.79 (m, 4H, PCH$_2$), 2.56 (m, 4H, PCH$_2$). ^{13}C{^{1}H} NMR (CD$_2$Cl$_2$, 125.8 MHz, 25°C): δ 251.6 (W≡C), 151.8, 140.9, 138.9, 134.1, 134.0, 130.7, 129.1, 128.7, 127.8, 127.4, 126.3, 123.0, 32.8. ^{31}P{^{1}H} NMR (161.9 MHz, CD$_2$Cl$_2$, 25°C): δ 45.3 (s, $^1J_{PW}$ = 281 Hz).

---

[*]The checkers used a glass-coated stir bar for this step, and noted that the reaction mixture may be stirred overnight without apparent reduction in the yield of product.

Acknowledgment

This research was primarily supported by the NSF MRSEC Program under grant DMR-0213745.

References

1. R. R. Schrock, *Chem. Rev.* **102**, 145 (2002).
2. (a) A. Mayr, M. P. Y. Yu, and V. W.-W. Yam, *J. Am. Chem. Soc.* **121**, 1760 (1999); (b) K. D. John and M. D. Hopkins, *Chem. Commun.*, 589 (1999); (c) M. P. Y. Yu, K.-K. Cheung, and A. Mayr, *J. Chem. Soc., Dalton Trans.* 2373 (1998); (d) R. E. Da Re and M. D. Hopkins, *Coord. Chem. Rev.* **249**, 1396 (2005).
3. (a) R. R. Schrock, J. Sancho, and S. F. Pedersen, *Inorg. Synth.* **26**, 44 (1989); (b) L. G. McCullough, R. R. Schrock, J. C. Dewan, and J. C. Murdzek, *J. Am. Chem. Soc.* **107**, 5987 (1985); (c) R. R. Schrock, D. N. Clark, J. Sancho, J. H. Wengrovius, S. M. Rocklage, and S. F. Pedersen, *Organometallics* **1**, 1645 (1982). (d) A. Mayr and G. A. McDermott, *J. Am. Chem. Soc.* **108**, 548 (1986).
4. M. A. Stevenson and M. D. Hopkins, *Organometallics* **16**, 3572 (1997).
5. A. B. Pangborn, M. A. Giardello, R. H. Grubbs, R. K. Rosen, and F. J. Timmers, *Organometallics* **15**, 1518 (1996).
6. M. L. Listemann and R. R. Schrock, *Organometallics* **4**, 74 (1985).
7. (a) M. Akiyama, M. H. Chisholm, F. A. Cotton, M. W. Extine, D. A. Haitko, D. Little, and P. E. Fanwick, *Inorg. Chem.* **18**, 2266 (1979). (b) M. H. Chisholm, B. W. Eichhorn, K. Folting, J. C. Huffman, C. D. Ontiveros, W. E. Streib, and W. G. Van Der Sluys, *Inorg. Chem.* **26**, 3182 (1987); (c) S. M. Broderick, S. C. Browne, and M. J. A. Johnson, *Inorg. Synth.* **36**, 95 (2014).
8. A. Mayr, A. M. Dorries, G. A. McDermott, and D. Van Engen, *Organometallics* **5**, 1504 (1986).

26. TUNGSTEN OXYTETRACHLORIDE AND (ACETONITRILE)TETRACHLOROTUNGSTEN IMIDO COMPLEXES

Submitted by LAUREL L. REITFORT-BAYSAL,[*] COREY B. WILDER,[*] and LISA MCELWEE-WHITE[*]
Checked by KRISTOPHER S. MIJARES,[†] JEFFREY D. KARCHER,[†] and ERIC A. MAATTA[†]

Alkyl- and arylimido complexes of the type $W(NR)Cl_4(NCMe)$, where R = phenyl, isopropyl, or allyl, have been used as precursors for the chemical vapor deposition (CVD) of tungsten nitride (WN_x) and tungsten carbide nitride (WN_xC_y), which are potential diffusion barrier materials for Cu metallized semiconductor devices.[1–3] Although the complexes do not have sufficient vapor pressure for

[*]Department of Chemistry, University of Florida, Gainesville, FL 32611.
[†]Department of Chemistry, Kansas State University, Manhattan, KS 66506.

volatilization by traditional CVD bubbler or sublimation techniques, CVD using these precursors is possible using a nebulizer assembly to generate an aerosol of solvent/precursor droplets, which is delivered to the reactor by a heated impinging jet. The syntheses of these complexes are based on reported preparations[4,5] for the tungsten imido dimers $[W(NPh)Cl_4]_2$ and $[W(N^iPr)Cl_4]_2$, but utilize a simpler experimental setup. These compounds can be converted in a straightforward fashion to the corresponding acetonitrile complexes.

General Procedures

All reactions and manipulations are performed in an inert atmosphere (N_2) glove box or using standard Schlenk and vacuum line techniques. All glassware is dried overnight in an oven at 150°C before use. WCl_6 is purchased from Strem and used without further purification. Hexamethyldisiloxane is purchased from Aldrich and distilled from sodium. Isocyanates (phenyl, isopropyl, and allyl) are purchased from Aldrich and used without further purification. Anhydrous heptane is purchased from Aldrich in a Sure-Seal® bottle and used as received. All other solvents are purchased from Fisher and passed through an M. Braun MB-SP solvent purification system before use. NMR solvents are degassed by three freeze–pump–thaw cycles and stored over 3 Å molecular sieves in an inert atmosphere glove box. ^{1}H and ^{13}C NMR spectra are recorded on Mercury 300, Gemini 300, VXR 300, or Inova 500 spectrometers. In cases where assignments of ^{1}H or ^{13}C NMR resonances are ambiguous, standard ^{1}H–^{13}C HMQC experiments are used. Elemental analyses are performed by Robertson Microlit (Madison, NJ).

A. TUNGSTEN OXYTETRACHLORIDE

$$WCl_6 + Me_3SiOSiMe_3 \rightarrow WOCl_4 + 2Me_3SiCl$$

Procedure

■ **Caution.** *When exposed to moisture, tungsten hexachloride and tungsten oxytetrachloride evolve hydrogen chloride, a toxic and corrosive gas.*

$WOCl_4$ is prepared by the following modification of a literature procedure (an alternative preparation, starting from sodium tungstate, appears elsewhere in this volume).[6] Inside a glove box, a 500 mL Schlenk flask equipped with an overhead stirrer is charged with WCl_6 (26.2 g, 0.066 mol) and methylene chloride (~350 mL). To the vigorously stirred suspension is added dropwise freshly distilled $Me_3SiOSiMe_3$ (10.7 g, 0.066 mol). The reaction mixture is stirred vigorously for 1 h during which time the solution changes color from deep red to bright orange. The flask is removed from the glove box and the solvent is removed in vacuum on a Schlenk line. The orange solid is washed several times with dry hexane, and dried in vacuum. Yield: 21.0 g (93%).[*]

Properties

Tungsten oxytetrachloride is an orange, air- and moisture-sensitive solid that is soluble in nonpolar solvents. It reacts with protic solvents such as alcohols and water and forms adducts with many Lewis bases.

B. (ACETONITRILE)TETRACHLORO(PHENYLIMIDO)- TUNGSTEN(VI)

$$2WOCl_4 + 2PhNCO \rightarrow [W(NPh)Cl_4]_2 + 2CO_2$$

$$[W(NPh)Cl_4]_2 + 2MeCN \rightarrow 2W(NPh)Cl_4(NCMe)$$

Procedure

W(NPh)Cl$_4$(NCMe) is prepared by the following modification of a literature procedure.[7] Inside a glove box, a Chemglass 350 mL heavy-walled pressure vessel with a Teflon® bushing and stirrer is charged with WOCl$_4$ (2.035 g, 5.96 mmol) and heptane (60 mL). To the vigorously stirring suspension is added dropwise via syringe phenyl isocyanate (0.851 g, 7.15 mmol). The sealed vessel is removed from the glove box and placed on a stir plate in a fume hood. (■ **Caution.** *Although we have had no problems with pressure buildup in the following step, a shield was placed in front of the stir plate.*) The vessel is warmed to 110°C, using either an oil bath or a heating mantle, and the suspension is stirred for 16 h. The reaction vessel is cooled to room temperature, and the solvent is removed from the greenish brown solution in vacuum. The greenish brown residue is dissolved in a minimal amount of acetonitrile (~10 mL) and stirred for 2 h at room temperature. The solvent is removed under reduced pressure. The resulting greenish black residue is washed with toluene (5 × 10 mL), and the resulting solution is filtered through glass wool inside a glove box. This solution is transferred to a Schlenk flask and concentrated to ~5 mL on a Schlenk line. Hexane is then added to precipitate the product. The green solid is filtered, washed with hexane, and dried in vacuum. Yield: 1.87 g (69%).[†]

---

[*]The checkers obtained a yield of 21.5 g (95%).
[†]The checkers obtained a yield of 1.67 g (61%).

Properties

W(NPh)Cl$_4$(NCMe) is a green, air- and moisture-sensitive, crystalline solid that is indefinitely stable when stored in an inert atmosphere at $-20°C$. It is soluble in ethers and other coordinating solvents. It is sparingly soluble in aromatic hydrocarbons and is insoluble in aliphatic hydrocarbons.

^{1}H NMR (C$_6$D$_6$): δ 0.22 (s, 3H, NC*Me*), 6.21 (t, 1H, $J = 7$ Hz, N*Ph*), 6.20–7.02 (m, 4H, Ph). ^{13}C{^{1}H} NMR (C$_6$D$_6$): δ 0.5 (NC*Me*), 120.3 (NCMe), 127.6, 132.1, 134.4, 140.2 (N*Ph*).

C. (ACETONITRILE)TETRACHLORO(2-PROPYLIMIDO)-TUNGSTEN(VI)

$$2WOCl_4 + 2^iPrNCO \rightarrow [W(N^iPr)Cl_4]_2 + 2CO_2$$

$$[W(N^iPr)Cl_4]_2 + 2MeCN \rightarrow 2W(N^iPr)Cl_4(NCMe)$$

Procedure

Inside a glove box, a Chemglass 350 mL heavy-walled pressure vessel with a Teflon$^®$ bushing is charged with WOCl$_4$ (3.186 g, 9.325 mmol) and heptanes (60 mL). To the vigorously stirring suspension is added dropwise via syringe *iso*-propyl isocyanate (1.215 g, 14.28 mmol). The sealed vessel is removed from the glove box and placed on a stir plate in a fume hood. (■ **Caution.** *Although we have had no problems with pressure buildup in the following step, a shield was placed in front of the stir plate.*) The vessel is warmed to 110°C, using either an oil bath or a heating mantle, and the suspension is stirred for 16 h. The reaction vessel is cooled to room temperature, and the solvent is removed in vacuum. The reddish orange residue is dissolved in a minimal amount of acetonitrile ($\sim$10 mL) and the mixture is stirred for 2 h at room temperature. The solvent is removed under reduced pressure. The resulting brown residue is extracted with toluene (5 × 10 mL), and the resulting extracts are combined and filtered through glass wool inside a glove box. This filtrate is transferred to a Schlenk flask and concentrated to $\sim$5 mL on a Schlenk line. Hexane (50 mL) is then added to precipitate the product. The orange solid is filtered, washed with hexane (5 mL), and dried in vacuum. Yield: 2.66 g (68%).*

Properties

W(N^{i}Pr)Cl$_4$(NCMe) is an orange, air- and moisture-sensitive, polycrystalline solid that is indefinitely stable when stored in an inert atmosphere at $-20°C$ in the

---

*The checkers obtained a yield of 2.15 g (55%).

absence of light. It is soluble in ethers and other coordinating solvents. It is also soluble in aromatic hydrocarbons, but insoluble in aliphatic hydrocarbons.

^{1}H NMR (CDCl$_3$): δ 1.67 (d, 6H, $J = 6$ Hz, NCH*Me*$_2$), 2.49 (s, 3H, NC*Me*), 7.07 (m, 1H, NC*H*Me$_2$). ^{13}C{^{1}H} NMR (CDCl$_3$): δ 3.4 (NC*Me*), 23.1 (NCH*Me*$_2$), 68.5 (NC*H*Me$_2$), 118.4 (N*C*Me). IR (CH$_2$Cl$_2$, cm^{-1}): 2319, 2282, 1462, 1452, 1386, 1366, 1310, 1271, 1154, 1108, 1075.

D. (ACETONITRILE)TETRACHLORO(2-PROPENYLIMIDO)-TUNGSTEN(VI)

$$2WOCl_4 + 2CH_2{=}CHCH_2NCO \rightarrow [W(NCH_2CH{=}CH_2)Cl_4]_2 + 2CO_2$$

$$[W(NCH_2CH{=}CH_2)Cl_4]_2 + 2MeCN \rightarrow 2W(NCH_2CH{=}CH_2)Cl_4(NCMe)$$

Procedure

Inside a glove box, a Chemglass 350 mL heavy-walled pressure vessel with a Teflon bushing is charged with WOCl$_4$ (1.23 g, 3.60 mmol) and heptane (60 mL). To the vigorously stirring suspension is added dropwise via syringe 2-propenyl isocyanate (0.366 g, 4.41 mmol). The sealed vessel is removed from the glove box and placed on a stir plate in a fume hood. (■ **Caution.** *Although we have had no problems with pressure buildup in the following step, a shield was placed in front of the stir plate.*) The vessel is warmed to 110 °C, using either an oil bath or a heating mantle, and the suspension is stirred for 36 h. The reaction vessel is then cooled to room temperature, and the solvent is removed from the dark red solution in vacuum. The reddish brown residue is dissolved in a minimal amount of acetonitrile (~10 mL) and stirred for 2 h at room temperature. The solvent is removed under reduced pressure. The resulting brown residue is extracted with toluene (5 × 10 mL), and the resulting extracts are combined and filtered through glass wool inside a glove box. This solution is transferred to a Schlenk flask and concentrated to ~5 mL on a Schlenk line. Hexane (50 mL) is added to precipitate the product. The orange-brown solid is filtered, washed with hexane, and dried in vacuum. Yield: 0.974 g (64%).[*]

Anal. Calcd. for C$_5$H$_8$N$_2$Cl$_4$W: C, 14.24; H, 1.91; N, 6.64. Found: C, 14.51; H, 1.86; N, 6.43.

Properties

W(NCH$_2$CH=CH$_2$)Cl$_4$(NCMe) is an orange-brown, air- and moisture-sensitive, polycrystalline solid that is indefinitely stable when stored in an inert atmosphere at

---

[*]The checkers obtained a yield of 0.92 g (61%).

−20°C in the absence of light. It is soluble in ethers and other coordinating solvents. It is also soluble in aromatic hydrocarbon solvents, but insoluble in aliphatic hydrocarbon solvents.

^{1}H NMR (CDCl$_3$): δ 2.50 (s, 3H, NCMe), 5.60 (dtd, J = 0.6, 1.4, 10.2 Hz, 1H, NCH$_2$CH=CH_2), 5.73 (dtd, J = 0.6, 1.5, 17.1 Hz, 1H, NCH$_2$CH=CH_2), 6.07 (tdd, J = 5.6, 10.2, 17.1 Hz, 1H, NCH$_2$CH=CH$_2$), 7.55 (ddd, J = 1.5, 1.4, 5.6 Hz, 2H, NCH_2CH=CH$_2$). ^{13}C{^{1}H} NMR (CDCl$_3$): δ 3.5 (NCMe), 68.3 (NCH$_2$CH=CH$_2$), 118.9 (NCMe), 121.9 (NCH$_2$CH=CH$_2$), 129.7 (NCH$_2$$C$H=CH$_2$). IR (KBr, cm^{-1}): 2929, 2314, 2287, 1649, 1409, 1364, 1317, 989, 919.

References

1. O. J. Bchir, K. M. Green, M. S. Hlad, T. J. Anderson, B. C. Brooks, C. B. Wilder, D. H. Powell, and L. McElwee-White, *J. Organomet. Chem.* **684**, 338 (2003).
2. O. J. Bchir, S. W. Johnston, A. C. Cuadra, T. J. Anderson, C. G. Ortiz, B. C. Brooks, D. H. Powell, and L. McElwee-White, *J. Cryst. Growth* **249**, 262 (2003).
3. O. J. Bchir, K. M. Green, H. M. Ajmera, E. A. Zapp, T. J. Anderson, B. C. Brooks, L. L. Reitfort, D. H. Powell, K. A. Abboud, and L. McElwee-White, *J. Am. Chem. Soc.* **127**, 7825 (2005).
4. A. J. Nielson, *Inorg. Synth.* **24**, 194 (1986).
5. B. R. Ashcroft, G. R. Clark, A. J. Nielson, and C. E. F. Rickard, *Polyhedron* **5**, 2081 (1986).
6. G. Gelbard, *Inorg. Synth.* **36**, 143 (2014).
7. A. J. Nielson and J. M. Waters, *Aust. J. Chem.* **36**, 243 (1983).

27. TUNGSTEN OXYTETRACHLORIDE AND SEVERAL TUNGSTATE SALTS

Submitted by GEORGES GELBARD[*]
Checked by JUSTIN L. MALLEK[†] and GREGORY S. GIROLAMI[†]

Tungsten oxytetrachloride is a good starting material to prepare other oxotungsten (VI) complexes and to incorporate tungsten species into silica-based materials.[1–3] The usual preparations of tungsten oxytetrachloride, which are based on tungstic acid, are somewhat tedious and require prolonged heating times.[4–7] The preparation of WOCl$_4$ from anhydrous sodium tungstate, reported here, is an improved and simplified method. An alternative preparation, starting from tungsten hexachloride, appears elsewhere in this volume.[8]

[*]Institut de Recherches sur la Catalyse et L'Evironment de Lyon, CNRS-Université de Lyon, Lyon 1, UMR 5256, 2 avenue Albert Einstein, F-69626 Villeurbanne, France.
[†]School of Chemical Sciences, University of Illinois at Urbana-Champaign, 600 South Mathews Avenue, Urbana, IL 61801.

The preparation of $(n\text{-}Bu_4N)_2W_6O_{19}$ reported here is a simplification of the procedure given in this series by Fournier[9] and Fuchs.[10,11] Peroxotungstic salts are catalysts for the epoxidation of alkenes with hydrogen peroxide.[12] In addition, the quaternary phosphonium[13] and ammonium[14,15] salts are of special interest due to their solubility in nonchlorinated solvents, but unfortunately these salts are difficult to isolate because their preparations involve the use of biphasic conditions. In contrast, the preparation of the cetylpyridinium cation requires only aqueous solutions.

Phosphoric, phosphinic, and phosphonic acids form complexes with both peroxotungstates and peroxomolybdates. In particular, phenylphosphonic acid forms a complex that is a very convenient alkene epoxidation catalyst.[16] In contrast, other analogous catalysts were not isolated and instead were used in situ.[17]

A. TUNGSTEN OXYTETRACHLORIDE

$$Na_2WO_4 + 3SOCl_2 \rightarrow WOCl_4 + 2NaCl + 3SO_2$$

Procedure

■ **Caution.** *When exposed to moisture, sulfinyl chloride evolves hydrogen chloride, a toxic and corrosive gas.*

Sodium tungstate, $Na_2WO_4 \cdot 2H_2O$ (3.9 g, 11.75 mmol), is placed in a 50 mL distillation flask, the outer diameter of which should be less than 45 mm so as to fit inside a Büchi glass oven (GRK-50 or B-580) or a similar horizontal electric oven.[*] The sodium tungstate is dried at 150°C for 3 h under the vacuum of a water pump and for an additional 2 h at 250°C under the vacuum of an oil pump. After being cooled, the flask is removed from the oven and a magnetic bar is introduced. Freshly distilled sulfinyl chloride (20 mL, 257 mmol) is added (the distilled $SOCl_2$ can be collected directly in the reaction flask). The flask is fitted with a reflux condenser and a calcium chloride moisture trap. The reaction mixture is stirred and progressively heated on an oil bath; an orange color appears as the temperature rises and SO_2 begins to evolve at about 60°C. Refluxing at 110°C and stirring are then continued overnight. The condenser is quickly removed and replaced by a small distillation still, and the excess of sulfinyl chloride is distilled off, the last traces being removed under reduced pressure. The flask is connected to a bulb-to-bulb distillation device and the product is sublimed at 140–150°C under vacuum (5 mmHg)[†] to give an orange red powder. Yield: 4.0 g (98%).

Anal. Calcd. for Cl_4OW: W, 53.8. Found: W, 57.0.

---

[*]The checkers carried out most of these operations on a Schlenk line.
[†]The checkers sublimed the product at 60–80°C and ~0.01 mmHg.

Properties

The orange-red solid is moisture sensitive and should be handled in a glove bag and stored in a Teflon®-stoppered flask under nitrogen. For long-term storage, the flask should be kept in a desiccator containing calcium chloride. The solid gives dark red solutions when soluble: 1,1,1-trichloroethane (soluble), 1,2-dichloro-ethane (soluble), toluene (low solubility), dichloromethane (insoluble), and *o*-dichlorobenzene (insoluble). The physical properties have been reviewed.[7]

B. BIS(TETRABUTYLAMMONIUM) HEXAPOLYTUNGSTATE

$$6Na_2WO_4 + 10(n\text{-}Bu_4N)HSO_4 \rightarrow (n\text{-}Bu_4N)_2W_6O_{19} + 6Na_2SO_4$$
$$+ 4(n\text{-}Bu_4N)_2SO_4 + 5H_2O$$

Procedure

A solution of sodium tungstate $Na_2WO_4 \cdot 2H_2O$ (13.2 g, 40 mmol) in water (150 mL) is stirred in a 500 mL conical flask; a solution of tetrabutylammonium hydrogensulfate, $(n\text{-}Bu_4N)HSO_4$ (27.2 g, 80 mmol), in water (150 mL) is added dropwise under efficient stirring. The solid begins to precipitate when about one half of the sulfate is added; stirring is continued 30 min after the end of the addition. The product is collected by filtration on a Büchner funnel, and the filtrate is washed with water and air dried overnight. Yield: 12.4 g (98%). Samples for spectroscopic characterization can be obtained by recrystallization from acetonitrile/methylene chloride.

Anal. Calcd. for $C_{32}H_{72}N_2O_{19}W_6$: C, 20.31; N, 1.48; W, 58.3. Found: C, 20.00; N, 1.51; W, 58.5.

Properties

The white flakes are soluble in most organic solvents, especially chlorinated hydrocarbons. FTIR (KBr, cm^{-1}): 1484, 1381, 955, 887, 796.* The crystal structure has been described by Fuchs.[10]

---

*The checkers carried out this preparation on one-third scale and obtained the same yield. *Anal.* Calcd.: C, 20.31; H, 3.84; N, 1.48. Found: C, 22.27; H, 3.97; N, 1.50. Their IR spectrum as a Nujol mull is somewhat different from that as a KBr pellet: IR (Nujol, cm^{-1}): 976 (s), 956 (s), 890 (m), 815 (vs), 785 (sh).

C. DI(CETYLPYRIDINIUM) PEROXODITUNGSTATE

$$H_2W_2O_{11} + 2C_{21}H_{38}NCl \rightarrow (C_{21}H_{38}N)_2W_2O_{11} + 2HCl$$

Procedure

A solution of peroxotungstic acid ($H_2W_2O_{11}$) is first prepared by stirring a mixture of "light" tungstic acid (15 g, 60 mmol)[*] and 35% hydrogen peroxide (60 mL) at 40°C for 3 h. The yellow suspension progressively gives a quite clear solution of peroxotungstic acid, which is centrifuged when cold; titration gives about 0.9 mol $W\,L^{-1}$. The solution is stable for months if kept in a refrigerator.

To a stirred solution of cetylpyridinium chloride (2.0 g, 5.56 mmol) in water (200 mL) is added dropwise peroxotungstic acid (5.7 mL of a 0.9 N solution, 5.1 mmol of tungsten); the flaky precipitate that forms immediately is stirred for additional 30 min, filtered on a Büchner funnel,[†] and washed with water until no frothing of the filtrate occurs. The cake is dried at 40°C under vacuum (heating above 60°C may induce decomposition of the peroxo salt). Yield: 3.1 g (96%).

Anal. Calcd. for $C_{42}H_{76}N_2O_{11}W_2$: C, 43.73; N, 2.43; W, 31.9. Found: C, 44.22; N, 2.45; W, 32.5.

Properties

White flakes. FTIR (KBr, cm^{-1}): 930 ν(W=O), 866 ν(O−O), 602 ν_{asym}(W−O), 534 ν_{sym}(W−O).[‡]

---

[*]Light tungstic acid, a bright yellow material having the formula $WO_3 \cdot H_2O$, dissolves easily in hydrogen peroxide, whereas many other forms of tungstic acid dissolve only after prolonged heating. The checkers did not have access to this material, so they made a solution of peroxotungstic acid by the following procedure, which was adapted from Ref. 18: A 500 mL Erlenmeyer flask charged with a stir bar, a thermometer, and 30% H_2O_2 (52 mL, 510 mmol) is heated to 40°C by immersion in a large ($\sim$1 L) warm water bath on a hot plate; the bath serves as a heat sink. Tungsten powder (12 µm particle size, 11.0 g, 60 mmol) is weighed out and separated into eight approximately equal portions. To the vigorously stirred H_2O_2 solution is added the tungsten powder over $\sim$4 h, one portion being added about every 30 min (i.e., after the solution temperature had returned to $\sim$40°C). (■ **Caution.** *The tungsten powder must not be added too quickly or in larger portions, otherwise, the exothermic reaction will cause violent boiling of the solution that can spew hot solution for large distances in the vicinity of the flask.*) The solution is filtered (a small amount of black powder remains on the filter paper) and the clear, pale yellow filtrate is diluted to 65 mL and used in the next step. The calculated normality of the peroxotungstic acid solution is 0.9 N.

[†]The checkers obtained a yield of 2.4 g (76%), having lost some material in the filtration step. They recommend using fine filter paper to prevent the loss.

[‡]The checkers find the following: *Anal.* Calcd.: C, 43.76; N, 2.43; H, 6.65. Found: C, 44.41; N, 2.73; H, 6.81. Their IR spectrum as a Nujol mull did not match well with that reported for a KBr pellet, possibly owing to reaction of the product with KBr during pellet formation. IR (Nujol, cm^{-1}): 3127 (w), 3057 (m), 1635 (m), 1582 (w), 1180 (m), 957 (sh), 946 (s), 896 (s), 849 (w), 831 (m), 799 (m), 779 (w), 736 (sh), 693 (m), 623 (m), 570 (m), 541 (m).

D. BIS(TETRABUTYLAMMONIUM) PHENYLPHOSPHONATODIPEROXOTUNGSTATE

$$C_6H_5PO_3H_2 + H_2W_2O_{11} + 2(n\text{-}Bu_4N)OH \rightarrow (n\text{-}Bu_4N)_2[C_6H_5PO_3(WO_5)_2] + 3H_2O$$

Procedure

Phenylphosphonic acid (1.58 g, 10 mmol) is added to a diluted solution of 40% tetrabutylammonium hydroxide (13 mL, 20 mmol) in water (100 mL); then a solution of $H_2W_2O_{11}$ (21 mL, 20 mmol of tungsten), prepared as above, is added dropwise to the clear solution. The flaky solid that forms is collected, rinsed with cold water, and dried under vacuum at 40°C (heating above 60°C may induce decomposition of the peroxo salt). Yield: 7.23 g (62%). An analytical sample was recrystallized from a mixture of 1,2-dichloroethane and ethyl acetate.

Anal. Calcd. for $C_{38}H_{77}N_2O_{13}PW_2$: C, 39.05; N, 2.4; W, 31.46; P, 2.65. Found: C, 38.93; N, 2.4; W, 31.45; P, 2.64. Iodometry: 4.1 active oxygen per mole.

Properties

IR (KBr pellet, cm^{-1}): 960 ν(W=O), 835 ν(O–O), 645.* ^{31}P NMR (1,2-dichloro-ethane/CD_2Cl_2): δ +11.34 (s with satellites, $^2J_{P\text{-}W} = 10.3$ Hz). ^{183}W NMR (1,2-dichloroethane/CD_2Cl_2): δ −624.6 (d, $^2J_{P\text{-}W} = 10.3$ Hz). The crystal structure is known.[16]

References

1. A. A. Eagle, E. R. T. Tiekink, and C. G. Young, *J. Chem. Soc., Chem. Commun.* 1746 (1991).
2. R. Neumann, M. Chava, and M. Levin, *J. Chem. Soc., Chem. Commun.* 1685 (1993).
3. J. Sundermayer, J. Putterlik, M. Foth, J. S. Field, and N. Ramesar, *Chem. Ber.* **127**, 1201 (1994).
4. J. Tillack, *Inorg. Synth.* **14**, 109 (1973).
5. A. J. Nielson, *Inorg. Synth.* **23**, 195 (1985); **28**, 323 (1990).
6. C. Crouch, G. W. A. Fowles, and R. A. Walton, *J. Inorg. Nucl. Chem.* **32**, 329 (1970).
7. V. V. Abramenko, A. D. Garnovski, and V. L. Abramenko, *Russ. J. Inorg. Chem.* **37**, 1389 (1992).
8. L. L. Reitfort-Baysal, C. B. Wilder, and L. McElwee-White, *Inorg. Synth.* **36**, 138 (2014).
9. M. Fournier, *Inorg. Synth.* **28**, 80 (1990).
10. K. F. Jahr, J. Fuchs, and R. Oberhauser, *Chem. Ber.* **101**, 477 (1968).
11. J. Fuchs, W. Freiwald, and H. Hartl, *Acta Crystallogr. B* **34**, 1764 (1978).
12. N. J. Campbell, A. C. Dengel, C. J. Edwards, and W. P. Griffith, *J. Chem. Soc., Dalton Trans.* 1203 (1989).

---

*The checkers find the following: *Anal.* Calcd.: C, 39.05; N, 2.40, H, 6.64. Found: C, 39.00; N, 2.54; H, 6.76. Their IR spectrum as a Nujol mull is somewhat different from that reported for a KBr pellet, but it is possible that the product reacts with the KBr during pellet formation. IR (Nujol, cm^{-1}): 1241 (m), 1128 (m), 1028 (w), 1003 (m), 991 (m), 982 (w), 964 (sh), 954 (s), 852 (w), 841 (m), 701 (w).

13. J. Prandi, H.B. Kagan, and H. Mimoun, *Tetrahedron Lett.* **27**, 2617 (1985).
14. T. Iwahama, S. Sakagushi, and Y. Ishii, *Tetrahedron Lett.* **36**, 1523 (1995).
15. C. Venturello, E. Alneri, and M. Ricci, *J. Org. Chem.* **48**, 3831 (1983).
16. G. Gelbard, F. Raison, E. Roditi-Lachter, R. Thouvenot, L. Ouahab, and D. Grandjean, *J. Mol. Catal.* **114**, 77 (1996).
17. K. Sato, M. Aoki, M. Ogawa, T. Hashimoto, D. Panyella, and R. Noyori, *Bull. Chem. Soc. Jpn.* **70**, 905 (1997).
18. K. Yamanaka, H. Oakamoto, H. Kidou, and T. Kudo, *Jpn. J. Appl. Phys.* **25**, 1420 (1986).

28. BROMOTRICARBONYLDI(PYRIDINE)MANGANESE(I)

Submitted by MARIO PONS[*] and GERHARD E. HERBERICH[*]
Checked by PATRICK N. KIRK,[†] MICHAEL P. CASTELLANI,[†]
MICHAEL J. RIZZO,[‡] and JIM D. ATWOOD[‡]

$$MnBr(CO)_5 + 2py \rightarrow MnBr(CO)_3(py)_2 + 2CO$$

Bromotricarbonyldi(pyridine)manganese(I) was first mentioned in the early 1960s in the context of infrared studies of metal carbonyl halide complexes.[1] Several other studies of this complex have appeared since then detailing its spectroscopic properties.[2] More recently, this complex has been found to be a useful synthon that efficiently transfers a $-Mn(CO)_3$ group,[3] primarily to prepare cymantrene complexes.[4]

Procedure

Under an inert atmosphere, a Schlenk flask equipped with a magnetic stir bar is flushed with nitrogen and charged with $MnBr(CO)_5$[5] (5.04 g, 25.8 mmol). The flask is stoppered and wrapped with a black cloth. Then degassed pyridine (10 mL, 124 mmol) is added to the flask. The reaction mixture is stirred for 15 min, by which time the initially rapid evolution of CO should have ceased. The flask is then placed in a water bath kept at 80°C and the reaction mixture is stirred for 1 h and then recooled to room temperature. To the flask is added hexane (20 mL) and the resulting precipitate is collected on a glass frit. The solid is redissolved in dichloromethane (~250 mL), and the resulting solution is filtered and taken to dryness.[§] The solid is then dried in vacuum. Yield: 6.62 g (96%).[¶]

[*]Institut für Anorganische Chemie, Technische Hochschule Aachen, Aachen, Germany.
[†]Department of Chemistry, Marshall University, Huntington, WV 25755.
[‡]Department of Chemistry, State University of New York, Buffalo, NY 14214.
[§]The checkers found that the dichloromethane extraction step was not necessary to produce high-quality product.
[¶]The checkers carried out the reaction successfully twice on smaller scale (0.5 and 1.5 g of product).

Properties

Bromotricarbonyldi(pyridine)manganese(I) is an orange-red, crystalline, light-sensitive solid that is soluble in pyridine, THF, and CH_2Cl_2. It is insoluble in hexane and toluene. The solution infrared spectrum displays $\nu(CO)$ bands at 2030, 1944, and 1910 cm^{-1} in CD_2Cl_2.[4d] The ^1H NMR spectrum shows multiplets at δ 8.74, 7.61, and 7.31 in CD_2Cl_2.[4d] Its UV–vis spectrum has λ_{max} in CH_2Cl_2 at 315 nm with a broad shoulder at $\sim$375 nm.

References

1. (a) R. J. Angelici, F. Basolo, and A. J. Poë, *J. Am. Chem. Soc.* **85**, 2215 (1963); (b) M. A. Bennett and R. J. H. Clark, *J. Chem. Soc. Suppl.* **1**, 5560 (1964).
2. (a) R. J. Angelici, *J. Inorg. Nucl. Chem.* **28**, 2627 (1966); (b) J. Dalton, I. Paul, J. G. Smith, and F. G. A. Stone, *J. Chem. Soc. A* 1208 (1968); (c) R. A. N. McLean, *J. Chem. Soc., Dalton Trans.* 1568, (1974).
3. M. Tachikawa, R. L. Geerts, and E. L. Muetterties, *J. Organomet. Chem.* **213**, 11 (1981).
4. (a) W. L. Bell, C. J. Curtis, A. Miedaner, C. W. Elgenbrot Jr., R. C. Haltiwanger, C. G. Pierpont, and J. C. Smart, *Organometallics* **7**, 691 (1988); (b) T. L. Lynch, M. C. Helvenston, A. L. Rheingold, and D. L. Staley, *Organometallics* **8**, 1959 (1989); (c) H. Plenio and D. Burth, *Organometallics* **15**, 1151 (1996); (d) G. E. Herberich, U. Englert, B. Ganter, and M. Pons, *Eur. J. Inorg. Chem.* 979 (2000); (e) P. M. Kirk and M. P. Castellani, *Inorg. Synth.* **36**, 62 (2014).
5. M. H. Quick and R. J. Angelici, *Inorg. Synth.* **28**, 156 (1990).

29. BIS(TETRAETHYLAMMONIUM) *fac*-TRIBROMOTRICARBONYLRHENATE(I) AND -TECHNETATE(I)

Submitted by STEFAN MUNDWILER,[*] HENRIK BRABAND,[*] and ROGER ALBERTO[*]
Checked by DONNA MCGREGOR,[†] ROBERTHA C. HOWELL,[†] and LYNN C. FRANCESCONI[†]

The organometallic chemistry of technetium is very poorly developed in comparison with that of its heavier congener rhenium. In part this is because high-pressure and high-temperature syntheses of radioactive compounds are almost impossible to perform nowadays due to radioactive safety regulations and other reasons. As a result, important precursors for low-valent organometallic chemistry such as $Tc_2(CO)_{10}$ and $TcX(CO)_5$ (X = Cl, Br, I) are not readily available.[1,2] It is therefore

[*]Institute of Inorganic Chemistry, University of Zürich, Zürich, Switzerland.
[†]Department of Chemistry, Hunter College of the City University of New York, New York, NY 10065.

important to develop ambient (or at least low-pressure) methods to obtain stable but substitutionally reactive low-valent compounds that can serve as starting materials for other technetium complexes. Furthermore, because the commercial driver in technetium chemistry is radiopharmacy (^{99m}Tc is the most widely used gamma-emitter in radiodiagnostics), an easily synthesized water (and air)-stable organometallic precursor would extend the portfolio of diagnostic agents by opening up new possibilities in labeling strategies.

The $(Et_4N)_2[MX_3(CO)_3]$ salts of technetium(I) and rhenium(I) described herein are almost ideal examples of such compounds. $(Et_4N)_2[^{99}TcCl_3(CO)_3]$ is a highly useful starting material to explore low-valent technetium chemistry, and $(Et_4N)_2$-$[ReBr_3(CO)_3]$ is playing a key role in the emerging field of bioorganometallic chemistry. The dissolution of $(Et_4N)_2[MX_3(CO)_3]$ in water readily yields the "semi" aquo ions $[M(OH_2)_3(CO)_3]^+$,[3–8] and additional important starting materials can be obtained by replacing the water ligands in $[M(OH_2)_3(CO)_3]^+$ by a wide variety of classical Werner-type or even typical organometallic ligands.[9–11] For example, treatment with suitable cyclopentadienyl reagents affords the piano stool complex $(\eta^5\text{-}C_5H_5)Tc(CO)_3$.[12]

The study of ^{99}Tc compounds is important not only for exploring the fundamental chemistry of technetium but also for developing clinical applications of its nuclear excited state ^{99m}Tc. Because ^{99m}Tc complexes are typically generated only in nanomole quantities, they cannot be characterized by standard analytical techniques. Instead, ^{99}Tc complexes are used as standards for the characterization of their ^{99m}Tc analogs. For example, an accepted way of identifying the composition of a ^{99m}Tc complex is by comparing its HPLC retention time with that of a fully characterized ^{99}Tc analog or a Re homolog. The identical chemistry of ^{99}Tc and ^{99m}Tc also means that a procedure to synthesize $[^{99m}Tc(OH_2)_3(CO)_3]^+$ directly from water in a kit formulation[13–16] must also work for $[^{99}Tc(OH_2)_3(CO)_3]^+$, and vice versa.

Salts of the *fac*-trihalogenotricarbonylrhenate anion were first synthesized by Abel et al.;[4,17] the procedure described herein is an improvement on that method. The procedure for the preparation of *fac*-trichlorotricarbonyltechnetate is an improved version of that originally described by us.[5,18]

General Procedures

Bromopentacarbonylrhenium is prepared by bromination of decacarbonyldirhenium,[19,20] but is also commercially available (e.g., from Strem Chemicals). Tetra(n-butyl)ammonium pertechnetate, (n-Bu$_4$N)[^{99}TcO$_4$], is prepared by precipitation from an aqueous (NH$_4$)[TcO$_4$] solution (Oak Ridge National Laboratories) with (n-Bu$_4$N)Cl. Tetra(n-butyl)ammonium oxotetrachlorotechnetate(V), (n-Bu$_4$N)[^{99}TcOCl$_4$], is prepared according to a literature procedure.[21]

A. BIS(TETRAETHYLAMMONIUM) *fac*-TRIBROMOTRICARBONYLRHENATE(I)

$$ReBr(CO)_5 + 2(Et_4N)Br \rightarrow (Et_4N)_2[ReBr_3(CO)_3] + 2CO$$

Procedure

■ **Caution.** *This reaction releases carbon monoxide, which is a toxic, odorless, and colorless gas. Work should be done in a well-ventilated fume hood.*

A 250 mL two-necked round-bottomed flask is fitted with a Teflon®-coated magnetic stir bar and a reflux condenser; the latter is topped with a gas inlet and is connected to a double Schlenk manifold, which serves as a source of vacuum and dinitrogen. A magnetic stirrer with a regulated oil bath is used for stirring and heating. Diethylene glycol dimethyl ether (diglyme, 2-methoxyethyl ether, 150 mL, Fluka purum grade) and tetraethylammonium bromide (5.25 g, 25.0 mmol, Fluka purum grade) are added to the flask. The NEt$_4$Br should be added as fine white crystals; lumps should be crushed before addition. The slurry is then stirred and heated to 80°C under a slight stream of dinitrogen, which enters through one neck of the flask and exits through the reflux condenser. The reflux condenser is temporarily replaced with a stopper, and the flask is *carefully* evacuated and backfilled with dinitrogen 10 times to remove dissolved dioxygen from the reaction mixture.[*] After the flask is refilled with N$_2$, the stopper is removed and bromopentacarbonylrhenium (5.00 g, 12.3 mmol) is added. A strong countercurrent of dinitrogen prevents the introduction of air. The reflux condenser is reattached, and the mixture is heated to 115°C under strong stirring and a slight dinitrogen stream to sweep out the evolved carbon monoxide through the reflux condenser and out to an oil/mercury bubbler. The mixture is stirred at 115°C overnight (15–20 h), during which time a white precipitation forms. The mixture is then cooled to room temperature and the solid is collected by filtration with a suction filter. The filtration step can be done in air because the products are not air sensitive. The solid is washed with diethylene glycol dimethyl ether (6 mL) and dried at 0.1 mmHg for 3 h to give a slightly off-white fine powder. This crude product is stirred in air with ice-cold absolute ethanol (20 mL) for 15 min to dissolve remaining traces of tetraethylammonium bromide. The solid is again collected on a suction filter, washed with ice-cold ethanol (6 mL), and dried at 0.1 mmHg for 5 h to give the product as a white to slightly off-white fine powder. Yield: 8.86 g (93%).[†]

---

[*]It is crucial to work under dioxygen-free conditions to prevent yellowing of the final product.
[†]A lower yield can occur due to the slight solubility of the product in ethanol. In this case, the ethanol washes should be worked up. The checkers note that the preparation works equally well on a 1/10 scale.

Anal. Calcd. for $C_{19}H_{40}Br_3N_2O_3Re$: C, 29.62; H, 5.23, N, 3.64. Found: C, 29.77; H, 5.39, N, 3.68.

Properties

Bis(tetraethylammonium) *fac*-tribromotricarbonylrhenate(I) is a white solid that is stable in air at room temperature. It is soluble in polar solvents such as water and methanol, but insoluble in less polar solvents such as dichloromethane or higher alcohols. In aqueous solution, the bromide ligands are partially exchanged by water. The compound is best analyzed by IR spectroscopy: the spectrum shows two characteristic strong peaks, one at $1868\,cm^{-1}$ (broad) and another at $1999\,cm^{-1}$ (sharp). IR (KBr, cm^{-1}): 3442 (w), 3007 (m), 2987 (m), 2948 (w), 1999 (s), 1868 (s), 1635 (w), 1458 (m), 1406 (m), 1372 (w), 1350 (w), 1307 (w), 1183 (m), 1136 (w), 1119 (w), 1079 (w), 1031 (m), 1004 (m), 893 (w), 798 (m), 653 (m), 638 (w), 516 (m), 507 (m).

B. BIS(TETRAETHYLAMMONIUM) *fac*-TRICHLOROTRICARBONYLTECHNETATE(I)

$$(n\text{-}Bu_4N)[^{99}TcO_4] + 3CO + 2BH_3 \cdot THF + 3(Et_4N)Cl + 8HOEt$$
$$\rightarrow (Et_4N)_2[^{99}TcCl_3(CO)_3] + 3H_2 + 2THF + 4H_2O + 2\text{``}(R_4N)[B(OEt)_4]\text{''}$$

■ **Caution.** *^{99}Tc is a weak β-emitter with a half-life of 212,000 years. According to safety regulations, it may be handled only in special radiolabs and with special equipment. It is of utmost importance to prevent any contamination, ingestion, or inhalation. A well-ventilated hood is crucial due to the evolution of toxic CO.*

Procedure

A 100 mL three-necked round-bottomed flask is equipped with a magnetic stir bar and a reflux condenser topped by a gas inlet adapter. The adapter is connected by a tube to a mercury bubbler to ensure a slight internal overpressure of about 10 mmHg. A second neck of the flask is fitted with a serum rubber stopper, through which is passed a Teflon tube (1–2 mm ID) that is connected to a source of CO. The third neck is initially sealed with another serum stopper. The reaction flask is evacuated and filled with CO three times (or flushed with CO for about 30 min at a flow of about 40 mL min^{-1}). Then BH$_3$ · THF (25 mL of a 1 M solution in THF, 25 mmol) is added to the flask by means of a glass syringe, and the CO flux slowed to about $1 \, mL \, min^{-1}$.[*] The tip of the Teflon tube is immersed in the solution, so that the CO bubbles through the reaction mixture ($1 \, mL \, min^{-1}$ corresponds to ~5–7 bubbles min^{-1}). A solution of $(n\text{-}Bu_4N)[^{99}TcO_4]$[†] (410 mg, 1.01 mmol) in THF (4–5 mL) is placed in a syringe pump[‡] connected to a Teflon tube (1 mm ID) that passes through a serum stopper fitted into the third neck of the reaction flask. The end of this tube should also be immersed in the solution. The $(n\text{-}Bu_4N)[^{99}TcO_4]$ solution is added at a constant rate of about $0.03 \, mL \, min^{-1}$ into the $BH_3 \cdot THF$ solution. As the addition proceeds, the solution turns yellow and eventually brown-black as a result of the formation of small amounts (<5%) of higher-valent technetium by-products. During the addition, the solution warms slightly and some precipitate (probably borates) is observed, which disappears in the course of the reaction. After the addition, stirring is continued for at least 1 h.[§] The THF is removed in vacuum or by passing a slow stream of N_2 through the reaction flask. Solid $(Et_4N)Cl$ (1.0 g, 6.0 mmol), ethanol (3 mL), and concentrated HCl (1 mL) are added to the brown residue, and the clear solution is stirred for 2 h. Some of the solid product may precipitate from the solution. The solvent is evaporated to dryness, ethanol (4 mL) is added, and the mixture is stirred for 1 h. The slightly yellowish powder is collected by filtration on a glass frit. More than 95% of this precipitate (crude yield ~450 mg) consists of $(Et_4N)_2[^{99}TcCl_3(CO)_3]$, the rest being $(Et_4N)_2[^{99}TcCl_6]$. The material is dissolved in water (10 mL); stirring overnight converts $(Et_4N)_2[^{99}TcCl_6]$ into TcO_2, which is filtered off.[¶] The filtrate is evaporated, and the residue is washed with CH_2Cl_2

[*]To obtain good yields of product, it is crucial to reduce the CO flow to an absolute minimum, ideally just keeping the internal pressure constant without causing the mercury bubbler to bubble. Possibly, a strong stream of CO sweeps the borane out of the flask as $BH_3 \cdot CO$ before the subsequent addition step is complete. To determine whether sufficient BH_3 is still present, the pertechnetate inlet is withdrawn briefly so that the tip is just above the surface of the solution. If the droplet of pertechnetate solution at the tip of the tube turns yellow, sufficient amounts of BH_3 are still present in the reaction flask.

[†]The technetium(V) salt $(n\text{-}Bu_4N)[^{99}TcOCl_4]$ is an alternative starting material. The procedure is exactly the same, and the yield of product from this starting material is often greater than 90%.

[‡]The best results are achieved with a syringe pump. Manual dropwise addition is also possible, but the yield is normally reduced. The checkers, who worked at 1/5 scale, added $(n\text{-}Bu_4N)[^{99}TcOCl_4]$ (116 mg) dissolved in THF (1.5 mL) to the borane solution (8 mL) dropwise at a rate of one drop per minute over the course of 1 h. The color changes, stirring times, and THF removal method (slow stream of N_2) were as stated in the larger-scale preparation. After the addition of Et_4NCl (283 mg), ethanol (0.85 mL), and concentrated HCl (0.29 mL), the clear yellow solution was stirred for 2 h and the solvent evaporated to dryness. The checkers used ethanol (1.2 mL) in the wash step and increased the stirring time to a few hours to remove the borates completely. The slightly yellow powder was stirred overnight in water to convert $(Et_4N)_2[^{99}TcCl_6]$ to $^{99}TcO_2$. The checkers obtained yields of $(Et_4N)_2[^{99}TcCl_3(CO)_3]$ of up to 67% by means of the dropwise addition of $(n\text{-}Bu_4N)[^{99}TcOCl_4]$.

[§]At this point, the success of the reaction can be assessed by withdrawing 0.05 mL of the solution, mixing it with methanol (0.2 mL) and concentrated HCl (0.05 mL), letting the mixture stand for 30 min, and analyzing the resulting solution by HPLC under conditions described elsewhere.[22]

[¶]The checkers note that the $^{99}TcO_2$ is a very fine powder and recommend filtering the solution through Celite® on a fine porosity sintered glass frit.

$(2 \times 5\,\text{mL})$ to afford white, analytically pure $(\text{Et}_4\text{N})_2[^{99}\text{TcCl}_3(\text{CO})_3]$. Yield: 425 mg (77%).

Anal. Calcd. for $\text{C}_{19}\text{H}_{40}\text{N}_2\text{Cl}_3\text{O}_3\text{Tc}$: Tc, 17.84; Cl, 19.38. Found: Tc, 17.92; Cl, 19.25.

Properties

Bis(tetraethylammonium)*fac*-trichlorotricarbonyltechnetate(I) is a white solid that is stable in air for months. It is highly soluble in water, but complete dissolution requires some time because the chloride ligands are substituted by water molecules. The compound is also soluble in methanol, DMSO, and DMF, but is insoluble in other common organic solvents. ^{99}Tc NMR (D_2O, 67.56 MHz): δ -868 relative to $[^{99}\text{TcO}_4]^-$. IR (KBr, cm^{-1}): 3435 (w), 2989 (m), 2025 (s), 1908 (s), 1460 (m), 1403 (w), 1261 (w), 1184 (m), 1137 (w), 1032 (m), 1005 (w), 798 (m), 667 (m), 635 (w), 500 (m).

The corresponding bromo complex $(\text{Et}_4\text{N})_2[^{99}\text{TcBr}_3(\text{CO})_3]$ can be synthesized by the same method by use of concentrated HBr and $(\text{Et}_4\text{N})\text{Br}$ in place of the corresponding chloride compounds. The halide ligands in $(\text{Et}_4\text{N})_2[^{99}\text{TcCl}_3(\text{CO})_3]$ can also be exchanged by addition of AgX, where X is nitrate, trifluoroacetate, hexafluorophosphate, and so on. These exchange products are soluble in common organic solvents such as dichloromethane, tetrahydrofuran, and higher alcohols. Solutions in these solvents are also stable in air, but at higher pH in water or in the presence of deprotonated alcohols, multinuclear hydroxo- or alcoxo-bridged species are reversibly formed.

References

1. W. Hieber, F. Lux, and C. Herget, *Z. Naturforsch. B* **20**, 1159 (1965).
2. J. C. Hileman, H. D. Kaesz, and D. K. Huggins, *Inorg. Chem.* **1**, 933 (1962).
3. R. Alberto, A. Egli, U. Abram, K. Hegetschweiler, V. Gramlich, and P. A. Schubiger, *J. Chem. Soc., Dalton Trans.* 2815 (1994).
4. R. Alberto, A. Egli, U. Abram, K. Hegetschweiler, and P. A. Schubiger, *J. Chem. Soc., Dalton Trans.* 2815 (1994).
5. R. Alberto, R. Schibli, A. Egli, P. A. Schubiger, W. A. Herrmann, G. Artus, U. Abram, and T. A. Kaden, *J. Organomet. Chem.* **493**, 119 (1995).
6. N. Aebischer, R. Schibli, R. Alberto, and A. E. Merbach, *Angew. Chem., Int. Ed.* **39**, 254 (2000).
7. R. Alberto, R. Schibli, U. Abram, D. Angst, T. A. Kaden, and P. A. Schubiger, *Transit. Met. Chem.* **22**, 597 (1997).
8. R. Alberto, R. Schibli, A. Egli, A. P. Schubiger, U. Abram, and T. A. Kaden, *J. Am. Chem. Soc.* **120**, 7987 (1998).
9. R. Alberto, R. Schibli, R. Waibel, U. Abram, and A. P. Schubiger, *Coord. Chem. Rev.* **190–192**, 901 (1999).
10. R. Alberto, *Top. Curr. Chem.* **176**, 149 (1996).

11. S. R. Banerjee, K. P. Maresca, L. Francesconi, J. Valliant, J. W. Babich, and J. Zubieta, *Nucl. Med. Biol.* **32**, 1 (2005).
12. J. Wald, R. Alberto, K. Ortner, and L. Candreia, *Angew. Chem., Int. Ed.* **40**, 3062 (2001).
13. R. Alberto, R. Schibli, A. P. Schubiger, U. Abram, H. J. Pietzsch, and B. Johannsen, *J. Am. Chem. Soc.* **121**, 6076 (1999).
14. R. Alberto, K. Ortner, N. Wheatley, R. Schibli, and A. P. Schubiger, *J. Am. Chem. Soc.* **123**, 3135 (2001).
15. N. Metzler-Nolte, *Angew. Chem., Int. Ed.* **40**, 1040 (2001).
16. R. Schibli and P. A. Schubiger, *Eur. J. Nucl. Med. Mol. Imaging* **29**, 1529 (2002).
17. E. W. Abel, I. S. Butler, M. C. Ganorkar, C. R. Jenkins, and M. H. Stiddard, *Inorg. Chem.* **5**, 25 (1966).
18. R. Alberto, R. Schibli, P. A. Schubiger, U. Abram, and T. A. Kaden, *Polyhedron* **15**, 1079 (1996).
19. S. P. Schmidt, W. C. Trogler, and F. Basolo, *Inorg. Synth.* **28**, 162 (1990).
20. C. Kutal, M. A. Weber, G. Ferraudi, and D. Geiger, *Organometallics* **4**, 2161 (1985).
21. (a) W. Preetz and G. Peters, *Z. Naturforsch. B* **35**, 1355 (1980); (b) A. Davison, H. S. Trop, B. V. Depamphilis, and A. G. Jones, *Inorg. Synth.* **21**, 160 (1982); (c) F. Poineau, E. V. Johnstone, and K. R. Czerwinski, *Inorg. Synth.* **36**, 110 (2014).
22. R. Schibli, R. Schwarzbach, R. Alberto, K. Ortner, H. Schmalle, C. Dumas, A. Egli, and P. A. Schubiger, *Bioconjug. Chem.* **13**, 750 (2002).

30. METHYL(OXO)RHENIUM(V) COMPLEXES WITH CHELATING LIGANDS

Submitted by XIAOPENG SHAN[*] and JAMES H. ESPENSON,[*]
Checked by SONGPING HUANG[†] and ROGER ALBERTO[‡]

Oxygen atom transfer reactions often do not occur in the absence of a catalyst even if there is a considerable driving force. For example, the reaction of pyridine *N*-oxide with PPh_3 to give pyridine and Ph_3PO is difficult to catalyze even though this reaction is characterized by a free energy change of about $-300 \, kJ \, mol^{-1}$. Methylrhenium(V) oxo compounds[1-13] with certain chelating ligands are often effective catalysts for oxygen atom transfer reactions.[8,10,11,14-16] Such rhenium compounds are also chemical relatives of the extensively studied molybdenum and tungsten oxotransferase enzymes.[17-19]

Many methylrhenium(V) oxo compounds are five-coordinate with a square pyramidal geometry about rhenium.[1-6,8,10,12,13] The equatorial positions are occupied by the methyl group, a heteroatom (usually S, P, N, or O), and two donor atoms of the bidentate chelate, which in many instances is a dithiolate.

[*]Ames Laboratory and Department of Chemistry, Iowa State University, Ames, IA 50011.
[†]Department of Chemistry and Biochemistry, Kent State University, Kent, OH 44242.
[‡]Institute of Inorganic Chemistry, University of Zürich, Winterthurerstrasse 190, CH-8057 Zürich, Switzerland.

Among the compounds in this class are dimeric compounds of stoichiometry [MeReO(SR)$_2$]$_2$ in which one of the sulfur atoms bridges between the two metal centers. Other methylrhenium(V) oxo compounds are six-coordinate, with the lower axial position occupied by one arm of a chelate ligand that is also connected to an equatorial position.[7,11,12]

All of these compounds can be prepared from methyltrioxorhenium(VII), abbreviated as MTO. Reduction to rhenium(V) is accomplished in these procedures with thiols, phosphanes, and sulfanes.[11,20,21] Suitable stabilizing ligands must be present to intercept the MeReO$_2$ intermediate, which is quite reactive: even perchlorate ions are reduced by it.[22,23] With neither a stabilizing ligand nor an oxidant, a black precipitate is formed from the polymerization of methyldioxorhenium(V).[23]

General Procedures

Chemicals and solvents are purchased from commercial sources and used as received. The rhenium products are stable toward O$_2$ and moisture, and so the synthetic procedures can be carried out in air. Methylrhenium trioxide is prepared by a literature route.[24] Abbreviations used: MTO = methyltrioxorhenium(VII); H$_2$EDT = 2-ethanedithiol; H(HQ) = 8-hydroxyquinoline.

A. METHYL(OXO)(1,2-ETHANEDITHIOLATO)RHENIUM(V) DIMER

Procedure

$$MeReO_3 + PhC{\equiv}CPh + PPh_3 \rightarrow MeReO_2(\eta^2\text{-}PhC{\equiv}CPh) + OPPh_3$$

$$2\ MeReO_2(\eta^2\text{-}PhC{\equiv}CPh) + 2\ H_2EDT \rightarrow \quad\quad\quad\quad + 2\ H_2O + 2\ PhC{\equiv}CPh$$

A flask is charged with MTO (250 mg, 1.0 mmol), diphenylacetylene (178 mg, 1.0 mmol), and toluene (20 mL). To the stirred mixture is added PPh$_3$ (262 mg, 1.0 mmol). After 10 min, the color of the resulting solution changes from colorless to yellow,[*] indicating the formation of the intermediate product MeReO$_2$(η^2-diphenylacetylene).[20] Then 1,2-ethanedithiol (94 mg, 84 μL, 1.0 mmol) is added

---

[*]The checkers noted that diphenylacetylene obtained from some suppliers is brown instead of white. In such cases, the solution color changes from yellow to brown when the PPh$_3$ is added.

dropwise. The mixture is stirred at room temperature for 6 h and then layered with hexane (20 mL). After 2 days, a brown-black solid is deposited, collected by filtration, and washed with hexane. Yield: 370 mg (60%).

Anal. Calcd. for $C_6H_{14}O_2Re_2S_4$: C, 11.82; H, 2.39. Found: C, 11.73; H, 2.27.

Properties

^{1}H NMR (C_6D_6, 25°C): δ 3.65 (m, 2H, SCH_2), 2.62 (s, 6H, Re$-$Me), 2.37 (m, 2H, SCH_2), 2.24 (m, 2H, SCH_2), 1.92 (m, 2H, SCH_2). $^{13}C\{^1H\}$ NMR (C_6D_6, 25°C): δ 47.1 (SCH_2), 36.6 (SCH_2), 13.2 (Re$-$Me). Analogous dithiolate complexes can be prepared in a variety of ways. For example, direct treatment of MTO with 2-(mercaptomethyl)thiophenol[25] gives the corresponding dithiolate complex.[1] In addition, MTO can be converted to $[MeReO(SPh)_2]_2$,[21] which can be treated with 1,3-propanedithiol to give the 1,3-propanedithiolate complex. These dinuclear rhenium(V) dithiolate complexes can be converted into monomeric complexes of stoichiometry MeReO(κ^2-dithiolate)L with Lewis bases such as pyridine, 2-ethyl-pyridine, 4-*tert*-butylpyridine, tetramethylthiourea, triphenylphosphine, or methyldiphenylphosphine.

B. METHYL(OXO)BIS(2-OXYQUINOLINE)RHENIUM(V)

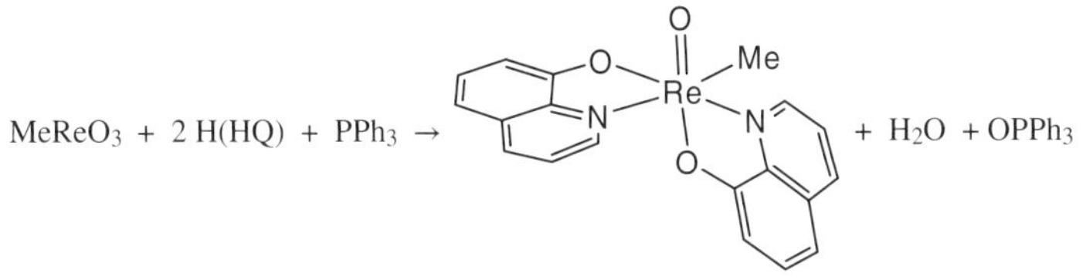

$$MeReO_3 + 2\ H(HQ) + PPh_3 \rightarrow \qquad\qquad + H_2O + OPPh_3$$

Procedure

To a solution containing MTO (50 mg, 0.2 mmol) and 2-hydroxyquinoline (58 mg, 0.4 mmol) in CH_2Cl_2 (20 mL) is added PPh_3 (53 mg, 0.2 mmol). The mixture is stirred for 12 h,[*] layered with hexanes (20 mL), and placed in a freezer at -12°C. After 24 h, a black powder is deposited. It is isolated from the solution by filtration, rinsed with hexanes, and dried. Yield: 81 mg (80%).

Anal. Calcd. for $C_{19}H_{15}N_2O_3Re$: C, 45.14; H, 2.99; N, 5.54. Found: C, 44.64; H, 2.92; N, 5.29.

---

[*]The checkers noted that during the synthesis of MeReO(hq)$_2$ a dark red solution forms immediately upon addition of PPh_3, which later turns slightly green.

Properties

^{1}H NMR (CDCl$_3$): δ 8.56 (d, 1H), 8.36 (m, 1H), 8.21 (m, 1H), 7.66 (m, 4H), 7.40 (m, 3H), 7.07 (d, 1H), 6.46 (d, 1H), 4.53 (s, 3H). IR (CHCl$_3$, cm^{-1}): 980 (Re=O). UV–vis (CHCl$_3$, λ_{max} (nm) (log ϵ (M^{-1} cm^{-1}))): 470 (sh), 417 (3.70), 360 (sh). The ^{13}C{^{1}H} NMR spectrum was difficult to obtain owing to the compound's low solubility in chloroform. The analogous compounds MeReO(PA)$_2$ and MeReO (MQ)$_2$ [7,13] can be synthesized from MTO and the corresponding protonated form of the bidentate ligand, 2-picoline or 8-mercaptoquinoline, with Re(VII) being reduced by triphenylphosphine or dimethylsulfide.[11] In the case of MeReO(DPPBA), where H(DPPBA) is 2-phenylphosphinobenzoic acid, the ligand itself acts as the reducing agent. Finally, the *N,N*-diethyldithiocarbamate complex MeReO(DDC)$_2$ can be made by treating MTO with NaDDC, acetic acid, and triphenylphosphine.

C. METHYL(OXO)(2,2′-THIODIACETATO)(TRIPHENYLPHOSPHINE)-RHENIUM(V)

$$\text{MeReO}_3 + \text{S(CH}_2\text{CO}_2\text{H)}_2 + \text{PPh}_3 \rightarrow$$

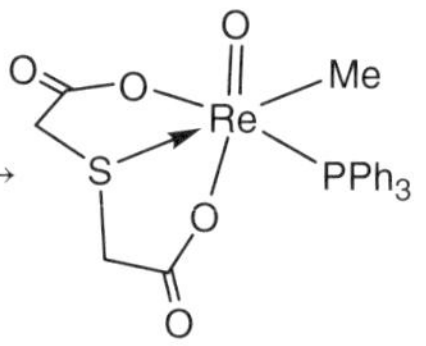

$$+ \text{H}_2\text{O} + \text{OPPh}_3$$

Procedure

MTO (250 mg, 1 mmol), 2,2′-thiodiacetic acid (150 mg, 1 mmol), and PPh$_3$ (525 mg, 2 mmol) are mixed in CH$_2$Cl$_2$ (20 mL). The blue mixture is stirred for 10 h, layered with hexane (20 mL), and placed in a freezer at $-12°$C. After 24 h, a blue powder is deposited. The solid is collected by filtration, rinsed with hexane, and dried. Yield: 585 mg (93%).[*]

Anal. Calcd. for C$_{23}$H$_{22}$O$_5$PReS: C, 44.01; H, 3.53; S, 5.11; P, 4.93. Found: C, 43.97; H, 3.51; S, 4.66; P, 4.30.

---

[*]The checkers noted that, when this product is examined with microscope, some colorless microcrystals are seen, which presumably are OPPh$_3$. They recrystallized the product from CH$_2$Cl$_2$ to give pure material. Yield: 577 mg (91%).

Properties

^{1}H NMR (CDCl$_3$): δ 7.47–7.70 (m, 15H), 4.58 (d, 3H), 3.61 (d, 1H), 3.36 (d, 1H), 2.89 (d, 1H), 1.38 (d, 1H). ^{13}C{^{1}H} NMR (CDCl$_3$): 185.7, 177.8, 134.1 (d), 132.1 (d), 131.5 (d), 129.2 (d), 128.6 (d), 37.9, 36.6, 15.4. ^{31}P{^{1}H} NMR (CDCl$_3$): δ 0.45. IR (CHCl$_3$, cm^{-1}): 1007 (Re=O). UV–vis (CHCl$_3$, λ_{max} (nm)): 265 (sh), 300 (sh). The same procedure is successful for other triarylphosphines such as P(C$_6$H$_4$-4-OMe)$_3$ and P(C$_6$H$_4$-4-F)$_3$. Ligand displacement from [MeReO(SPh)$_2$]$_2$ using the tridentate ligands 2,2′-thiodiethanethiol (H$_2$SSS) or 2-mercaptoethyl ether (H$_2$SOS) affords the five-coordinate compounds MeReO(SSS) and MeReO (SOS). Treatment of MTO with the tridentate ligand 2-(salicylideneamino)benzoic acid (H$_2$ONO)[26] in the presence of PPh$_3$ affords the six-coordinate compound MeReO(ONO)PPh$_3$.

References

1. J. Jacob, I. A. Guzei, and J. H. Espenson, *Inorg. Chem.* **38**, 1040 (1999).
2. J. Jacob, G. Lente, I. A. Guzei, and J. H. Espenson, *Inorg. Chem.* **38**, 3762 (1999).
3. J. Jacob, I. A. Guzei, and J. H. Espenson, *Inorg. Chem.* **38**, 3266 (1999).
4. G. Lente, I. A. Guzei, and J. H. Espenson, *Inorg. Chem.* **39**, 1311 (2000).
5. G. Lente, J. Jacob, I. A. Guzei, and J. H. Espenson, *Inorg. React. Mech.* **2**, 169 (2000).
6. G. Lente, X. Shan, I. A. Guzei, and J. H. Espenson, *Inorg. Chem.* **39**, 3572 (2000).
7. J. H. Espenson, X. Shan, D. W. Lahti, T. M. Rockey, B. Saha, and A. Ellern, *Inorg. Chem.* **40**, 6717 (2001).
8. R. Huang and J. H. Espenson, *Inorg. Chem.* **40**, 994 (2001).
9. D. W. Lahti and J. H. Espenson, *J. Am. Chem. Soc.* **123**, 6014 (2001).
10. J. H. Espenson, X. Shan, Y. Wang, R. Huang, D. W. Lahti, J. Dixon, G. Lente, A. Ellern, and I. A. Guzei, *Inorg. Chem.* **41**, 2583 (2002).
11. X. Shan, A. Ellern, and J. H. Espenson, *Inorg. Chem.* **41**, 7136 (2002).
12. X. Shan, A. Ellern, I. A. Guzei, and J. H. Espenson, *Inorg. Chem.* **42**, 2362 (2003).
13. X. Shan, A. Ellern, I. A. Guzei, and J. H. Espenson, *Inorg. Chem.* **43**, 3854 (2004).
14. Y. Wang and J. H. Espenson, *Org. Lett.* **2**, 3525 (2000).
15. Y. Wang, G. Lente, and J. H. Espenson, *Inorg. Chem.* **41**, 1272 (2002).
16. J. H. Espenson, *Coord. Chem. Rev.* **249**, 329 (2005).
17. R. H. Holm, *Coord. Chem. Rev.* **100**, 183 (1990).
18. C. G. Young and A. G. Wedd, *Chem. Commun.* 1251 (1997).
19. J. H. Enemark, J. J. A. Cooney, J.-J. Wang, and R. H. Holm, *Chem. Rev.* **104**, 1175 (2004).
20. J. K. Felixberger, J. G. Kuchler, E. Herdtweck, R. A. Paciello, and W. A. Herrmann, *Angew. Chem., Int. Ed. Engl.* **100**, 975 (1988).
21. J. Takacs, M. R. Cook, P. Kiprof, J. G. Kuchler, and W. A. Herrmann, *Organometallics* **10**, 316 (1991).
22. M. M. Abu-Omar and J. H. Espenson, *Inorg. Chem.* **34**, 6239 (1995).
23. M. M. Abu-Omar, E. H. Appelman, and J. H. Espenson, *Inorg. Chem.* **35**, 7751 (1996).
24. W. A. Herrmann and R. M. Kratzer, *Inorg. Synth.* **33**, 111 (2002).
25. A. G. Hortmann, A. J. Aron, and A. K. Bhattacharya, *J. Org. Chem.* **43**, 3374 (1978).
26. A. D. Westland and M. T. H. Tarafder, *Inorg. Chem.* **20**, 3992 (1981).

31. HEXAHYDRIDOFERRATE(II) SALTS

Submitted by DONALD E. LINN JR. [*] **and GABRIEL M. SKIDD** [*]
Checked by JEFFREY L. CROSS [†] **and GREGORY J. KUBAS** [†]

Homoleptic complex hydrides of the transition metals are rare species and solution synthetic procedures for their preparation are virtually nonexistent. The hexahydrido iron tetraanion, $[FeH_6]^{4-}$, was apparently unwittingly prepared by Theodore Weichselfelder, a student of Wilhelm Schlenk.[1] Much later, Sidney Gibbins was able to separate and analyze the yellow crystalline solid and arrive at its correct formulation.[2] Gibbins' procedure was hampered by the many difficult manipulations required for the separation and purification of the title complex. Our preparation is a single-step procedure that can be done conveniently in a modern synthetic laboratory, all with readily available equipment and materials. Here, iron halide (either the bromide or chloride) is treated with phenylmagnesium bromide in tetrahydrofuran (THF), starting at low temperature and under hydrogen pressure. This modification avoids decomposition reactions that yield a black tar containing biphenyl and metallic iron. Furthermore, the reaction time and workup is reduced from days to just hours.

The result, a salt of the $[FeH_6]^{4-}$ anion with four $[MgBr]^+$ cations, is a crystalline yellow solid that is soluble in tetrahydrofuran to the extent of 0.006 M. The bromide atoms in the $[MgBr]^+$ cations can be displaced by other good nucleophiles. Substitution of the bromide with *tert*-butoxide enhances the solubility in tetrahydrofuran to about 0.5 M and also increases solubility in other less polar organic solvents.[3,4] The complex hydride is capable of hydrogenating olefins and arenes under hydrogen.[4] Recent interest has focused on the role of $[FeH_6]^{4-}$ as a precursor for iron nanoparticles.[5]

General Procedures

Peroxide-free THF and diethyl ether are distilled under argon from sodium benzophenone ketyl. Phenylmagnesium bromide (3 M solution, Aldrich) is obtained commercially. Anhydrous iron(II) halides (99.9+%) can also be obtained from Aldrich or prepared in the laboratory.[6] Lithium *tert*-butoxide (Aldrich) is used as received.

[*]Department of Chemistry, Indiana University Purdue University, Fort Wayne, IN 46805.
[†]Chemistry Division, Los Alamos National Laboratory, Los Alamos, NM 87545.

A. TETRAKIS[BROMOBIS(TETRAHYDROFURAN)MAGNESIUM] HEXAHYDRIDOFERRATE(II)

$$FeBr_2 + 6PhMgBr + 6H_2 + 10THF \rightarrow [MgBr(THF)_2]_4[FeH_6] + 2MgBr_2 \cdot THF + 6PhH$$

Procedure

■   **Caution.**   *Hydrogen is explosive when mixed with air, and the apparatus for the following experiment should be kept in a well-ventilated space behind a safety shield. Benzene, a by-product of this synthesis, is a suspected carcinogen; the waste should be discarded in a well-ventilated hood. The drying of tetrahydrofuran can cause explosions if peroxides are present. See Inorg. Synth., 12, 111, 317 (1970).*

A pressure bottle[*] with a 15 mm threaded opening (120 mL, see Fig. 1) containing a magnetic stir bar (oblong, 25 mm × 10 mm × 5 mm) is charged with powdered

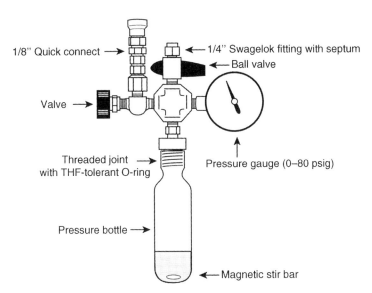

Figure 1.   The pressure reactor used in the hydrogenation reaction to prepare bis(tetrahydrofuran)bromomagnesium hexahydridoferrate(II). A useful modification according to the checkers employs a three-way valve rather than the quick-connect to enable rapid switching between vacuum and hydrogen supplies. (Figure used with permission from Ref. 4. © Elsevier.)

---

[*]The pressure tube (Ace Glass, Vineland, NJ) is equipped with the #15 Ace Thred and CHEMRAZ® O-ring (10.8 mm ID).

FeCl$_2$ (0.23 g, 1.81 mmol)* and THF (10 mL) in a dry box. This bottle is then fitted to a pressure head while in the dry box. The pressure head consists of 15 mm threaded joint to 1/4 in. NPT fitting to which is attached an X-connector; the latter is connected to a ball valve with 1/4 in. Swagelok fitting and rubber septum, a pressure gauge, and a packless valve with a quick connect fitting. This assembly is sealed by tightening the threaded O-ring fitting with a tetrahydrofuran-tolerant O-ring and closing the valves. The unit is then attached to a vacuum line using the quick connect fitting. The flask is cooled with liquid nitrogen ($-196°C$) and evacuated to $\sim 10^{-3}$ mmHg. The sealed flask is then warmed to $-30°C$ (an *o*-xylene/liquid nitrogen bath works well). The reactor is pressurized with hydrogen (2.7 bar) and stirred for 5 min. The pressure is vented to leave slightly above 1 bar of hydrogen in the reactor. The ball valve (whose opening is sealed with a septum) is turned to the open position.

■ **Caution.** *In the next step, slightly elevated hydrogen pressure inside the flask requires that a firm grip must be exercised to hold the syringe plunger.*

Phenyl magnesium bromide (4 mL of a 3.0 M solution,† 12 mmol) is added dropwise to the reactor with continuous efficient stirring over about 15 min; the addition is done using a gas-tight (Hamilton) syringe equipped with a long needle (30 cm, 20 or 22 gauge) by passing the needle through the septum and ball valve. After the addition is complete, the syringe needle is withdrawn and the ball valve is closed. The reactor is kept at $-30°C$, and the hydrogen pressure is raised to 4 bar for 2 h. For the next 4 h, the hydrogen pressure is monitored using either the dial gauge or a digital pressure manometer attached to the manifold and maintained between 3 and 5 bar. Within 4 h of stirring, the mixture changes from a murky red-brown solution with a tan precipitate to a dark brown solution with a yellow precipitate. The reactor is left in the bath and allowed to warm to room temperature, and then is stirred for an additional 4 h at room temperature. The reactor is then transferred to a glove box, the hydrogen is vented, the pressure head is detached, and the mixture in the bottle is filtered‡ to obtain a light yellow powder. The crude product is washed§ with THF (1 × 10 mL and then 4 × 2 mL) and diethyl ether (2 × 2 mL) to obtain [FeH$_6$]-

---

*The product yield depends on the starting halide. FeCl$_2$, as used in this preparation, gave somewhat better yields than FeBr$_2$. These materials can be prepared in the laboratory (see Ref. 6).

†The PhMgBr is best added as a solution in 4:1 (v/v) tetrahydrofuran/diethyl ether. The diethyl ether improves the solubility of PhMgBr in THF. If the commercially available 3 M diethyl ether solution is employed instead, the product purity will suffer due to the coprecipitation of MgBr$_2$. The high solubility of MgBr$_2$ (~ 0.4 M) in THF permits facile separation from the desired product.

‡The checkers report that the solution should be cooled in a dry box cold well or freezer before the filtration step, in order to obtain the reported yield.

§The checkers report that the wash steps should be done with cold THF and diethyl ether, in order to obtain the reported yield.

$[MgBr(THF)_2]_4$ as a bright yellow powder. Yield: 0.80 g (40%). The reaction can be performed on larger scales with modifications.[*]

Anal. Calcd. for $C_{32}H_{70}Br_4FeMg_4O_8$: C, 36.41; H, 6.68; hydride, 0.59; Fe, 5.3. Found: C, 36.75; H, 6.59; hydride, 0.56; Fe, 5.1.

Properties

This compound darkens upon loss of tetrahydrofuran, which can occur even at near room temperature. The compound is sensitive to air and moisture in the solid state and in solution. Reaction with water results in quantitative hydrogen evolution according to the equation $[FeH_6]^{4-} + 6H_2O \rightarrow Fe^{2+} + 6H_2 + 6HO^-$. This hydride titer is conveniently performed in a small septum-capped vial, and the hydrogen evolution measured by an oil-lubricated glass syringe.[†]

The ^{1}H NMR spectrum in tetrahydrofuran-d_8 shows a broad singlet at $\delta -20.3$ (FWHM = 5 Hz) due to the Fe−H groups. The deuteride $[FeD_6]^{4-}$ is readily prepared by using deuterium in the above procedure and shows the analogous Fe−D resonance in its ^{2}H NMR spectrum. The ^{57}Fe-labeled compound exhibits a $^1J_{Fe-H}$ coupling constant of 15.7 Hz. The infrared spectrum shows a doublet, $\nu(Fe-H) = 1569$, $1523\ cm^{-1}$, both in KBr and in a Nujol mull. The UV–vis spectrum in tetrahydrofuran shows ligand field absorptions at 355 and 414 nm.[2]

B. TETRAKIS[2-METHYL-2-PROPOXOMAGNESIUM] HEXAHYDRIDOFERRATE(II)

$$[MgBr(THF)_2]_4[FeH_6] + 4Li[OC(CH_3)_3]$$
$$\rightarrow [MgOC(CH_3)_3]_4[FeH_6] \cdot 4LiBr \cdot 3THF + 5THF$$

Procedure

In an inert atmosphere glove box, a 50 mL Schlenk flask is charged with [MgBr-(THF)$_2$]$_4$[FeH$_6$] (0.200 g, 0.195 mmol) and a magnetic stirring bar. To this flask is added diethyl ether (12 mL) and lithium *tert*-butoxide (0.062 g, 0.78 mmol). A reaction occurs within 1 h of stirring; after 2 h, the yellow-brown mixture is

---

[*]A larger reaction (40 mmol) required mechanically stirring for 4 h under hydrogen.
[†]Percentage purity is calculated using the reaction stoichiometry and the ideal gas law, where R is the gas constant, T is temperature, and P is pressure: % purity = $100\% \times [1071.7 \times (mL\ H_2)]/[6 \times (RT/P) \times (mg$ sample)]. Here a 10.3 mg sample of hydride evolves 1.8 mL of hydrogen under our conditions, where $P_{tot} = 730$ mmHg, $P_{solvent} = 183$ mmHg, and $T = 295$ K. The solubility of hydrogen in the aqueous solution can be neglected. This method is accurate to within 10% and helps to determine whether the MgBr$_2$ · solvent by-product has been removed.

filtered. The filtrate is evaporated and the resulting solid is triturated with hexane and then dried under vacuum to yield tan crystals of $[MgOC(CH_3)_3]_4[FeH_6] \cdot 4LiBr \cdot 3THF$. Yield: 0.17 g (64%).

Anal. Calcd. for $C_{28}H_{66}Br_4FeLi_4Mg_4O_7$: C, 33.13; H, 6.55; hydride, 0.59; Fe, 5.5. Found: C, 33.36; H, 6.42; hydride, 0.59; Fe, 5.1.

Properties

The hexahydridoferrate anion as the magnesium *tert*-butoxide salt is several orders of magnitude more soluble than the magnesium bromide salt due to a distinctive ion-pairing behavior.[3] The ^1H NMR spectrum in C_6D_6 shows[*] a multiplet at δ 3.75 (12H, α-CH$_2$ from THF), a broad singlet at δ 1.34 (48H, β-CH$_2$ from THF and CH$_3$ from the alkoxide), and a broad singlet at δ −20.1 (6H, FWHM = 500 Hz, FeH). The hydride resonance is broadened by van Vleck paramagnetism.[3,4] The ^6Li NMR spectrum shows a singlet at δ 0.78. The infrared spectrum as a Nujol mull shows a prominent Fe−H stretching absorption at $1460 \, cm^{-1}$. EXAFS data report that the Fe−Mg distance increases from 2.47 to 2.60 Å upon adding the alkoxide ion.[7,8]

Acknowledgments

We acknowledge support from the Army Research Office (under DAAH04-94-0312) and partially from the Petroleum Research Fund, administered through the American Chemical Society for support of this work.

References

1. T. Weichselfelder, *Justus Liebigs Ann. Chem.* **447**, 64 (1926).
2. (a) S. G. Gibbins, *Inorg. Chem.* **16**, 2571 (1977); (b) D. E. Linn and S. G. Gibbins, *Inorg. Chem.* **36**, 3461 (1997).
3. D. E. Linn Jr., and S. G. Gibbins, *J. Organomet. Chem.* **554**, 171 (1998).
4. D. E. Linn Jr., G. M. Skidd, and E. M. Tippmann, *Inorg. Chim. Acta* **291**, 142 (1999).
5. D. E. Linn Jr., Y. Guo, and S. P. Cramer, *Inorg. Chim. Acta* **361**, 1552 (2008).
6. G. Winter, *Inorg. Synth.* **7**, 101 (1964).
7. D. E. Linn Jr., G. M. Skidd, and S. N. McVay, *Inorg. Chem.* **41**, 5320 (2002).
8. A soluble nonahydridodicobaltate species is found to be accessible, $[(CoH_3)_2H_3]^{5-}$., see D. Linn, J. Kieser, J. Shearer, *Dalton Trans.*, (2014), DOI: 10.1039/C3DT53117A.

[*]The checkers reported that when the product is dissolved in C_6D_6, a small amount of white solid appears, which is readily removed by filtration. This material is likely LiBr and should not affect the use of the hydride.

32. TRIS(ALLYL)IRIDIUM AND -RHODIUM

Submitted by KEVIN D. JOHN,[*] JUDITH L. EGLIN,[*] KENNETH V. SALAZAR,[*]
R. THOMAS BAKER,[*] and ALFRED P. SATTELBERGER[*]
Checked by DANIEL SERRA[†] and LISA MCELWEE WHITE[†]

Recent efforts aimed at preparing and characterizing "single-site" catalysts consisting of well-defined, metal oxide surface-bound, ligated transition metals have yielded exciting new catalytic systems for alkane metathesis and polyolefin depolymerization.[1,2] One particularly well-characterized class of catalytic sites is derived from reactions of tris(allyl)rhodium, $Rh(C_3H_5)_3$, with carefully prepared, high-purity metal oxides.[3] The use of (allyl)rhodium-based systems for catalysis is hampered by weak $Rh-O$ bonds to the metal oxide surface and the proclivity of trivalent rhodium intermediates to undergo reductive elimination.[4]

The investigation of tris(allyl)iridium for the preparation of single-site catalysts[5] with better thermal and reductive stability resulted in the development of improved synthetic routes to both the iridium and rhodium analogs. The syntheses of $M(allyl)_3$, modified syntheses of $MCl_3(THT)_3$ precursors (THT = tetrahydrothiophene),[6] and an improved synthesis of allyl lithium are described here.

General Procedures

Reactions involving air-sensitive reagents and products are carried out at atmospheric pressure (600 mmHg in Los Alamos, elevation 2200 m) in an inert atmosphere glove box or by using vacuum line and Schlenk techniques, unless specified otherwise. $IrCl_3 \cdot 3H_2O$ (Pressure Chemical), $RhCl_3 \cdot 3H_2O$ (Pressure Chemical), 2-methoxyethanol (Aldrich), ethanol (Aldrich), tetraallyltin (Aldrich), and tetrahydrothiophene (Aldrich) are used as received. Hexane and ether are dried by elution from columns of activated alumina and copper oxide BTS catalyst according to the procedure described by Grubbs and coworkers.[7] Salicylaldehyde phenylhydrazone is prepared according to the literature method and used as an indicator to titrate the butyllithium (Aldrich, 2.5 M solution in hexanes) and allyllithium reagents.[8] NMR spectra are recorded with a Varian UNITY series 300 MHz spectrometer at room temperature. NMR solvents are dried over activated 4 Å molecular sieves and freeze–pump–thaw–degassed prior to use. Elemental analyses are performed on a Perkin Elmer 2400 system with a Series II CHNS/O analyzer equipped with an AD-4 autobalance.

[*]Los Alamos National Laboratory, Los Alamos, NM 87545.
[†]Department of Chemistry, University of Florida, Gainesville, FL 32611.

A. ALLYLLITHIUM

$$Sn(allyl)_4 + 2Li(n\text{-}Bu) \xrightarrow[25°C]{Et_2O/hexanes} 2Li(allyl) + Sn(allyl)_2(n\text{-}Bu)_2$$

Procedure[9]

Inside a glove box, n-butyllithium (7.1 mL of a 2.5 M solution in hexanes, 17.7 mmol) is reduced to a thick oil under vacuum at room temperature. The pale yellow n-butyllithium oil is then dissolved in diethyl ether (5 mL). This solution is added dropwise over 5 min to a stirred solution of tetra(allyl)tin (2.5 g, 8.83 mmol) in diethyl ether (40 mL) in a 100 mL Erlenmeyer flask. The solution is stirred for 15 min before most of the solvent is removed under vacuum. To the resultant paste, hexane (25 mL) is added and the resultant suspension is stirred vigorously for 30 min. If the trituration fails to provide a white solid, vigorous scraping of the flask walls with a spatula usually affords the desired powder. Alternatively, the hexane can be decanted and fresh hexane added and the agitation repeated. The product is filtered and rinsed with hexane (2 × 10 mL) to give a white powder. Yield: 0.74 g (87%).

Properties

Allyllithium is a pyrophoric white solid. Solid samples stored under an inert atmosphere at room temperature become "yellow" over time (days) and lose their effectiveness. Samples should be stored at −20°C or prepared immediately before use. Fresh samples can be dissolved readily in ether solvents to give pale yellow solutions. Allyllithium is insoluble in aliphatic hydrocarbons, slightly soluble in aromatic solvents, and reacts with methylene chloride, chloroform, and acetonitrile. The product can be dissolved in pyridine to provide orange solutions for NMR analyses. ^{1}H NMR (NC_5D_5, 25°C): δ 6.25 (s, 1H, $CH_2-CH=CH_2$), 5.19–5.11 (m, 2H, $CH_2-CH=CH_2$), 2.56 (s, 2H, $CH_2-CH=CH_2$). ^{13}C{^{1}H} NMR (NC_5D_5, 25°C): δ 138.99 (1C, $CH_2-CH=CH_2$), 114.44 (1C, $CH_2-CH=CH_2$), 50.74 (1C, $CH_2-CH=CH_2$). Note that the chemical shifts in pyridine solvent indicate sigma coordination of the allyl anion to the lithium cation, unlike the π-coordination observed in THF.[10]

B. *mer*-TRICHLOROTRIS(TETRAHYDROTHIOPHENE)IRIDIUM(III)

$$IrCl_3 \cdot 3H_2O + 3THT \xrightarrow[\Delta]{CH_3O(CH_2)_2OH} mer\text{-}IrCl_3(THT)_3 + 3H_2O$$

Procedure[6]

■ **Caution.** *Tetrahydrothiophene is a foul-smelling liquid and should only be opened in a good fume hood; all subsequent transfers and filtrations should also be performed in a hood.*

$IrCl_3 \cdot 3H_2O$ (2.00 g, 5.67 mmol) and 2-methoxyethanol (100 mL) are placed in a 250 mL one-necked round-bottomed flask fitted with a reflux condenser and a magnetic stir bar. Tetrahydrothiophene (2.50 mL, 28.4 mmol) is added in one portion with stirring. The initial suspension is refluxed for 12 h, typically providing a clear yellow orange solution. After being cooled to room temperature, the solution (if cloudy) is filtered through a 60 mL medium-porosity sintered glass frit and the solvent volume is reduced by ~50% under vacuum.[*] Then water (150 mL) is added to the room temperature solution, and the suspension is cooled overnight (0°C), and filtered using a 60 mL medium-porosity sintered glass frit to provide the crude product as a yellow powder. The powder is washed with water (2 × 100 mL) and allowed to air dry overnight.[†] The product is further purified by recrystallization from boiling ethanol (~400 mL) to yield yellow microcrystals. Yield: 2.92 g (91%).[‡]

Anal. Calcd. for $C_{12}H_{24}Cl_3S_3Ir$: C, 25.60; H, 4.30. Found: C, 25.70; H, 4.39.

Properties

mer-$IrCl_3(THT)_3$ is a yellow solid that is indefinitely air stable.[§] It is sparingly soluble in aromatic hydrocarbons, alcohols, methylene chloride, chloroform, and ethers. Yields do not vary significantly upon addition of larger quantities of tetrahydrothiophene or scaling of the reaction; for example, a second preparation with $IrCl_3 \cdot 3H_2O$ (5.00 g, 14.18 mmol), 2-methoxyethanol (250 mL), and tetrahydrothiophene (9.0 mL, 102.2 mmol) provided *mer*-$IrCl_3(THT)_3$ in a yield of 7.12 g (89%). ^1H NMR (CD_2Cl_2, 25°C): δ 3.60 (m, 4H, SCH_2CH$_2$), 3.23 (m, 2H, SCH_2CH$_2$), 2.9 (overlapping m, 6H, SCH_2CH$_2$), 2.4–2.1 (overlapping m, 12H, SCH$_2$CH_2). ^{13}C{^{1}H} NMR (CD_2Cl_2, 25°C): δ 36.92 (2C, SCH_2CH$_2$), 36.51 (4C, SCH_2CH$_2$), 30.49 (4C, SCH$_2$CH_2), 30.40 (2C, SCH$_2$CH_2).

C. *mer*-TRICHLOROTRIS(TETRAHYDROTHIOPHENE)-RHODIUM(III)

$$RhCl_3 \cdot 3H_2O + 3THT \xrightarrow[\Delta]{CH_3O(CH_2)_2OH} \textit{mer-}RhCl_3(THT)_3 + 3H_2O$$

---

[*]The checkers report that 2-methoxyethanol easily condenses in hoses and elsewhere, and that the solvent reduction step is best conducted with a vacuum transfer system or distillation head. They also report that some yellow solid forms when the volume is reduced, and cooling may not be necessary after the addition of water in the next step if the solution is colorless.

[†]The checkers dried the material in vacuum.

[‡]The checkers obtained a 56% yield, but the $IrCl_3(H_2O)_3$ starting material they used was later found to be impure.

[§]The checkers report that the product is light sensitive, and for long-term storage the material should be kept in the dark.

Procedure[6]

■   **Caution.**   *Tetrahydrothiophene is a foul-smelling liquid and should only be opened in a good fume hood; all subsequent transfers and filtrations should also be performed in a hood.*

To a room-temperature suspension of $RhCl_3 \cdot 3H_2O$ (2.00 g, 7.60 mmol) in 2-methoxyethanol (100 mL) in a 250 mL one-necked round-bottomed flask fitted with a condenser and a magnetic stir bar is added tetrahydrothiophene (4.00 mL, 45.4 mmol) with stirring. The solution is refluxed in air for 12 h to provide a clear orange solution. After the mixture is cooled to room temperature, the solvent volume is reduced by ~50% in vacuum[*] and water (150 mL) is added to the room-temperature solution. The solution is cooled overnight (0°C) and filtered using a 60 mL medium-porosity sintered glass frit to provide the crude product as an orange powder. The powder is washed with water (2 × 100 mL) and allowed to air dry overnight.[†] The product is further purified by recrystallization from boiling ethanol (~400 mL) to yield orange microcrystals. Yield: 3.59 g (99%).

Anal. Calcd. for $C_{12}H_{24}Cl_3S_3Rh$: C, 30.42; H, 5.11. Found: C, 30.43; H, 5.44.

Properties

mer-$RhCl_3(THT)_3$ is an air-stable orange crystalline solid. It is sparingly soluble in aromatic hydrocarbons, alcohols, chloroform, and ethers. ^1H NMR (CD_2Cl_2, 25°C): δ 3.70 (m, 4H, SCH_2CH_2), 3.28 (m, 2H, SCH_2CH_2), 2.90 (overlapping m, 6H, SCH_2CH_2), 2.3–2.0 (overlapping m, 12H, SCH_2CH_2). $^{13}C\{^1H\}$ NMR (CD_2Cl_2, 25°C): δ 37.69 (2C, SCH_2CH_2), 37.31 (4C, SCH_2CH_2), 30.38 (4C, SCH_2CH_2), 30.23 (2C, SCH_2CH_2).

D. TRIS(ALLYL)IRIDIUM(III)

$$mer\text{-}IrCl_3(THT)_3 + 3Li(allyl) \xrightarrow[25°C]{Et_2O/hexanes} Ir(allyl)_3 + 3THT + 3LiCl$$

Procedure[11]

Inside a glove box, $IrCl_3(THT)_3$ (0.509 g, 0.90 mmol) and Et_2O (40 mL) are placed in a stoppered 150 mL Erlenmeyer flask equipped with magnetic stir bar. A

---

[*]The checkers report that 2-methoxyethanol easily condenses in hoses and elsewhere, and that the solvent reduction step is best conducted with a vacuum transfer system or distillation head. Cooling may not be necessary after the addition of water in the next step if the solution is colorless.
[†]The checkers dried the product in vacuum.

solution of allyllithium (0.130 g, 2.70 mmol) in Et_2O (10 mL) is added dropwise over the course of 20 min with vigorous stirring. Imperative to the success of the reaction is the slow addition of the allyllithium solution; copious amounts of white/beige-colored precipitate are observed if the allyllithium is added too quickly, and this observation is coupled with little or no product formation. The reaction mixture is stirred for 14 h at room temperature during which time the color of the suspension changes from pale yellow to brown similar to the appearance of coffee with cream. A yellow tinge remains in the solution without the development of the coffee color if insufficient allyllithium is used in the preparation.

After 14 h, the reaction mixture is filtered through a 60 mL medium-porosity sintered glass frit and the residue is rinsed with Et_2O (2×10 mL). The filtrate is then reduced to dryness under vacuum. Note that the $Ir(C_3H_5)_3$ is volatile at room temperature and excessive drying of the solid will result in loss of the product. The solid residue is extracted with hexane (2×25 mL) and the extract is filtered using a 30 mL medium-porosity sintered glass frit. The brown hexane extract is reduced in volume to about 3 mL and transferred to a sublimator where the remaining hexane is carefully removed and the brown residue is sublimed onto a cold finger ($-78°C$) at 10^{-5} mmHg with mild heating (oil bath; $<60°C$).* Sublimation affords colorless microcrystalline $Ir(allyl)_3$. Yield: 0.126 g (44%).

Anal. Calcd. for $C_9H_{15}Ir$: C, 34.26; H, 4.79. Found: C, 34.57; H, 5.05.

Properties

Tris(allyl)iridium is a moderately air-sensitive, white microcrystalline solid that can be sublimed under dynamic vacuum at room temperature. Samples exposed to air slowly transform to a tan material.[†] Tris(allyl)iridium can be recovered from this material via sublimation or by extracting the solid with pentane, filtering, and stripping the solvent. Reactions performed on a larger scale (above 2.0 g of $IrCl_3(THT)_3$) result in significantly reduced yields. A solution of $Ir(allyl)_3$ can be prepared in over 80% yield (based on NMR analysis compared to an internal standard) from the reaction of $IrCl_3(THT)_3$ and allyllithium in benzene. However, the product can only be isolated in ~20% yield after sublimation. The poor yield is presumably due to loss of $Ir(allyl)_3$ upon removal of the benzene solvent under vacuum prior to sublimation. Benzene solutions prepared in this manner have been used to generate a variety of complexes with no effect on the product quality or yields.[12] ^{1}H NMR (C_6D_6, 25°C): δ 5.00 (m, 1H, CH_2CHCH_2), 3.30 (m, 2H, CH_2CHCH_2), 2.86 (d, 4H, *syn*-CH_2CHCH_2), 2.75 (d, 2H, *syn*-CH_2CHCH_2), 2.66

---

*The checkers carried out the sublimation at 80–90°C/0.005 mmHg.
†The checkers report that the product is light sensitive, and for long-term storage the material should be kept in the dark.

(d, 2H, *anti-CH$_2$CHCH$_2$*), 1.72(d, 4H, *anti-CH$_2$CHCH$_2$*). ^{13}C{^{1}H} NMR (C$_6$D$_6$, 25°C): δ 91.98 (s, 1C, CH$_2$CHCH$_2$), 82.17 (s, 2C, CH$_2$CHCH$_2$), 38.07 (s, 2C, CH$_2$CHCH$_2$), 28.33 (s, 4C, CH$_2$CHCH$_2$).

E. TRIS(ALLYL)RHODIUM(III)

$$mer\text{-}RhCl_3(THT)_3 + 3Li(allyl) \xrightarrow[25°C]{Et_2O/hexanes} Rh(allyl)_3 + 3THT + 3LiCl$$

Procedure[13]

Inside a glove box, RhCl$_3$(THT)$_3$ (0.500 g, 1.06 mmol) and Et$_2$O (40 mL) are placed in a stoppered 150 mL Erlenmeyer flask equipped with magnetic stir bar. A solution of allyllithium (0.153 g, 3.18 mmol) in Et$_2$O (10 mL) is added dropwise over 20 min with vigorous stirring. Imperative to the success of the reaction is the slow addition of allyllithium solution. The reaction mixture is stirred for 14 h at room temperature during which time the color of the solution changes from orange to dark brown. The reaction mixture is then filtered through a 60 mL medium-porosity sintered glass frit and the residue rinsed with Et$_2$O (2 × 10 mL). The ether solution is then reduced to dryness under vacuum. As in the case of Ir(allyl)$_3$, Rh(allyl)$_3$ sublimes at room temperature and excess vacuum drying of the solid will result in loss of the product. The solid residue is extracted with hexane (50 mL) and the extract is filtered. The hexane extract is reduced in volume to about 3 mL and transferred to a sublimator where the remaining hexane is "carefully" removed and the brown residue is sublimed onto a cold finger (−78°C) at 10^{-5} mmHg with mild heating (oil bath; <60°C). Sublimation affords bright yellow microcrystalline Rh(allyl)$_3$. Yield: 0.243 g (86%).[*]

Anal. Calcd. for C$_9$H$_{15}$Rh: C, 47.81; H, 6.69. Found: C, 47.76; H, 6.72.

Properties

Tris(allyl)rhodium is a moderately air-sensitive, bright yellow microcrystalline solid that can be sublimed at room temperature.[†] ^{1}H NMR (C$_6$D$_6$, 25°C): δ 5.11 (m, 1H, CH$_2$CHCH$_2$), 3.68 (m, 2H, CH$_2$CHCH$_2$), 2.83 (d, 2H, *syn-CH$_2$CHCH$_2$*), 2.65 (d, 4H, *syn-CH$_2$CHCH$_2$*), 2.53 (d, 2H, *anti-CH$_2$CHCH$_2$*), 1.53 (d, 4H, *anti-CH$_2$CHCH$_2$*). ^{13}C{^{1}H} NMR (C$_6$D$_6$, 25°C): δ 100.18 (s, 1C, $^1J_{RhC}$ = 4.3 Hz,

---

[*]The checkers carried out the sublimation at 0.01 mmHg and a bath temperature of 80–90°C and obtained a yield of 55%.
[†]The checkers suggest that the compound may also be light sensitive.

CH_2CHCH_2), 95.06 (s, 2C, $^1J_{RhC} = 4.0\,Hz$, CH_2CHCH_2), 48.23 (s, 2C, $^1J_{RhC} = 9.2\,Hz$, CH_2CHCH_2), 41.22 (s, 4C, $^1J_{RhC} = 9.1\,Hz$, CH_2CHCH_2).

References

1. F. Lefebvre, J. Thivolle-Cazat, V. Dufaud, G. P. Niccolai, and J. M. Basset, *Appl. Catal. A* **182**, 1 (1999).
2. V. R. Dufaud and J. M. Basset, *Angew. Chem., Int. Ed.* **37**, 806 (1998).
3. C. C. Santini, S. L. Scott, and J. M. Basset, *J. Mol. Catal. A* **107**, 263 (1996), and references therein.
4. J. M. Basset, F. Lefebvre, and C. Santini, *Coord. Chem. Rev.* **178–180**, 1703 (1998).
5. R. J. Trovich, N. Guo, M. T. Janicke, H. Li, C. L. Marshall, J. T. Miller, A. P. Sattelberger, K. D. John, and R. T. Baker, *Inorg. Chem.* **49**, 2247 (2010).
6. E. A. Allen and G. Wilkinson, *J. Chem. Soc., Dalton Trans.* 613 (1972).
7. A. B. Pangborn, M. A. Giardellow, R. H. Grubbs, R. K. Rosen, and F. J. Timmer, *Organometallics* **15**, 1518 (1996).
8. B. E. Love and E. G. Jones, *J. Org. Chem.* **64**, 3755 (1999).
9. (a) D. Seyferth and M. A. Weiner, *J. Org. Chem.* **26**, 4797 (1961); (b) J. J. Eisch, *Nontransition Metal Compounds: Organometallic Syntheses*, Academic Press, New York, 1981, Vol. **2**; (c) Y. Horikawa and T. Takeda, *J. Organomet. Chem.* **523**, 99 (1996).
10. T. B. Thompson and W. T. Ford, *J. Am. Chem. Soc.* **101**, 5459 (1979).
11. P. Chini and S. Martinengo, *Inorg. Chem.* **6**, 837 (1967).
12. K. D. John, K. V. Salazar, B. L. Scott, R. T. Baker, and A. P. Sattelberger, *Organometallics* **20**, 296 (2001).
13. J. Powell and B. L. Shaw, *J. Chem. Soc. A* 583 (1968).

33. TRINUCLEAR PALLADIUM(II) ACETATE

Submitted by JOHN F. BERRY,[*] **F. ALBERT COTTON,**[*†]
SERGEY IBRAGIMOV,[*] **and CARLOS A. MURILLO**[*]
Checked by DAREN J. TIMMONS[‡] **and KYLE A. FRICKE**[‡]

$$PdCl_2 + NaO_2CH + NaOH \rightarrow Pd + CO_2 + 2NaCl + H_2O$$

$$3Pd + 6HNO_3 + 6HO_2CCH_3 \rightarrow Pd_3(O_2CCH_3)_6 + 6NO_2 + 6H_2O$$

One of the most versatile starting materials for the preparation of palladium(II) complexes is the trinuclear $Pd_3(OAc)_6$, which has the unusual but well-known D_{3h} structure in which six acetate groups bridge three nearly square planar metal atoms.[1–4] Recently, several syntheses have been reported[5–7] for this compound,

[*]Laboratory for Molecular Structure and Bonding, Department of Chemistry, Texas A&M University, College Station, TX 77842.
[†]Deceased, February 20, 2007.
[‡]Department of Chemistry, Virginia Military Institute, Lexington, VA 24450.

which is extensively used to make catalysts for organic synthesis.[8-14] An insoluble polymeric form has also been characterized structurally.[15]

There are various reports suggesting the hypothesis that various $[Pd(OAc)_2]_n$ aggregates also form in solution because the ^{1}H NMR spectra usually showed a complex pattern in the acetate region where only a single peak would be expected from the highly symmetrical D_{3h} structure.[16,17] Recently, we found that a complex pattern in the NMR spectrum in C_6D_6 and $CDCl_3$ may be attributed to hydrolysis or to the formation of a nitrite derivative that is formed following some of the recently reported preparations for $Pd_3(OAc)_6$.[18]

Here we report a reproducible and dependable two-step procedure for the synthesis of trinuclear $Pd_3(OAc)_6$ from $PdCl_2$. The method proceeds in high yield to produce a crystalline material that is soluble in various organic solvents and that gives a ^{1}H NMR spectrum with only one acetate resonance.

Procedure

■ **Caution.** *Concentrated nitric acid is very reactive and when used as an oxidant it produces toxic nitrogen oxides. All operations should be carried out in a well-ventilated fume hood; protective gloves and safety glasses should be worn. PdCl₂ should be handled with caution because of its toxicity.*

A mixture of sodium hydroxide (1.0 g, 25 mmol) and sodium formate (0.80 g, 11.7 mmol) is added to a suspension of $PdCl_2$ (0.500 g, 2.86 mmol) in water (50 mL). Formation of elemental palladium, as a fine black powder, occurs immediately. The suspension is stirred for 30 min to allow the palladium particles to coagulate. The palladium metal is separated by filtration, washed with acetone, and dried under vacuum. Yield: 0.302 g (98%).

The palladium powder is then suspended in glacial acetic acid (20 mL), and concentrated nitric acid (0.3 mL) is slowly added with stirring. The resulting solution is heated to reflux for 30 min while N_2 is continuously bubbled through the reaction mixture to avoid formation of $Pd_3(OAc)_5(NO_2)$. The volume of the solution is then reduced to a third of the original volume by slow evaporation using mild heating. After the solution is transferred to a Petri dish and cooled to room temperature, an orange powder forms, which is isolated the following day by filtration. Yield: 0.605 g (94.4%).[*]

Properties

The orange $Pd_3(OAc)_6$ is readily soluble in dichloromethane, toluene, benzene, and tetrahydrofuran, and is stable in air. Crystals can be easily obtained by slow

---

[*]The checkers report a yield of 70%.

evaporation under a N_2 stream from solutions containing the trinuclear complex in a mixed solvent consisting of equal volumes of CH_2Cl_2 and hexanes. The ^1H NMR spectrum (300 MHz) in thoroughly dried solvents shows one resonance at δ 2.006 (s, 12H) in $CDCl_3$ and at δ 1.622 (s, 12H) in C_6D_6. In solvents that have not been thoroughly dried, several additional resonances appear. IR (KBr, cm^{-1}): 1600 (vs), 1430 (vs), 1350 (w), 1157 (vw), 1047 (vw), 951 (vw), 696 (m), 625 (vw).

Acknowledgments

We thank the Robert A. Welch Foundation and NSF for support. J.F.B. thanks the NSF for a predoctoral fellowship.

References

1. A. C. Skapski and M. L. Smart, *J. Chem. Soc., Chem. Commun.* 658 (1970).
2. F. A. Cotton and S. Han, *Rev. Chim. Minér.* **20**, 496 (1983).
3. F. A. Cotton and S. Han, *Rev. Chim. Minér.* **22**, 277 (1985).
4. N. N. Lyalina, S. V. Dargina, A. N. Sobolev, T. M. Buslaeva, and I. P. Romm, *Koord. Khim.* **19**, 57 (1993); *Russ. J. Coord. Chem.* **19**, 50, (1993).
5. R. Zhang, C. Ma, and H. Yin, *Huaxue Shiji* **16**, 383 (1994); *Chem. Abstr.* **122**, 203713 (1995).
6. I. P. Romm, Yu. G. Noskov, T. I. Perepelkova, S. V. Kravtsova, and T. M. Buslaeva, *Russ. J. Gen. Chem.* **68**, 681 (1998).
7. D. Jiang, J. Wang, B. Tang, and W. Wang, *Guijinshu* **18**, 16 (1997); *Chem. Abstr.* **128**, 225136 (1998).
8. N. E. Leadbeater and M. Marco, *J. Org. Chem.* **68**, 888 (2003).
9. F. Y. Zhao, M. Shirai, and M. Arai, *J. Mol. Catal. A: Chem.* **154**, 39 (2000).
10. C. G. Jia, W. J. Lu, T. Kitamura, and Y. Fujiwara, *Org. Lett.* **1**, 2097 (1999).
11. R. B. Bedford, M. E. Blake, C. P. Butts, and D. Holder, *Chem. Commun.* 466 (2003).
12. Q. W. Yao, E. P. Kinney, and Z. Yang, *J. Org. Chem.* **68**, 7528 (2003).
13. H. M. Lee and S. P. Nolan, *Org. Lett.* **2**, 2053 (2000).
14. J. P. Wolfe, R. A. Singer, B. H. Yang, and S. L. Buchwald, *J. Am. Chem. Soc.* **121**, 9550 (1999).
15. S. D. Kirik, R. F. Mulagaleev, and A. I. Blokhin, *Acta Crystallogr. C* **60**, m449 (2004).
16. A. Marson, A. B. van Oort, and W. P. Mul, *Eur. J. Inorg. Chem.* 3028 (2002).
17. C. Bianchini, A. Meli, and W. Oberhauser, *Organometallics* **22**, 4281 (2003).
18. V. I. Bakhmutov, J. F. Berry, F. A. Cotton, S. Ibragimov, and C. A. Murillo, *Dalton Trans.* 1989 (2005).

Chapter Five

MAIN GROUP COMPOUNDS AND LIGANDS

34. MONOCARBABORANE ANIONS WITH 10 OR 12 VERTICES

Submitted by ANDREAS FRANKEN[*] and JOHN D. KENNEDY[*]
Checked by JUDE CLAPPER[†] and LARRY G. SNEDDON[†]

Monocarbaborane cluster chemistry is sparsely investigated compared to its flanking fields of binary boron hydride cluster chemistry and dicarbaborane cluster chemistry.[1–4] Dicarbaborane compounds can be synthesized relatively easily by addition of alkynes to boron hydrides such as decaborane, but there are no simple routes to monocarbaborane compounds. Traditionally, the principal methods involve removal of one carbon atom from dicarbaboranes,[5] or addition of cyanide or organic nitriles to borane clusters to give C-aminated monocarbaboranes,[6] which can subsequently be deaminated.[3]

There is considerable contemporary interest in monocarbaborane chemistry:[7] (a) as synthons for metallamonocarbaborane compounds,[3,8–10] (b) as components in ionic liquids,[11] (c) in liquid crystal chemistry and other rheological applications,[12] (d) as rigid zwitterions of high dipolarities and hyperpolarizabilities,[13,14] (e) as building blocks for "molecular meccano" or "molecular tinkertoy" nanoarchitectural chemistries,[15,16] and (f) for tailoring of compounds for host–guest molecular recognition and supramolecular assembly.[17] Of particular interest are the *closo* 10-vertex $[CB_9H_{10}]^-$ and *closo* 12-vertex $[CB_{11}H_{12}]^-$ anions and their alkylated or halogenated derivatives.[18–20] These anions are chemically, electrochemically, and

[*]Department of Chemistry, University of Leeds, Leeds LS2 9JT, UK.
[†]Department of Chemistry, University of Pennsylvania, Philadelphia, PA 19104-6323.

Inorganic Syntheses, Volume 36, First Edition. Edited by Gregory S. Girolami and Alfred P. Sattelberger.
© 2014 John Wiley & Sons, Inc. Published 2014 by John Wiley & Sons, Inc.

thermally stable, and the anionic charge is highly delocalized, so that they are useful as "least-coordinating" counterions for cations of high Lewis acidity[3,18] such as cationic metallocenes (which are useful for olefin polymerization catalysis),[21] coordinatively unsaturated cations,[22] and strong Brønsted acids.[23]

One excellent and high-yield entry into monocarbaborane chemistry is the Brellochs reaction[24] of decaborane with aldehydes. For example, formaldehyde gives the $[arachno\text{-}6\text{-}CB_9H_{14}]^-$ anion, aliphatic aldehydes give the $[6\text{-}R\text{-}arachno\text{-}6\text{-}CB_9H_{13}]^-$ anions, and aromatic aldehydes give the $[6\text{-}Ar\text{-}nido\text{-}6\text{-}CB_9H_{11}]^-$ anions.[8,25] Functionalized aldehydes can be used to generate series of C-functionalized analogs.[26–30] Here we describe the use of this reaction to prepare two key monocarbaborane anions: the $[arachno\text{-}6\text{-}CB_9H_{14}]^-$ anion **1** and the $[6\text{-}Ph\text{-}nido\text{-}6\text{-}CB_9H_{11}]^-$ anion **5**. We also describe the conversion of these starting materials to various *closo* 10-vertex and 12-vertex species, the $[closo\text{-}2\text{-}CB_9H_{10}]^-$ anion **2**, the $[closo\text{-}1\text{-}CB_9H_{10}]^-$ anion **3**, the $[closo\text{-}1\text{-}CB_{11}H_{12}]^-$ anion **4**, the $[2\text{-}Ph\text{-}closo\text{-}2\text{-}CB_9H_9]^-$ anion **6**, the $[1\text{-}Ph\text{-}closo\text{-}1\text{-}CB_9H_9]^-$ anion **7**, and the $[1\text{-}Ph\text{-}closo\text{-}1\text{-}CB_{11}H_{11}]^-$ anion **8**. For all these species, compound identity and purity are determined by ^{11}B and ^{1}H NMR spectroscopy. In each case, crystals can be obtained by overlayering a saturated acetone solution with diethyl ether, and then allowing the mixture to stand at room temperature. Drawings of the structures of the products are given in Fig. 1.

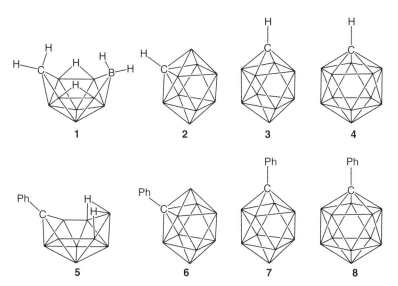

Figure 1. Schematic molecular structures of the anions $[arachno\text{-}6\text{-}CB_9H_{14}]^-$ **1**, $[closo\text{-}2\text{-}CB_9H_{10}]^-$ **2**, $[closo\text{-}1\text{-}CB_9H_{10}]^-$ **3**, $[closo\text{-}1\text{-}CB_{11}H_{12}]^-$ **4**, $[6\text{-}Ph\text{-}nido\text{-}6\text{-}CB_9H_{11}]^-$ **5**, $[2\text{-}Ph\text{-}closo\text{-}2\text{-}CB_9H_9]^-$ **6**, $[1\text{-}Ph\text{-}closo\text{-}1\text{-}CB_9H_9]^-$ **7**, and $[1\text{-}Ph\text{-}closo\text{-}1\text{-}CB_{11}H_{11}]^-$ **8**. Unlabeled vertices are BH(*exo*) units.

General Procedures

All reactions are carried out under an inert atmosphere (dry N_2) using standard Schlenk techniques. Subsequent manipulations are carried out in air. No special techniques or apparatus are required. Evaporations designated "under reduced pressure" are carried out using a standard rotary film evaporator and water pump pressure. Solids designated as dried in vacuum are dried using a standard vacuum oil pump. Organic solvents are dried and distilled before use. Crystallographic data as given for the $[NEt_4]^+$ salts of anions **5**, **6**, **7**, and **8** were measured at 150(2) K. ^{11}B NMR chemical shifts are referenced to $BF_3(OEt_2)$ in $CDCl_3$.

A. TETRAETHYLAMMONIUM *arachno*-6-CARBA-DECABORANATE(14)

$$B_{10}H_{14} + KOH + 2H_2O + HCHO \rightarrow [K][arachno\text{-}6\text{-}CB_9H_{14}] + B(OH)_3 + 2H_2$$

$$[K][arachno\text{-}6\text{-}CB_9H_{14}] + [NEt_4]Cl \rightarrow [NEt_4][arachno\text{-}6\text{-}CB_9H_{14}] + KCl$$

Procedure

■   **Caution.**   *$B_{10}H_{14}$ is toxic and should be used in a well-ventilated fume hood.*

A 500 mL flask is charged with $B_{10}H_{14}$ (6.0 g, 49.1 mmol; Katchem, Prague). The flask is cooled to 0°C and a solution of KOH (10 g, 0.18 mol) in water (150 mL) is added. Over 6 h, aqueous 37% formaldehyde (stabilized with 10% methanol; 40 mL, 0.51 mol HCHO; Aldrich) is added in five portions at approximately equal intervals. The reaction mixture is stirred overnight,[*] the methanol is removed under reduced pressure, and water (100 mL) is added. The alkaline solution is extracted with Et_2O (3 × 100 mL). Water (300 mL) is added to the combined ether extracts. The Et_2O is removed under reduced pressure and NEt_4Cl (10 g, 60 mmol) is added to the resulting aqueous solution. A white precipitate forms, which is collected by filtration, washed with water (3 × 50 mL), and dried in vacuum to yield $[NEt_4]$ $[arachno\text{-}6\text{-}CB_9H_{14}]$, **1**. Yield: 9.2 g (74%).

Anal. Calcd. for $C_9H_{34}NB_9$: C, 42.6; H, 13.5; N, 5.5. Found: C, 42.7; H, 13.7; N, 5.5.

---

[*]The checkers stirred the mixture for 16 h.

Properties

[NEt$_4$][*arachno*-6-CB$_9$H$_{14}$] is a white air-stable solid, mp >280°C. It is soluble in acetone, tetrahydrofuran, acetonitrile, and CH$_2$Cl$_2$. IR (KBr, cm^{-1}): 2989 (m), 2522 (s), 1642 (m), 1435 (m), 1305 (m), 1002 (m), 784 (m), 709 (m). ^{11}B{^{1}H} or ^{13}C{^{1}H} NMR [^{1}H NMR data in brackets], in CD$_3$CN at 294–299 K: BH(4) −1.5 [+2.14], BH(2) −10.3 [+1.89], BH(5,7) −13.0 [+2.10], BH$_2$(9) −23.2 [+0.74, −0.78], BH(8.10) −28.8 [+0.64], BH(1,3) −39.8 [+0.22], CH$_2$(cluster 6-position) −4.5 [−0.15, −0.78], α-Et +47.6 [+3.49 (8H, quartet)], β-Et +7.9 [+1.41 (12H, triplet)], μH(7,8)/μH(5,10) [−3.96].

B. TETRAETHYLAMMONIUM *closo*-2-CARBA-DECABORANATE(10)

[NEt$_4$][*arachno*-6-CB$_9$H$_{14}$] + HCl + H$_2$O → [H$_3$O][*arachno*-6-CB$_9$H$_{14}$] + [NEt$_4$]Cl

[H$_3$O][*arachno*-6-CB$_9$H$_{14}$] + KOH → [K][*arachno*-6-CB$_9$H$_{14}$] + 2H$_2$O

[K][*arachno*-6-CB$_9$H$_{14}$] + 2I$_2$ + 4KOH → [K][*closo*-2-CB$_9$H$_{10}$] + 4KI + 4H$_2$O

[K][*closo*-2-CB$_9$H$_{10}$] + NEt$_4$Cl → [NEt$_4$][*closo*-2-CB$_9$H$_{10}$] + KCl

Procedure

A solution of [NEt$_4$][*arachno*-6-CB$_9$H$_{14}$] (**1**, 3.04 g, 12.0 mmol) in aqueous 5% HCl (160 mL) is extracted several times with Et$_2$O (3 × 80 mL). The separated organic layers are shaken with aqueous 1.0 M KOH (400 mL), and the Et$_2$O is removed under reduced pressure. To the resulting alkaline aqueous solution, elemental I$_2$ (8.0 g, 31.6 mmol)* is added. The solution is stirred for 2 h at ambient temperature, and then NEt$_4$Cl (3.32 g, 20 mmol) is added. The mixture is stirred briefly, and then the precipitate is collected by filtration, washed with H$_2$O (3 × 50 mL), and dried in vacuum to afford [NEt$_4$][*closo*-2-CB$_9$H$_{10}$], **2**. Yield: 2.76 g (92%).

Anal. Calcd. for C$_9$H$_{30}$NB$_9$: C, 43.3; H, 12.1; N, 5.6. Found: C, 43.1; H, 12.3; N, 5.7.

---

*The original recipe submitted by the authors called for just over 1 equiv of I$_2$ per carborane unit, but since the procedure was checked, independent workers have reported that the reaction proceeds more reliably if more than 2 equiv are added; the above recipe includes this modification.

Properties

[NEt$_4$][*closo*-2-CB$_9$H$_{10}$] is a white air-stable solid, mp >280°C. It is soluble in acetone, tetrahydrofuran, acetonitrile, and CH$_2$Cl$_2$. IR (KBr, cm^{-1}): 2992 (m), 2534 (s), 1649 (m), 1485 (m), 1375 (m), 1002 (m), 785 (m), 619 (m). ^{11}B{^{1}H} or ^{13}C{^{1}H} NMR [^{1}H NMR in brackets], in (CD$_3$)$_2$CO at 299 K: BH(10) −0.0 [+3.74], BH(1) −7.4 [+3.53], BH(4) −23.0 [+1.08], BH(6,9) −27.3 [+0.61], BH (3,5) −32.2 [+1.05], BH(7,8) −32.2 [+0.32], CH(2) +26.6 [+2.05], α-Et +47.6 [+3.48 (8H, quartet)], β-Et +7.8 [+1.39 (12H, triplet)].

C. TETRAETHYLAMMONIUM *closo*-1-CARBA-DECABORANATE(10)

$$[NEt_4][closo\text{-}2\text{-}CB_9H_{10}] \xrightarrow{\Delta} [NEt_4][closo\text{-}1\text{-}CB_9H_{10}]$$

$$[NEt_4][closo\text{-}1\text{-}CB_9H_{10}] + HCl + H_2O \rightarrow [H_3O][closo\text{-}1\text{-}CB_9H_{10}] + NEt_4Cl$$

$$[H_3O][closo\text{-}1\text{-}CB_9H_{10}] + NEt_4Cl \rightarrow [NEt_4][closo\text{-}1\text{-}CB_9H_{10}] + HCl + H_2O$$

Procedure

A flask is charged with [NEt$_4$][*closo*-2-CB$_9$H$_{10}$] (**2**, 2.0 g, 8.0 mmol) and 1,2-dimethoxyethane (DME) (30 mL), and the mixture is heated to reflux for 24 h. The solution is allowed to cool to room temperature and the DME is removed under reduced pressure. The residue is treated with H$_2$O (50 mL) and aqueous 10% HCl (150 mL), and the mixture is extracted with Et$_2$O (3 × 80 mL). H$_2$O (300 mL) is added to the combined ether extracts, and the Et$_2$O is removed under reduced pressure. The aqueous solution is filtered, and the filtrate is treated with NEt$_4$Cl (2.0 g, 12 mmol). The resulting white precipitate is collected by filtration and dried in vacuum to afford [NEt$_4$][*closo*-1-CB$_9$H$_{10}$], **3**. Yield: 1.9 g (94%).

Anal. Calcd. for C$_9$H$_{30}$NB$_9$: C, 43.3; H, 12.1; N, 5.6. Found: C, 43.4; H, 11.9; N, 5.4.

Properties

[NEt$_4$][*closo*-1-CB$_9$H$_{10}$] is a white air-stable solid, mp >280°C. It is soluble in acetone, tetrahydrofuran, acetonitrile, and CH$_2$Cl$_2$. IR (KBr, cm^{-1}): 2993 (m), 2541 (s), 2497 (s), 1650 (w), 1483 (m), 1394 (m), 1308 (m), 1174 (m), 1002 (m), 784 (m). ^{11}B{^{1}H} or ^{13}C{^{1}H} NMR [^{1}H NMR in brackets], in (CD$_3$)$_2$CO at 299 K: BH(10) +29.9 [+5.87], BH(2,3,4,5) −19.5 [+1.73], BH(6,7,8,9) −25.0 [+0.90], CH(1) +52.8 [+4.63], α-Et +52.5 [+3.50 (8H, quartet)], β-Et +7.1 [+1.41 (12H, triplet)].

D. TETRAETHYLAMMONIUM *closo*-1-CARBA-
DODECABORANATE(12)

$$[NEt_4][arachno\text{-}6\text{-}CB_9H_{14}] + 2BH_3(SMe_2) \rightarrow [NEt_4][closo\text{-}1\text{-}CB_{11}H_{12}] + 2SMe_2 + 3H_2$$

$$[NEt_4][closo\text{-}1\text{-}CB_{11}H_{12}] + HCl + H_2O \rightarrow [H_3O][closo\text{-}1\text{-}CB_{11}H_{12}] + NEt_4Cl$$

$$[H_3O][closo\text{-}1\text{-}CB_{11}H_{12}] + NEt_4Cl \rightarrow [NEt_4][closo\text{-}1\text{-}CB_{11}H_{12}] + HCl + H_2O$$

Procedure

■ **Caution.** *BH$_3$(SMe$_2$) is toxic and should be used in a well-ventilated fume hood.*

A solution of [NEt$_4$][*arachno*-6-CB$_9$H$_{14}$] (**1**, 4.6 g, 18.2 mmol) dissolved in 1,2-Cl$_2$C$_2$H$_4$ (50 mL) is treated with BH$_3$(SMe$_2$) (40 mL, 430 mmol; Lancaster Synthesis). The mixture is heated to reflux for 4 days, and then cooled to 0°C. The mixture is diluted slowly with H$_2$O (50 mL), and then with aqueous 10% HCl (150 mL). The 1,2-Cl$_2$C$_2$H$_4$ is removed under reduced pressure, and the resulting precipitate is collected by filtration and dried in vacuum.* Aqueous 10% HCl (500 mL) is added to the colorless solid, and the mixture is extracted with Et$_2$O (3 × 150 mL). H$_2$O (300 mL) is added to the combined ether extracts, and the Et$_2$O is then removed under reduced pressure. The remaining aqueous solution is filtered, and the filtrate is treated with NEt$_4$Cl (4.14 g, 25 mmol). The colorless precipitate is collected by filtration and dried in vacuum. To remove traces of SMe$_2$-containing by-products (evident from mass spectrometry), the crude product is dissolved in the minimum volume of hot ethanol, and then cooled to room temperature to afford a white precipitate, which is collected by filtration and dried in vacuum to afford [NEt$_4$][*closo*-1-CB$_{11}$H$_{12}$], **4**. Yield: 4.41 g (89%).

Anal. Calcd. for C$_9$H$_{22}$NB$_{11}$: C, 39.6; H, 11.8; N, 5.1. Found: C, 39.6; H, 11.6; N, 5.0.

Properties

[NEt$_4$][*closo*-1-CB$_{11}$H$_{12}$] is a white air-stable solid, mp >280°C. It is soluble in acetone, tetrahydrofuran, acetonitrile, and CH$_2$Cl$_2$. IR (KBr, cm^{-1}): 2990 (m), 2540 (s), 2503 (s), 1637 (w), 1483 (m), 1395 (m), 1017 (m), 786 (m), 717 (w). ^{11}B{^{1}H} or ^{13}C{^{1}H} NMR [^{1}H NMR in brackets], in (CD$_3$)$_2$CO at 294–299 K:

---

*Impurities in the BH$_3$(SMe$_2$) and solvent can sometimes result in a gummy yellow material at this stage. If this is the case, then an intermediate crystallization may be effected by dissolution in a minimum of hot ethanol and cooling to room temperature.

BH(12) −6.8 [+2.36], BH(7,8,9,10,11) −13.2 [+1.61], BH(2,3,4,5,6) −16.3 [+1.45], CH(1) +50.9 [+2.81], α-Et +52.47 [+3.49 (8H, quartet)], β-Et +7.07 [+1.33 (12H, triplet)].

CARBON-SUBSTITUTED ANALOGS

As implied in the introduction, the above routes can often be adapted to afford C-alkylated species by the use of the appropriate aliphatic aldehydes RCHO. For example, acetaldehyde affords [NEt$_4$][6-Me-*arachno*-6-CB$_9$H$_{14}$],[25] although the general route has not yet been significantly investigated for the synthesis of an extensive series of C-alkylated *closo* species. The reaction can also be adapted to prepare the corresponding C-arylated *closo* species from aromatic aldehydes ArCHO, except that the initial reaction of aromatic aldehydes with B$_{10}$H$_{14}$ generally gives the *nido* 10-vertex anions [6-Ar-*nido*-6-CB$_9$H$_{11}$]$^-$ rather than their *arachno* cousins.[31,32] In some cases, however, for example, with 2-acetamidobenzaldehyde,[26] the *arachno* species can be formed. Otherwise, the procedures for all the C-substituted monocarbaboranes are similar. Those for the C-phenyl series are given here. The purity of the starting aldehydes is probably critical for optimum yields.

E. TETRAETHYLAMMONIUM *nido*-6-PHENYL-6-CARBA-DECABORANATE(12)

B$_{10}$H$_{14}$ + KOH + 2H$_2$O + PhCHO → [K][6-Ph-*nido*-6-CB$_9$H$_{11}$] + B(OH)$_3$ + 3H$_2$

[K][6-Ph-*nido*-6-CB$_9$H$_{11}$] + [NEt$_4$]Cl → [NEt$_4$][6-Ph-*nido*-6-CB$_9$H$_{11}$] + KCl

Procedure

A 500 mL flask is charged with B$_{10}$H$_{14}$ (6.0 g, 49 mmol). The flask is cooled to 0°C and a solution of KOH (15 g, 0.27 mol) in H$_2$O (150 mL) and ethanol (70 mL) is added. Over 6 h, PhCHO (40 mL, 400 mmol) is added in five portions at approximately equal intervals. The ethanol is removed under reduced pressure and the remaining mixture is diluted with H$_2$O (100 mL). The alkaline solution is extracted with Et$_2$O (3 × 50 mL), and the combined ether extracts are diluted with H$_2$O (150 mL). The Et$_2$O is removed under reduced pressure and the remaining solution is treated with NEt$_4$Cl (10 g, 60 mmol) to afford a yellow oil. After addition of EtOH (100 mL), the resulting colorless precipitate is filtered off, washed with water (3 × 40 mL), and dried in vacuum to afford [NEt$_4$][6-Ph-*nido*-6-CB$_9$H$_{11}$], **5**. Yield: 10.3 g (64 %).

Anal. Calcd. for $C_{15}H_{36}NB_9$: C, 55.0; H, 11.1; N, 4.3. Found: C, 55.2; H, 11.1; N, 4.1.

Properties

$[NEt_4][6\text{-Ph-}nido\text{-}6\text{-}CB_9H_{11}]$ is a white air-stable solid, mp 101–104°C. It is soluble in acetone, tetrahydrofuran, acetonitrile, and CH_2Cl_2. IR (KBr, cm^{-1}): 2987 (m), 2530 (s), 2510 (s), 2484 (s), 1650 (w), 1479 (m), 1393 (m), 1171 (m), 999 (w), 781 (m), 701 (w). $^{11}B\{^1H\}$ or $^{13}C\{^1H\}$ NMR [^1H NMR in brackets], in $(CD_3)_2CO$ at 294–299 K: BH(5,7) +1.8 [+3.38], BH(9) −1.9 [+2.94], BH(1,3) −4.4 [+2.50], BH(8,10) −12.2 [+2.00], BH(2) −26.0 [+0.61], BH(4) −37.7 [+0.42], C(cluster 6-position) +63.2, Ph +124.3 (1C), +127.2 (2C), +127.7 (2C), +141.5 (1C) [+7.29 (5H, compact overlapping multiplet)], α-Et +52.5 [+3.49 (8H, quartet)], β-Et +7.1 [+1.33 (12H, triplet)], μH(8,9)(9,10) [−3.32]. Crystals are monoclinic, space group $C2/c$, $a = 18.6342(4)$ Å, $b = 10.2871(2)$ Å, $c = 23.4709(5)$ Å, $\beta = 109.019(1)°$, CCDC No. 184220.

F. TETRAETHYLAMMONIUM *closo*-2-PHENYL-2-CARBA-DECABORANATE(10)

$$[NEt_4][6\text{-Ph-}nido\text{-}6\text{-}CB_9H_{11}] + HCl + H_2O \rightarrow [H_3O][6\text{-Ph-}nido\text{-}6\text{-}CB_9H_{11}] + NEt_4Cl$$

$$[H_3O][6\text{-Ph-}nido\text{-}6\text{-}CB_9H_{11}] + I_2 + 3KOH \rightarrow [K][2\text{-Ph-}closo\text{-}2\text{-}CB_9H_9] + 2KI + 4H_2O$$

$$[K][2\text{-Ph-}closo\text{-}2\text{-}CB_9H_9] + [NEt_4]Cl \rightarrow [NEt_4][2\text{-Ph-}closo\text{-}2\text{-}CB_9H_9] + KCl$$

Procedure

A solution of $[NEt_4][6\text{-Ph-}nido\text{-}6\text{-}CB_9H_{11}]$ (**5**, 3.92 g, 12.0 mmol) in aqueous 5% HCl (160 mL) is extracted with Et_2O (3 × 80 mL). The separated organic layers are shaken with aqueous 1 M KOH (400 mL), and the Et_2O is evaporated under reduced pressure. To the resulting alkaline aqueous solution, elemental I_2 (4.0 g, 15.8 mmol) is added. The solution is stirred for 1 h at ambient temperature, and then NEt_4Cl (3.31 g, 20.0 mmol) is added. The mixture is stirred briefly, and then the colorless precipitate is filtered off, washed with H_2O (3 × 50 mL), and dried in vacuum to afford $[NEt_4][2\text{-Ph-}closo\text{-}2\text{-}CB_9H_9]$, **6**. Yield: 3.72 g (92 %).

Anal. Calcd. for $C_{15}H_{34}NB_9$: C, 55.3; H, 10.5; N, 4.3. Found: C, 55.1; H, 10.3; N, 4.2.

Properties

$[NEt_4][2\text{-Ph-}closo\text{-}2\text{-}CB_9H_9]$ is a white air-stable solid, mp 128–131°C. It is soluble in acetone, tetrahydrofuran, acetonitrile, and CH_2Cl_2. IR (KBr, cm^{-1}):

2985 (m), 2601 (s), 2542 (s), 2504 (s), 1650 (w), 1484 (m), 1395 (m), 1171 (m), 998 (m), 779 (m), 706 (w). ^{11}B$\{^1$H$\}$ or ^{13}C$\{^1$H$\}$ NMR [^{1}H NMR in brackets], (CD$_3$)$_2$CO at 294–299 K: BH(10) +1.7 [+3.84], BH(1) −2.8 [1–3.41], BH(4) −21.1 [+1.27], BH(6,7) −25.6 [+0.93], BH(3,5) −28.6 [+1.72], BH(8,9) −28.6 [+0.63], C(cluster 2-position) +49.2, Ph +144.2 (1C), +127.0 (2C), +126.9 (2C), +124.2 (1C) [+6.96–6.85 (unresolved multiplets, 5H)], α-Et +52.5 [+3.49 (8H, quartet)], β-Et +7.1 [+1.33 (12H, triplet)]. Crystals are orthorhombic, space group $P2_12_12_1$, $a = 10.0208(2)$ Å, $b = 13.6557(3)$ Å, $c = 15.4382(3)$ Å, $U = 2112.58$ (7) Å^3, CCDC No. 184221.

G. TETRAETHYLAMMONIUM *closo*-1-PHENYL-1-CARBA-DECABORANATE(10)

$$[NEt_4][2\text{-}Ph\text{-}closo\text{-}2\text{-}CB_9H_9] \xrightarrow{\Delta} [NEt_4][1\text{-}Ph\text{-}closo\text{-}1\text{-}CB_9H_9]$$

$$[NEt_4][1\text{-}Ph\text{-}closo\text{-}1\text{-}CB_9H_9] + HCl + H_2O \rightarrow [H_3O][1\text{-}Ph\text{-}closo\text{-}1\text{-}CB_9H_9] + NEt_4Cl$$

$$[H_3O][1\text{-}Ph\text{-}closo\text{-}1\text{-}CB_9H_9] + NEt_4Cl \rightarrow [NEt_4][1\text{-}Ph\text{-}closo\text{-}1\text{-}CB_9H_9] + HCl + H_2O$$

Procedure

A solution of [NEt$_4$][2-Ph-*closo*-2-CB$_9$H$_9$] (2.61 g, 8 0 mmol) in DME (30 mL) is heated to reflux for 24 h. The solution is allowed to cool to room temperature and the DME is removed under reduced pressure. After adding H$_2$O (50 mL) and aqueous HCl (5%, 150 mL), the mixture is extracted with Et$_2$O (3 × 80 mL). H$_2$O (100 mL) is added to the combined Et$_2$O extracts, and the Et$_2$O is removed under reduced pressure. The aqueous solution is filtered, and [NEt$_4$]Cl (2.0 g, 12 mmol) is added to the filtrate. The resulting precipitate is collected by filtration and dried in vacuum to afford [NEt$_4$][1-Ph-*closo*-1-CB$_9$H$_9$], **7**. Yield: 2.48 g (94%).

Anal. Calcd. for C$_{15}$H$_{34}$NB$_9$: C, 55.3; H, 10.5; N, 4.3. Found: C, 55.4; H, 10.4; N, 4.2.

Properties

[NEt$_4$][1-Ph-*closo*-l-CB$_9$H$_9$] is a white air-stable solid, mp 154–157°C. It is soluble in acetone, tetrahydrofuran, acetonitrile, and CH$_2$Cl$_2$. IR (KBr, cm^{-1}): 2992 (m), 2540 (s), 2502 (s), 1650 (w), 1484 (m), 1395 (m), 1173 (m), 1001 (m), 766 (m), 703 (w). ^{11}B$\{^1$H$\}$ or ^{13}C$\{^1$H$\}$ NMR [^{1}H NMR in brackets], CD$_3$CN at 294–299 K: BH(10) +27.1 [+5.49], BH(2,3,4,5) −16.1 [+1.79], BH(6,7,8,9) −24.4 [+0.81], C(cluster 1-position) +74.3, α-Et +52.7 [+3.49 (8H, quartet)], β-Et +7.4 [+1.33 (12H, triplet)], Ph +144.6 (1C), +130.6 (2C), +127.7 (2C),

+125.7 (1C) [+7.29 (5H, compact overlapping multiplet)]. Crystals are mono-clinic, space group $C2/c$, $a = 24.4404(16)$ Å, $b = 10.5641(6)$ Å, $c = 16.7478(14)$ Å, $\beta = 99.099(3)°$, CCDC No. 164851.

H. TETRAETHYLAMMONIUM *closo*-1-PHENYL-1-CARBA-DODECABORANATE(12)

$$[NEt_4][6\text{-Ph-}nido\text{-}6\text{-CB}_9H_{11}] + 2BH_3(SMe_2) \rightarrow$$

$$[NEt_4][1\text{-Ph-}closo\text{-}1\text{-B}_{11}H_{11}] + 2SMe_2 + H_2$$

$$[NEt_4][1\text{-Ph-}closo\text{-}1\text{-B}_{11}H_{11}] \rightarrow [H_3O][1\text{-Ph-}closo\text{-}1\text{-B}_{11}H_{11}] + NEt_4Cl$$

$$[H_3O][1\text{-Ph-}closo\text{-}1\text{-B}_{11}H_{11}] + NEt_4Cl \rightarrow [H_3O][1\text{-Ph-}closo\text{-}1\text{-B}_{11}H_{11}] + HCl + H_2O$$

Procedure

A sample of $[NEt_4][6\text{-Ph-}nido\text{-}6\text{-CB}_9H_{11}]$ (**5**, 6.0 g, 18.4 mmol) is dissolved in 1,2-$Cl_2C_2H_4$ (50 mL). $BH_3(SMe_2)$ (40 mL, 430 mmol) is added and the mixture is heated to reflux for 4 days. The mixture is cooled to 0°C, and H_2O (50 mL) is added slowly, followed by aqueous HCl (10 %, 150 mL).[*] The 1,2-$Cl_2C_2H_4$ is removed under reduced pressure, and the remaining precipitate is filtered off and dried in vacuum. Aqueous HCl (10%, 500 mL) is added to the colorless solid, and the mixture is extracted with Et_2O (3×150 mL). H_2O (400 mL) is added to the combined ethereal extracts, and the Et_2O is then evaporated under reduced pressure. The remaining aqueous solution is filtered, and $[NEt_4]Cl$ (4.14 g, 25 mmol) is added to the filtrate. The resulting precipitate is filtered off and dried in vacuum. To remove traces of SMe_2-containing by-products (evident from mass spectrometry), the crude product is dissolved in the minimum volume of a hot ethanol–acetone mixture (75:25); on cooling, a white precipitate forms, which is collected by filtration and dried in vacuum to afford $[NEt_4][1\text{-Ph-}closo\text{-}1\text{-}CB_{11}H_{11}]$, **8**. Yield: 5.9 g (94%).

Anal. Calcd. for $C_{15}H_{36}NB_{11}$: C, 51.6; H, 10.4; N, 4.0. Found: C, 51.4; H, 10.1; N, 3.8.

Properties

$[NEt_4][1\text{-Ph-}closo\text{-}1\text{-}CB_{11}H_{11}]$ is a white air-stable solid, and, as prepared above, has mp (decomp.) 187–190°C. It is soluble in acetone, tetrahydrofuran, acetoni-trile, and CH_2Cl_2. IR (KBr, cm^{-1}): 2992 (m), 2530 (s), 2513 (s), 1650 (w),

---

[*]The checker stirred the mixture for 10 min at this point.

1474 (m), 1397 (m), 1172 (w), 1001 (w), 783 (w). ^{11}B{^{1}H} or ^{13}C{^{1}H} NMR [^{1}H NMR in brackets], CD_3CN at 294–299 K: BH(12) -7.7 [+1.80], BH(2,3,4,5,6) -12.8 [+1.80], BH(7,8,9,10,11) -12.8 [+1.74], C(cluster 1-position) +72.6, Ph +141.5 (1C), +127.7 (2C), +127.2 (2C), +124.35 (1C) [+7.29 (5H, compact overlapping multiplet)], α-Et +52.5 [+3.49 (8H, quartet)], β-Et +7.1 [+1.33 (12H, triplet)]. The compound seems prone to form a variety of crystal morphologies and solvates, and so the melting point will vary: crystals from acetone–diethyl ether can be monoclinic, space group $P2_1/c$, $a = 13.0151(3)$ Å, $b = 10.2266(3)$ Å, $c = 16.9186(5)$ Å, $\beta = 100.5580(13)°$, CCDC No. 194221, or orthorhombic, space group *Pbca*, $a = 10.2352(2)$ Å, $b = 16.9027(5)$ Å, $c = 25.6191(9)$ Å, CCDC No. 194222; from dichloromethane–hexane they are a monoclinic CH_2Cl_2 monosolvate, space group $P2_1/c$, $a = 15.7334(5)$ Å, $b = 10.8812(4)$ Å, $c = 15.7928(6)$ Å, $\beta = 110.9350(14)°$, CCDC No. 164850.

Acknowledgments

We thank the UK EPSRC and the UK DTI for support, and Professor Dwayne Heard for his good offices.

References

1. T. Onak, in *Boron Hydride Chemistry*, E. L. Muetterties, ed., Academic Press, New York, 1973, pp. 349–382.
2. T. Onak, in *Comprehensive Organometallic Chemistry*, E. W. Abel, G. Wilkinson, and F. G. A. Stone, eds., Pergamon Press, Oxford, 1982, Vol. **1**, Chapter 5.4, pp. 411–457.
3. (a) B. Štíbr, *Chem. Rev.* **92**, 225 (1992); (b) B. Štíbr, *Pure Appl. Chem.* **75**, 1295 (2003).
4. L. J. Todd, in *Comprehensive Organometallic Chemistry*, E. W. Abel, G. Wilkinson, and F.G.A. Stone, eds., Pergamon Press, Oxford, 1982, Vol. **1**, Chapter 5.6, pp. 543–553.
5. (a) J. Plešek, T. Jelínek, B. Štíbr, and S. Heřmánek, *J. Chem. Soc., Chem. Commun.* 348 (1988); (b) J. Plešek, B. Štíbr, X. L. R. Fontaine, J. D. Kennedy, S. Heřmánek, and T. Jelínek, *Collect. Czech. Chem. Commun.* **56**, 1618 (1991); (c) A. E. Wille, J. Plešek, J. Holub, B. Štíbr, P. J. Carrol, and L. G. Sneddon, *Inorg. Chem.* **35**, 5342 (1996).
6. W. H. Knoth, *J. Am. Chem. Soc.* **89**, 1274 (1967); *Inorg. Chem.* **10**, 598 (1971).
7. (a) T. Jelínek, B. Štíbr, J. Plešek, and S. Heřmánek, *J. Organomet. Chem.* **307**, C13 (1986); (b) K. Baše, B. Štíbr, J. Dolanský, and J. Duben, *Collect. Czech. Chem. Commun.* **46**, 2345 (1981).
8. (a) J. A. Kautz, D. A. Kisounko, N. S. Kissounko, and F. G. A. Stone, *J. Organomet. Chem.* **651**, 34 (2002); (b) D. Shaowu, J. A. Kautz, T. D. McGrath, and F. G. A. Stone, *Inorg. Chem.* **41**, 3202 (2002); (c) J. A. Kautz, D. A. Kissounko, N. S. Kissounko, and F. G. A. Stone, *Organometallics* **21**, 2547 (2002); (d) D. Shaowu, A. Franken, P. A. Jelliss, J. A. Kautz, F. G. A. Stone, and P.-Y. Yu, *J. Chem. Soc., Dalton Trans.* 1846 (2001).
9. B. Štíbr and B. Wrackmeyer, *J. Organomet. Chem.* **657**, 3 (2002); (b) A. Franken, C. A. Kilner, M. Thornton-Pett, and J. D. Kennedy, *Collect. Czech. Chem. Commun.* **67**, 869 (2002).
10. (a) I. V. Pisareva, V. E. Konoplev, E. V. Balagurova, A. I. Yanovsky, F. M. Dolgushin, D. N. Cheredilin, I. T. Chizhevsky, A. Franken, M. J. Carr, M. Thornton-Pett, and J. D. Kennedy, in *Boron Chemistry at the Beginning of the 21st Century (Proceedings of the 11th International Conference on the Chemistry of Boron, Moscow, July 28–August 1, 2002)*, Y. N. Bubnov, ed., Nauka, Moscow, 2003, p. 271; (b) I. V. Pisareva, I. T. Chizhevsky, P. V. Petrovskii, V. I. Bregadze,

F. M. Dolgushin, and A. I. Yanovsky, *Organometallics* **16**, 5598 (1997); (c) I. V. Pisareva, F. M. Dolgushin, A. I. Yanovsky, E. V. Balagurova, P. V. Petrovskii, and I. T. Chizhevsky, *Inorg. Chem.* **40**, 5318 (2001).

11. A. S. Larsen, J. D. Holbrey, F. S. Tham, and C. A. Reed, *J. Am. Chem. Soc.* **122**, 7264 (2000).

12. (a) P. Kaszynski, *Collect. Czech. Chem. Commun.* **64**, 895 (1999); (b) P. Kaszynski, S. Pakhomov, K. F. Tesh, and V. G. Young, *Inorg. Chem.* **40**, 6622 (2001).

13. S. Pakhomov, W. Piecek, P. Kaszynski, Y. N. Bubnov, M. E. Gurskii, and V. G. Young, *Abstracts, Boron Americas 8, Death Valley, California, USA, January 2–6*, 2002, Abstract No. P15, p. 71.

14. B. Grüner, Z. Janoušek, B. T. King, J. N. Woodford, C. H. Wang, V. Všetečka, and J. Michl, *J. Am. Chem. Soc.* **121**, 3122 (1999).

15. A. Franken, B. T. King, J. Rudolph, P. Rao, P. C. Noll, and J. Michl, *Collect. Czech. Chem. Commun.* **66**, 1238 (2001).

16. (a) X. Yang, W. Jiang, C. B. Knobler, and M. F. Hawthorne, *J. Am. Chem. Soc.* **114**, 9719 (1992); (b) J. Muller, K. Baše, T. F. Magnera, and J. Michl, *J. Am. Chem. Soc.* **114**, 9721 (1992); (c) S. J. Cantrill, A. R. Pease, and J. F. Stoddart, *J. Chem. Soc., Dalton Trans.* 3715 (2000).

17. (a) M. J. Hardie and C. L. Raston, *Chem. Commun.* 905 (2001); (b) M. J. Hardie, N. Malic, B. Roberts, and C. L. Raston, *Chem. Commun.* 865 (2001); (c) A. Franken, C. A. Kilner, M. Thornton-Pett, and J. D. Kennedy, *Chem. Commun.* 2048 (2002); (d) S. L. Renard, A. Franken, C. A. Kilner, J. D. Kennedy, and M. A. Halcrow, *New J. Chem.* **26**, 1634 (2002).

18. (a) C. A. Reed, *Acc. Chem. Res.* **31**, 133 (1998); (b) S. H. Strauss, *Chem. Rev.* **93**, 927 (1993); (c) see also S. H. Strauss, *Spec. Publ. R. Soc. Chem.* **253**, 44 (2000).

19. (a) B. T. King, Z. Janoušek, B. Grüner, M. Trammell, B. C. Noll, and J. Michl, *J. Am. Chem. Soc.* **118**, 3313 (1996); (b) C. W. Tsang and Z. W. Xie, *Chem. Commun.* 1839 (2000); (c) B. T. King and J. Michl, *J. Am. Chem. Soc.* **122**, 10255 (2000).

20. (a) S. V. Ivanov, J. J. Rockwell, O. G. Polyakov, C. M. Gaudinski, O. P. Anderson, K. A. Solntsev, and S. H. Strauss, *J. Am. Chem. Soc.* **120**, 4224 (1998); (b) Z. W. Xie, C. W. Tsang, E. T. P. Sze, Q. C. Yang, D. T. W. Chan, and T. C. W. Mak, *Inorg. Chem.* **37**, 6444 (1998); (c) S. V. Ivanov, A. J. Lupinetti, S. M. Miller, O. P. Anderson, K. A. Solntsev, and S. H. Strauss, *Inorg. Chem.* **34**, 6419 (1995).

21. H. W. Turner and G. G. Hlatky, European Patent Appl. 277003 (1988); U.S. Patent 5,278,119 (1994); U.S. Patent 5,407,884 (1995); U.S. Patent 5,483,014 (1996).

22. (a) Z. Xie, R. Bau, and C. A. Reed, *Angew. Chem., Int. Ed. Engl.* **33**, 2433 (1994); (b) Z. Xie, J. Manning, R. W. Reed, R. Mathur, P. D. W. Boyd, A. Benesi, and C. A. Reed, *J. Am. Chem. Soc.* **118**, 2922 (1996); (c) N. J. Patmore, M. F. Mahon, J. W. Steed, and A. S. Weller, *J. Chem. Soc., Dalton Trans.* 277 (2001).

23. (a) C. A. Reed, N. L. P. Fackler, K.-C. Kim, D. Stasko, D. R. Evans, P. D. W. Boyd, and C. E. F. Rickard, *J. Am. Chem. Soc.* **121**, 6314 (1999); (b) I. A. Koppel, P. Burk, I. Koppel, I. Leito, T. Sonoda, and M. Mishima, *J. Am. Chem. Soc.* **122**, 5114 (2000).

24. B. Brellochs, in *Contemporary Boron Chemistry*, M. G. Davidson, A. K. Hughes, T. B. Marder, and K. Wade, eds., Royal Society of Chemistry, Cambridge, UK, 2000, pp. 212–214.

25. T. Jelínek, M. Thornton-Pett, and J. D. Kennedy, *Collect. Czech. Chem. Commun.* **67**, 1035 (2002).

26. I. Sivaev, S. Sjoberg, and V. Bregadze, *Abstracts, 11th International Meeting on Boron Chemistry (IMEBORON XI), Moscow, Russia, July 28–August 2*, 2002, Abstract No. CA-5, p. 69.

27. N. J. Bullen, A. Franken, C. A. Kilner, and J. D. Kennedy, *Chem. Commun.* 1684 (2003).

28. A. Franken, C. A. Kilner, and J. D. Kennedy, *Inorg. Chem. Commun.* **6**, 1104 (2003).

29. A. Franken, C. A. Kilner, and J. D. Kennedy, *Chem. Commun.* 328 (2004).

30. A. Franken, M. J. Carr, W. Clegg, C. A. Kilner, and J. D. Kennedy, *Dalton Trans.* 3552 (2004).

31. A. Franken, C. A. Kilner, M. Thornton-Pett, and J. D. Kennedy, *Collect. Czech. Chem. Commun. (J. Plešek, Special Edition: Boron)* **67**, 869 (2002).

32. A. Franken, T. Jelínek, R. G. Taylor, D. L. Ormsby, C. A. Kilner, W. Clegg, and J. D. Kennedy, *Dalton Trans.* 5753 (2006).

35. TETRAKIS(5-*tert*-BUTYL-2-HYDROXYPHENYL)ETHENE

Submitted by MEE-KYUNG CHUNG[*] and JEFFREY M. STRYKER[*]
Checked by MORGANE OLLIVAULT-SHIFLETT,[†] CLIFFORD J. UNKEFER,[†] and LOUIS A. SILKS III[†]

Tetrakis(2-hydroxyphenyl)ethene and the more soluble analog tetrakis(5-*tert*-butyl-2-hydroxyphenyl)ethene[1] are preorganized tetradentate ligands that support the construction of polymetallic coordination complexes of potential importance to coordination chemistry, supramolecular chemistry, and catalysis.[2] These ligands, which contain aryloxy groups organized around an ethylene core, provide a topological alternative to the extensively investigated calixarene ligands, in which four or more aryloxy rings are connected into a ring by *ortho*-methylene bridges.[3] Although the ethylene core imposes some conformational rigidity, the rotational degrees of freedom of the aryl groups enable tetrakis(2-hydroxyphenyl)ethene to adjust to the coordination requirements of a wide array of metal atoms. Preliminary investigations using both main group and transition metals demonstrate that these ligands are versatile and afford complexes with distinctive structures and properties, as reflected by the formation of higher-order bridged polymetallic coordination compounds of unusual robustness.[1] The roughly square binding platform, although similar to that of the more familiar calixarenes, can be more easily modified by incorporating sterically isolating alkyl[1] or trialkylsilyl[4] substituents at the *ortho* positions of the aryl rings.

The synthetic challenge presented by the sterically congested *ortho*-substituted tetraphenylethylene structure was initially addressed through the adaptation of a known[5] acid-catalyzed homocoupling of substituted diaryl diazomethanes.[1] This procedure, although reasonably efficient, nonetheless requires the synthesis of a diaryl diazomethane precursor in two steps from 2,2′-dimethoxybenzophenone. Recently, however, a more efficient procedure for the synthesis of tetrakis(2-methoxyphenyl)ethene and tetrakis(5-*tert*-butyl-2-methoxyphenyl)ethene has been developed that involves the direct coupling of a suitably substituted benzophenone,[6] thus rendering the original synthesis obsolete. In this report, the full experimental details of the synthetic procedures are reported, along with the subsequent dealkylation of the methoxy groups[1] to give the corresponding tetrakis (2-hydroxyphenyl)ethene ligands.

The new protocol involves the McMurry-type reductive coupling[7,8] of 2,2′-dialkoxybenzophenones, sterically hindered substrates that have often been identified as unsuited for this procedure. For example, McMurry reactions of

[*]Department of Chemistry, University of Alberta, Edmonton, Alberta, Canada T6G 2G2.
[†]Bioscience Division, Los Alamos National Laboratory, Los Alamos, NM 87545.

2,2′-substituted diaryl ketones have invariably afforded overreduced 1,1,2,2-tetraarylethane products in preference to the desired 1,1,2,2-tetraarylalkenes.[9] Despite the application of a range of standard McMurry reaction conditions (e.g., TiCl$_3$/LiAlH$_4$, TiCl$_4$/Zn/pyridine), the isolated yields of alkene remain marginal, at best. This observation has generally been attributed to the high degree of steric congestion around the carbonyl group, although there is no rational mechanistic basis for this assertion.

After extensive optimization, the use of either TiCl$_3$ or TiCl$_3$ · 1.5DME and Zn–Cu couple in controlled stoichiometry reproducibly provides the desired tetrakis-(2-methoxyphenyl)ethene in excellent isolated yield after separation from relatively minor amounts of the corresponding 1,1,2,2-tetrakis(2-methoxyphenyl)ethane and reduced but uncoupled benzophenone. These procedures thus have distinct advantages in efficiency and overall yield over our previously published synthesis. The use of TiCl$_3$ · 1.5DME as the titanium source is, in particular, a good alternative to commercial TiCl$_3$. The experimental conditions for coupling 2,2′-dimethoxybenzophenone and 5,5′-di-*tert*-butyl-2,2′-dimethoxybenzophenone are identical, except for the reaction time to complete the McMurry reaction; the procedure for the preparation of tetrakis(5-*tert*-butyl-2-hydroxyphenyl)ethene is detailed below.

General Procedures

Commercial titanium trichloride (Alfa Aesar), titanium powder (Strem Chemicals), and titanium tetrachloride (Strem Chemicals) are used as received. All titanium reagents are handled under an inert atmosphere in a glove box. Dimethylcarbamoyl chloride, *p-tert*-butylanisole, *n*-butyllithium (10.0 M in hexanes), and boron tribromide (99%) are purchased from Aldrich and used as received. *N,N, N′,N′*-Tetramethylethylenediamine is dried over sodium. 1,2-Dimethoxyethane (DME) and diethyl ether are dried and deoxygenated by distillation from sodium/benzophenone ketyl. Dichloromethane is dried by distillation from calcium hydride. Zn–Cu couple was prepared by a literature procedure.[8b,c] All synthetic operations are performed in an inert atmosphere unless otherwise specified. A Vacuum Atmospheres He-553-2 Dri-lab equipped with a Mo-41-1 inert gas purifier is a suitable dry box for air-sensitive operations.

A. 5,5′-DI-*tert*-BUTYL-2,2′-DIMETHOXYBENZOPHENONE

Procedure[10]

An oven-dried, 2 L three-necked round-bottomed flask equipped with an argon inlet, a dropping funnel, and a gas adaptor connected to a gas bubbler is flushed with argon and charged with commercial *p-tert*-butylanisole (134.5 g, 0.819 mol), anhydrous *N,N,N',N'*-tetramethylethylenediamine (6.18 mL, 4.76 g, 0.041 mol), and diethyl ether (1 L). The dropping funnel is charged with *n*-butyllithium (172.0 mL of a 5.0 M solution in hexanes, 0.860 mol).* The anisole solution is cooled to −78°C (dry ice/acetone bath) and the *n*-butyllithium is added dropwise. After the addition is complete and ensuring that the flask is open to the gas bubbler to release the butane evolved, the cooling bath is removed, and the solution is allowed to warm to room temperature and stirred for 24 h. The reaction mixture is recooled to −78°C (dry ice/acetone bath). The dropping funnel is charged under an argon flow with dimethylcarbamoyl chloride (37.96 mL, 44.03 g, 0.409 mol), which is subsequently added dropwise with rapid stirring. The reaction mixture is slowly warmed to room temperature with stirring by removal of the cooling bath. The progress of the reaction is monitored by TLC,† and after 24 h water (10 mL) is added slowly; the remaining steps can be carried out in air. The resulting sludge is poured onto ice (500 g) and the mixture is acidified to pH 6 using 10% aqueous (w/w) sulfuric acid solution. The aqueous phase is extracted three times with diethyl ether (100 mL). The combined organic phases are dried over $MgSO_4$ and reduced to dryness in a rotary evaporator. Recrystallization from hot hexanes (more than one recrystallization may be required to attain sufficient purity) followed by drying under vacuum gives the product. Yield: 107 g (74%).

HRMS (EI). Calcd. for $C_{23}H_{30}O_3$: *m/z* 354.21948. Found: *m/z* 354.21973 (24%).

Properties

5,5'-Di-*tert*-butyl-2,2'-dimethoxybenzophenone is an air-stable white solid that is very soluble in pentane. IR (neat): 1646 (ν_{CO}) cm^{-1}. ^1H NMR (CDCl$_3$, 360 MHz): δ 7.52 (d, *J* = 2.6 Hz, 2H), 7.44 (dd, *J* = 8.7, 2.6 Hz, 2H), 6.85 (d, *J* = 8.7 Hz, 2H), 3.64 (s, 6H), 1.29 (s, 18H). $^{13}C\{^1H\}$ NMR (75 MHz, CDCl$_3$): δ 195.8, 156.2, 142.9, 129.6, 129.2, 127.4, 111.3, 55.8, 34.1, 31.3. TLC: R_f = 0.82 (50% CH$_2$Cl$_2$/hexanes).

---

*The 5 M solution of *n*-butyllithium in hexanes is made by diluting the commercially available 10 M *n*-butyllithium in hexanes with anhydrous hexanes. The reaction can also be carried out with 2.5 M *n*-butyllithium in hexanes with or without concentration to ~5 M.

†The progress of the reaction between (5-*tert*-butyl-2-methoxyphenyl)lithium and dimethylcarbamoyl chloride should be monitored by analytical TLC because the reaction times vary with concentration and the composition of the solvent.

B. TITANIUM TRICHLORIDE 1,2-DIMETHOXYETHANE (1:1.5)

$$Ti + 3TiCl_4 + 6DME \rightarrow 4TiCl_3 \cdot 1.5DME$$

Procedure[11]

In a glove box, an oven-dried 1 L three-necked round-bottomed flask equipped with an adaptor and two stoppers is charged with ~100 mesh titanium powder (2.98 g, 62.3 mmol). After the flask is removed from the glove box, anhydrous DME (580 mL) is added under an argon flow, the stoppers are replaced with a thermometer, and a condenser connected to an argon-flushed gas bubbler. This flask is cooled in ice. Separately, an oven-dried 100 mL Schlenk flask is charged with titanium tetrachloride (36.6 g, 21.2 mL, 192.9 mmol) in the glove box. After this flask is removed from the glove box, the titanium tetrachloride is slowly added under a flow of argon to the ice-cooled titanium suspension by cannula, maintaining the pot temperature between 4 and 7°C. After the addition is complete and the thermometer is replaced with a stopper, the reaction mixture is vigorously stirred and slowly heated to reflux in an oil bath. The temperature is maintained at 100–110°C for 30 h. The completion of the reaction is indicated by the rich greenish blue color of the slurry and the disappearance of the gray particulate titanium. The slurry is cooled to room temperature and filtered under an atmosphere of argon. The light greenish blue solid product is dried in vacuum overnight and then dried further on a high-vacuum line for several days (ultimate vacuum 6×10^{-5} mmHg). Yield: 63.8 g (88% based on the titanium powder).

Anal. Calcd. for $C_6H_{15}Cl_3O_3Ti$: Ti, 16.6; Cl, 36.8. Found: Ti, 16.6; Cl, 36.6.

Properties

Titanium trichloride 1,2-dimethoxyethane (1:1.5) is an air- and moisture-sensitive light greenish blue solid.

C. TETRAKIS(5-*tert*-BUTYL-2-METHOXYPHENYL)ETHENE

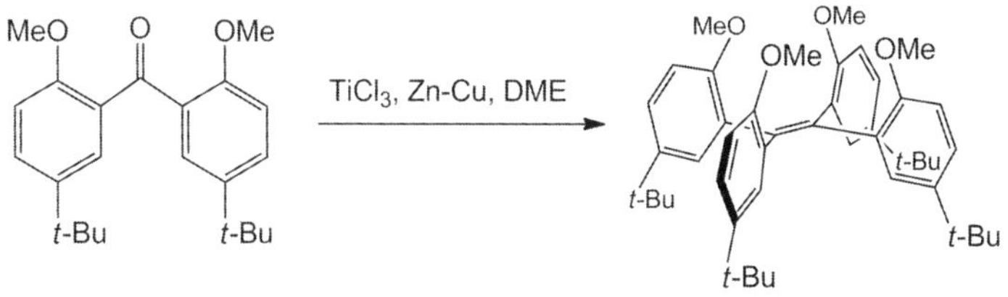

Procedure

Method 1. In an argon-filled dry box, an oven-dried 1 L Schlenk flask is charged with titanium trichloride (24.69 g, 160.0 mmol) and Zn–Cu couple (5.16 g, 79.0 mmol). After the flask is removed from the glove box, anhydrous and deoxygenated DME (350 mL)* is added under an argon flow and the flask is equipped with a condenser connected to gas bubbler that is kept under argon in case of suckback (Fig. 1). The resulting slurry is heated to reflux for 14 h in an oil bath maintained at 100–110°C. After 10 min, the solution phase of the slurry is light blue, and after 14 h the solution color is dark green. The reaction mixture is cooled to room temperature (typical cooling time ~30 min).

Separately, an oven-dried 250 mL Schlenk flask is charged with 5,5′-di-*tert*-butyl-2,2′-dimethoxybenzophenone (9.72 g, 27.4 mmol) and anhydrous 1,2-dimethoxyethane (130 mL) under an atmosphere of argon. The resulting solution is transferred by cannula under a flow of argon to the cooled slurry of low-valent titanium, and the reaction mixture is stirred at room temperature for 24 h. The progress of the reaction is monitored by analytical TLC. After approximately 1 h, the slurry becomes dark brownish black, which is a

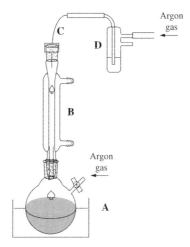

Figure 1. Reaction assembly for the McMurry carbonyl coupling: (A) 1 L reaction flask under argon and oil bath, (B) water-cooled condenser, (C) a rubber septum punctured by a needle that is connected to a gas bubbler, and (D) a mineral oil gas bubbler that is also connected to a countercurrent of argon.

---

*The checkers report that the TiCl₃/Zn(Cu) mixture clumps badly if the 1,2-dimethoxyethane is used without being purified and deoxygenated.

qualitative indication of a functioning McMurry reaction. After completion of the reaction, the volume of the solution is reduced to two-thirds in vacuum and the reaction quenched slowly by adding 1.0 M aqueous HCl (4 mL) and then water (200 mL). The resulting mixture is extracted three times with diethyl ether (100 mL). The combined organic layers are dried over MgSO₄ and the volatile material is removed on a rotary evaporator. The crude product is crystallized from 10% CH_2Cl_2 in pentane, isolated by filtration, and dried under vacuum to give 6.20 g of crystals. The residue from the mother liquor is purified by column chromatography on silica gel using a gradient solvent system of CH_2Cl_2/hexanes,[*] providing an additional 1.56 g of purified product. Total yield: 7.76 g (84%).[†]

Method 2. In an argon-filled glove box, an oven-dried 1 L Schlenk flask is charged with $TiCl_3 \cdot 1.5DME$ (13.27 g, 45.9 mmol)[‡] and Zn–Cu (1.50 g, 23.0 mmol).[§] After the flask is removed from the glove box, anhydrous DME (105 mL) is added under an argon flow at ambient temperature. Separately, an oven-dried 250 mL Schlenk flask is charged with 5,5′-di-*tert*-butyl-2,2′-dimethoxybenzophenone (6.0 g, 16.9 mmol) and anhydrous DME (80 mL) under an atmosphere of argon. The resulting solution is transferred to the slurry of $TiCl_3 \cdot 1.5DME$/Zn–Cu by cannula under a flow of argon, and the reaction mixture is stirred at ambient temperature for 2 days. The progress of the reaction is monitored by analytical TLC. After approximately 1 h, the slurry becomes dark brownish black, which is a qualitative indication of a functioning McMurry reaction. After completion of the reaction, in the same way as described in Section 35.B, the volume of solution is reduced to two-thirds in vacuum and the reaction is quenched by slowly adding 1.0 M aqueous HCl (3 mL) and water (100 mL). The resulting mixture is extracted with diethyl ether (3 × 100 mL). The combined organic layers are dried over MgSO₄ and the volatile material is removed on a rotary evaporator. The crude product is crystallized from 10% CH_2Cl_2 in pentane to afford 3.42 g of material, which is isolated by filtration and dried in vacuum. Purification of the mother liquors by column chromatography on silica gel

---

[*]Column chromatography elution progression: (1) column packed with hexanes; (2) 50% CH_2Cl_2/hexanes (100 mL); (3) 80% CH_2Cl_2/hexanes (600 mL); (4) 100% CH_2Cl_2 (800 mL). TLC: R_f of tetralos-(5-*tert*-butyl-2-methoxyphenyl)ethene = 0.34 (100% CH_2Cl_2), 0.23 (50% CH_2Cl_2/hexanes). The chromatographic separations work much more poorly if the crude product is not crystallized first or the significantly less polar solvent pentane is substituted for hexane.

[†]In smaller scale experiments using 240 mg (0.68 mmol) of 5,5′-di-*tert*-butyl-2,2′-dimethoxybenzophenone, the product tetrakis(5-*tert*-butyl-2-methoxyphenyl)ethene is obtained in 74% yield.

[‡]Note that the molar excess of titanium is about half that needed when anhydrous $TiCl_3$ is the titanium source.

[§]The McMurry reagent preparation step, that is, the reaction of the Ti[III] reagent and Zn–Cu in DME at reflux, is unnecessary for optimum results and is omitted from this procedure. This improvement was introduced by Fürstner et al.[12]

(same conditions as in method 1) provides an additional 1.13 g of purified product. Total yield: 4.55 g (79%).

HRMS (EI). Calcd. for $C_{46}H_{60}O_4$: m/z 676.44916. Found: m/z 676.45063 (100%).

Properties

Tetrakis(5-*tert*-butyl-2-methoxyphenyl)ethene is an air-stable white solid. IR (neat): 3032 (vinylic ν_{CH}), 1603 ($\nu_{C=C}$) cm^{-1}. ^{1}H NMR (CDCl$_3$, 360 MHz): δ 7.09 (d, $J = 2.1$ Hz, 4H), 6.93 (dd, $J = 8.5$, 2.5 Hz, 4H), 6.56 (d, $J = 8.5$ Hz, 4H), 3.48 (s, 12H, OMe), 1.06 (s, 36H, *t*-Bu). ^{13}C{^{1}H} NMR (75 MHz, CDCl$_3$): δ 155.5, 142.0, 131.8, 130.0, 123.1, 110.9, 55.9, 33.7, 31.3. One carbon (presumably the quaternary olefin) was not located.

The parent tetrakis(2-methoxyphenyl)ethene can be prepared in 73% yield from 2,2′-dimethoxybenzophenone (Rieke Metals, Inc., Lincoln, Nebraska) in the same way, except the McMurry reduction is typically complete within 10–11 h.

D. TETRAKIS(5-*tert*-BUTYL-2-HYDROXYPHENYL)ETHENE

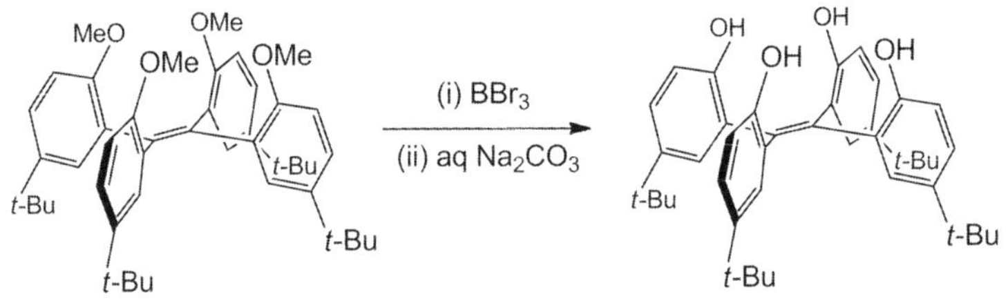

Procedure[13]

An oven-dried 100 mL Schlenk flask is charged with tetrakis(5-*tert*-butyl-2-methoxyphenyl)ethene (2.38 g, 3.52 mmol) and anhydrous dichloromethane (45 mL) under an argon atmosphere. The solution is cooled to −78°C in a dry ice/acetone bath, and boron tribromide (2.00 mL, 5.30 g, 21.1 mmol) is added slowly under an argon flow.* After the addition is complete, the flask is fitted with a condenser protected from moisture by a calcium chloride drying tube (Fig. 2). The reaction mixture is heated to reflux for 2 h in an oil bath maintained at 45°C, and then is allowed to stir overnight at room temperature. The solution is concentrated in vacuum to half its volume. Saturated aqueous NaHCO$_3$ (25 mL) is slowly added

---

*The checkers report that addition of BBr$_3$ causes the solution to turn black, and the yields are lowered, if the dichloromethane is used without being purified and deoxygenated.

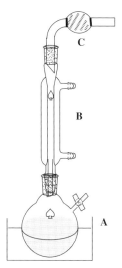

Figure 2. Demethylation reaction assembly: (A) 100 mL reaction flask with an oil bath, (B) water-cooled condenser, and (C) a drying tube containing anhydrous $CaCl_2$ powder between two plugs of glass wool.

and the resultant aqueous phase is extracted with diethyl ether (3×50 mL). The combined organic phases are washed with water (4×30 mL), dried over $MgSO_4$, and taken to dryness in a rotary evaporator. The product is recrystallized from diethyl ether/pentane followed by drying in vacuum. Yield: 2.04 g (93%).

HRMS (EI). Calcd. for $C_{42}H_{52}O_4$: *m/z* 620.38654. Found: *m/z* 620.38676 (100%).

Properties

Tetrakis(5-*tert*-butyl-2-hydroxyphenyl)ethene is an air-stable white solid. IR (neat): 3215 (ν_{OH}) cm^{-1}. ^{1}H NMR (CDCl$_3$, 360 MHz): δ 7.23 (br s, 4H), 6.87 (dd, $J = 8.5$, 2.4 Hz, 4H), 6.53 (d, $J = 8.5$ Hz, 4H), 3.9–3.5 (br s, 4H, OH), 1.12 (s, 36H, *t*-Bu). ^{13}C{^{1}H} NMR (75 MHz, acetone-d_6): δ 151.6, 142.0, 140.1, 128.9, 127.7, 125.0, 115.1, 34.1, 31.6.

The demethylation of tetrakis(2-methoxyphenyl)ethene proceeds under identical reaction conditions and workup procedures. At a scale of 0.47 g (1.0 mmol) of tetrakis(2-methoxyphenyl)ethene, the reaction with BBr$_3$ provides tetrakis(2-hydroxyphenyl)ethene in 90% yield after recrystallization from diethyl ether followed by drying in vacuum. ^{1}H NMR (CD$_2$Cl$_2$, 200 MHz): δ 7.14 (dd, $J = 7.4$, 1.8 Hz, 4H), 7.04 (ddd, $J = 8.0$, 7.4, 1.8 Hz, 4H), 6.77 (dt, $J = 7.4$, 1.0 Hz, 4H), 6.69 (br d, $J = 8.0$ Hz, 4H), 5.91 (br s, 4H, OH).

Acknowledgments

The authors thank the Natural Sciences and Engineering Research Council of Canada (Collaborative Research Opportunities), the University of Alberta, the Killam Trusts, and the Province of Alberta (Alberta Ingenuity) for research funding and fellowships (to M.-K.C.).

References

1. U. Verkerk, M. Fujita, T. D. Dzwiniel, R. McDonald, and J. M. Stryker, *J. Am. Chem. Soc.* **124**, 9988 (2002).
2. (a) S. Leininger, B. Olenyuk, and P. J. Stang, *Chem. Rev.* **100**, 853 (2000); (b) G. F. Swiegers and T. J. Malefetse, *Chem. Rev.* **100**, 3483 (2000); (c) M. Eddaoudi, D. Moler, H. Li, B. Chen, T. Reineke, M. O'Keefe, and O. M. Yaghi, *Acc. Chem. Res.* **34**, 319 (2001).
3. (a) C. D. Gutsche, *Calixarenes Revisited*, Royal Society of Chemistry, Cambridge, UK, 1998; (b) C. Wieser, C. B. Dielman, and D. Matt, *Coord. Chem. Rev.* **165**, 93 (1997); (c) M. P. Oude Wolbers, F. C. van Veggel, F. G. A. Peters, E. S. E. van Beelen, J. W. Hofstraat, F. A. J. Geurts, and D. N. Reinhoudt, *Chem. Eur. J.* **4**, 772 (1998).
4. M.-K. Chung, O. C. Lightbody, and J. M. Stryker, *Org. Lett.* **10**, 3825 (2008).
5. (a) J. D. Roberts and W. Watanabe, *J. Am. Chem. Soc.* **72**, 4869 (1950); (b) D. Bethell and J. D. Callister, *J. Chem. Soc.* **3801**, 3808 (1963).
6. M.-K. Chung, G. Qi, and J. M. Stryker, *Org. Lett.* **8**, 1491 (2006).
7. Original references: (a) T. Mukaiyama, T. Sato, and J. J. Hanna, *Chem. Lett.* 1041 (1973); (b) S. Tyrlik and I. Wolochowicz, *Bull. Soc. Chem. Fr.* 2147 (1973); (c) J. E. McMurry and M. P. Fleming, *J. Am. Chem. Soc.* **96**, 4708 (1974).
8. Reviews: (a) J. E. McMurry, *Chem. Rev.* **89**, 1513 (1989); (b) J. E. McMurry, M. P. Fleming, K. L. Kees, and L. R. Krepski, *J. Org. Chem.* **43**, 3255 (1978); (c) T. Lectka, in *Active Metals: Preparation, Characterization, Applications*, A. Fürstner, ed., VCH, New York, 1996, Chapter 3.
9. (a) F.-A. von Itter and F. Vögtle, *Chem. Ber.* **118**, 2300 (1985); (b) R. Willem, H. Pepermans, K. Hallenga, M. Gielen, R. Dams, and H. J. Geise, *J. Org. Chem.* **48**, 1890 (1983); (c) F. A. Bottino, P. Finocchiaro, E. Libertini, A. Reale, and A. Recca, *J. Chem. Soc., Perkin Trans. II* 77 (1982).
10. P. Lucas, N. E. Mehdi, H. A. Ho, D. Belanger, and L. Breau, *Synthesis* 1253 (2000).
11. J. M. Sullivan (Boulder Scientific Company, USA), U.S. Patent 6,307,063 (2001).
12. A. Fürstner, A. Hupperts, A. Ptock, and E. Janssen, *J. Org. Chem.* **59**, 5215 (1994).
13. (a) A. M. Felix, *J. Org. Chem.* **39**, 1427 (1974); (b) M. V. Bhatt and S. U. Kulkarni, *Synthesis* **4**, 249 (1983).

36. ELECTROCHEMICAL SYNTHESIS OF TETRAETHYLAMMONIUM TETRATHIOOXALATE

Submitted by JONATHAN G. BREITZER,[**] GEOFFREY A. HOLLOWAY,[*]
THOMAS B. RAUCHFUSS,[*] and MICHAEL R. SALATA[*]
Checked by CHEE LEONG KEE and YAW KAI YAN[†]

$$2CS_2 + 2e^- + 2Et_4N^+ \rightarrow (Et_4N)_2C_2S_4$$

Tetrathiooxalate was originally claimed to result from the reduction of CS_2 with sodium.[1] It was eventually shown that the reaction of concentrated solutions of CS_2 with Na metal gives the heterocycle $C_3S_5^{2-}$, commonly referred to as dmit^{2-}.[2] Isotopic labeling studies have since shown that $C_3S_5^{2-}$ arises from the condensation of $C_2S_4^{2-}$ with CS_2 according to the equation $C_2S_4^{2-} + CS_2 \rightarrow C_3S_5^{2-} + S$.[3]

Jeroschewski demonstrated a practical synthesis of $C_2S_4^{2-}$ by suppressing its condensation with CS_2.[4] His method involves the electrolytic reduction of CS_2 in the presence of a cation that forms a poorly soluble salt with tetrathiooxalate. Additional features of the Jeroschewski method are the use of low temperatures and dilute solutions, both of which minimize the condensation side reaction. This method opened the door to more extensive studies of $(Et_4N)_2C_2S_4$. Salts of $C_2S_4^{2-}$ can also be generated by the reaction of C_2Cl_4 with sodium sulfide, but this method gives erratic results.[3] The present procedure is an adaptation of Jeroschewski's method,[5] but uses simpler apparatus. It affords gram quantities of $(Et_4N)_2C_2S_4$ in a few hours work.

Procedure

■ **Caution.** *CS_2 is a suspected neurotoxin and should be handled in a well-functioning hood. Mercury is toxic and precautions should be observed to avoid spills and for its recovery at the end of the reaction. The use of nitrile gloves and ready access to a mercury spill kit is recommended.*

To a 150 mL beaker are added a Teflon®-coated stir bar and enough mercury to cover the bottom of the beaker (about 100 g). A saturated solution of Et_4NBr (20 g) in acetonitrile (200 mL) is prepared in advance, and 110 mL of this solution is added to the beaker followed by CS_2 (3.3 mL, 0.055 mol). A paper Soxhlet shell

[*]Department of Chemistry, University of Illinois at Urbana-Champaign, Urbana, IL 61801.
[**]Fayetteville State University, Fayetteville, NC 28301.
[†]Natural Sciences and Science Education, Nanyang Technological University, Singapore 637616.

(2.5 cm diameter, 7.5 cm length) is mounted to the inside of the beaker with a plastic clip so that the bottom of the shell does not touch the mercury pool (Fig. 1). A platinum wire (at least 10 cm long) is inserted in the beaker to contact the mercury pool and then secured to the side of the beaker with tape. A piece of platinum foil (6 cm × 1 cm × 0.1 mm)* is fitted into the Soxhlet shell. The electrodes are connected via alligator clips to a 12 V DC power supply. The negative terminal is connected to the wire that is inserted in the mercury pool and the positive terminal is connected to the foil inside the Soxhlet thimble. The beaker is placed in a plastic tub of $\sim$500 mL volume and placed on a magnetic stirring plate. The tub is packed with ice around the beaker, and the solution is stirred magnetically gently so as not to disturb the mercury pool or the platinum wire. The power supply is switched on, whereupon the solution color immediately becomes orange. The current is maintained between 0.3 and 0.5 A. After a few minutes, yellow-orange crystals of $(Et_4N)_2C_2S_4$ begin to appear. The electrolysis is continued for 4 h while monitoring the current. If the current drops below 0.3 A, several milliliters of solution inside the Soxhlet shell are removed by pipette and discarded, and the solution outside the Soxhlet thimble is allowed to flow into the shell. The amount of solution discarded is replaced with fresh electrolyte solution added *outside* the Soxhlet shell. If the current remains low, the power supply is turned off, and the platinum foil inside the Soxhlet shell is removed and cleaned. If this still did not cause the current to increase, $\sim$1 mL of CS_2 is added to the solution *outside* the Soxhlet shell.† After 4 h, the current is discontinued, and the Soxhlet shell is removed quickly and placed in another beaker to minimize contamination of the cathode solution with the anode solution. The wire and stir bar are removed.

At this stage, the mixture consists of a dark orange solution and yellow-orange crystals suspended above the pool of mercury. The product should be collected immediately upon completion of the electrolysis to minimize further reaction of the tetrathiooxalate with CS_2. The entire contents of the beaker are poured into a 500 mL separatory funnel, and the mercury is carefully drained out, leaving an orange slurry. The crude product (2.4–2.7 g) is collected by filtration and is washed consecutively with CH_3CN, CH_2Cl_2, and ether (about 15 mL portions of each). Elemental analysis of typical crude product shows that it is a mixture of about 73% $(Et_4N)_2C_2S_4$ and 27% Et_4NBr by weight. The mercury can be reused in subsequent preparations.

The crude product is extracted into degassed, deionized water (about 16–20 mL). This deep orange extract is filtered to remove a small quantity of brown solid, and the solid is washed with a minimal volume of water and the washings added to the filtrate. The solution is transferred to a 250 mL round-bottomed flask (alternatively, and even better, the filtration can be done directly *into* the round-

---

*The checkers used a 2.5 cm × 2.5 cm × 0.1 mm piece of Pt foil instead.
†The checkers cleaned the Pt anode once every 15 min and replaced the electrolyte in the thimble about once every 30 min or when it becomes cloudy.

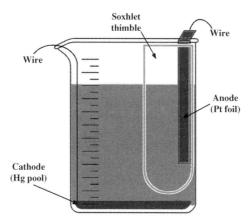

Figure 1. Apparatus for the electrochemical synthesis of $(Et_4N)_2C_2S_4$.

bottomed flask). The flask is attached to a vacuum. The solution volume of the filtrate is then reduced while stirring and gentle heating with a $\sim40°C$ water bath. When orange crystals first appear, the vacuum is disconnected, and the mixture is heated to $\sim60°C$ with the water bath (it is helpful to have this warm bath ready in advance) to redissolve the crystals. The *homogeneous* solution is then diluted with degassed acetone (200 mL) to precipitate the crystalline product. The slurry of $(Et_4N)_2C_2S_4$ is cooled in ice before being filtered. If conducted in air, the filtration should be done quickly to minimize air oxidation. The solid is washed with acetone (40 mL) and stored under nitrogen. Yield: 1.77 g (24%).

Properties

Solid $(Et_4N)_2C_2S_4$ can be stored under N_2 or in a vacuum for months without decomposition. The most useful check of the purity is by optical spectroscopy. It is soluble in highly polar solvents including water. It forms insoluble polymers with simple metal cations, but soluble mixed ligand complexes have been isolated.[6] IR (KBr, cm^{-1}): 2976 (m), 1456 (m), 1397 (m), 1308 (w), 1182 (m), 1129 (m), 988 (s, $\nu_{C=S}$), 798 (m), 760 (m), 740 (m). UV–vis (MeOH, λ_{max} (ϵ ($cm^{-1} \cdot M^{-1}$))): 282 (12,000), 344 (17,000), 380 (8000) nm.

References

1. E. Hoyer, *Comments Inorg. Chem.* **2**, 261 (1983).
2. (a) G. Steimecke, H.-J. Sieler, R. Kirmse, and E. Hoyer, *Phosphorus Sulfur Silicon Relat. Elem.* **7**, 49 (1979); (b) N, Svenstrup and J. Becher, *Synthesis* 215 (1995); (c) J. G. Breitzer and T. B. Rauchfuss, *Polyhedron* **19**, 1283 (2000).

3. J. G. Breitzer, J.-H. Chou, and T. B. Rauchfuss, *Inorg. Chem.* **37**, 2080 (1998).
4. (a) P. Jeroschewski, *Z. Chem.* **21**, 412 (1981); (b) P. Jeroschewski, P. Hansen, *Sulfur Rep.* **7**, 1 (1986).
5. H. Lund, E. Hoyer, and R. G. Hazell, *Acta Chem. Scand. B* **36**, 207 (1982).
6. (a) G. A. Holloway, K. K. Klausmeyer, and T. B. Rauchfuss, *Organometallics* **19**, 5370 (2000); (b) K. Kubo, A. Nakao, H. M. Yamamoto, and R. Kato, *J. Am. Chem. Soc.* **128**, 12358 (2006).

37. MID-INFRARED EMITTING LEAD SELENIDE NANOCRYSTAL QUANTUM DOTS

Submitted by JEFFREY M. PIETRYGA[*] and JENNIFER A. HOLLINGSWORTH[**]
Checked by FUDONG WANG[†] and WILLIAM E. BUHRO[†]

$$Pb(O_2CMe)_2 \cdot 3H_2O + Se=P(C_8H_{17})_3 \rightarrow PbSe + 2HO_2CMe + 2H_2O + O=P(C_8H_{17})_3$$

Colloidal nanocrystal quantum dots (NQDs) couple size-tunable optical properties with molecule-like chemical properties such as solubility, processability, and chemical reactivity (the latter depending on the identity of the surface-bound ligands).[1] Photoluminescent NQDs are potentially useful as molecular probes in biology and as materials for the construction of solid-state light-emitting devices. Synthetic routes have previously been reported for NQDs that emit in the ultraviolet (e.g., ZnSe),[2] in the visible (e.g., CdSe),[3] and in the near-IR (e.g., small-size PbSe <8 nm in diameter).[4] Because bulk PbSe has a bandgap of 0.26 eV (4.7 μm) at room temperature, PbSe NQDs have the potential to exhibit photoluminescence (PL) in the mid-IR energy range (≥ 2.5 μm).

The synthetic routes described here are useful for obtaining large-size PbSe NQDs (8–17 nm) with demonstrated emission in the mid-IR (2.5–3.5 μm).[5] The key feature of the synthetic approach described in detail here is the use of a two-step growth process. The first step involves nucleation and initial growth by rapid injection of lead and selenium precursors into a hot surfactant/solvent mixture. Under these conditions, growth is sufficiently fast that narrow size dispersions can be obtained; the resulting PbSe NQDs are small and emit at wavelengths less than 2.5 μm. Additional particle growth to larger NQD sizes is accomplished in the second step, which involves slow addition of precursors to the solution containing the "seed" particles. In this way, large-size PbSe NQDs are obtained that possess optimized optical properties and minimal size dispersions.

[*]Chemistry Division, Los Alamos National Laboratory, Los Alamos, NM 87545.
[**]Materials Physics & Applications Division, Los Alamos National Laboratory, Los Alamos, NM 87545.
[†]Department of Chemistry, Washington University, St. Louis, MO 63130.

General Procedures

Operations are conducted under nitrogen gas in a glove box (O_2 <1 ppm) or by using standard Schlenk manifold techniques. Trioctylphosphine (TOP, 90%, Fluka) is heated to 200°C under an inert atmosphere for 2 h before use. Selenium shot (99.999%, Alfa Aesar), diphenyl ether (99%, Acros), lead(II) acetate trihydrate (99.999%, Aldrich), and oleic acid (90%, Aldrich) are used without further purification. Oleic acid is stored in the glove box freezer (−30°C). The lead(II) acetate trihydrate slowly dehydrates in the glove box and becomes less soluble; therefore, use of fresh material is recommended. The 1 M TOPSe in TOP stock solution is prepared by covering selenium shot (1.58 g, 20 mmol) with TOP (20 mL) and stirring vigorously under inert atmosphere until the selenium dissolves (~12 h). Temperature regulation is achieved using a temperature controller (Barnant Standard Model) fitted with a heating mantle and thermocouple probe (Type K). PL measurements are performed on dilute NQD solutions in hexane or $CDCl_3$; samples are excited at 808 nm, and emission is analyzed using a grating monochromator and a chilled InSb detector.

A. LEAD SELENIDE NQDs EMITTING AT 2.5 µm (0.50 eV)

Procedure

A three-necked flask is fitted with a condenser (topped with a valved gas adapter), a rubber septum, and a glass stopper. In a glove box, the flask is loaded with a stir bar, lead(II) acetate trihydrate (0.87 g, 2.3 mmol), and oleic acid (2.3 mL, 7.2 mmol). The reactants are covered with diphenyl ether (2 mL) and TOP (8 mL). The flask is sealed, removed from the glove box, and connected to a Schlenk manifold through the valve at the top of the condenser. While keeping the contents of the flask under an inert atmosphere, a thermocouple probe is passed through the previously glass-stoppered neck using a PTFE 1/8″ thermocouple adapter (available from Chemglass or J-Kem Scientific). With stirring, the mixture is heated to 65°C under dynamic vacuum for 1 h, forming a clear solution. In the same manner, diphenyl ether (10 mL) is heated to 45°C under dynamic vacuum for 1 h in a four-necked flask fitted with a similarly valved condenser, septum, thermocouple probe, and an addition funnel. Both flasks are then placed under inert atmosphere. The three-necked flask is allowed to cool to room temperature, while the four-necked flask is heated to 185°C. TOPSe stock solution (2.3 mL) is then injected into the lead-containing flask, and an aliquot of this mixture (3.5 mL) is removed and placed in the addition funnel on the four-necked flask. The remaining reactant solution is then quickly injected into the rapidly stirring hot diphenyl ether. The resulting mixture turns brown, indicating nucleation of PbSe NQDs. After 4 min (as the solution returns to 185°C), the contents of the addition funnel are added slowly over ~5 min. Eleven minutes after the initial injection, the flask is removed from the heat and the mixture is quenched by addition of cold (−90°C) toluene (15 mL), and then allowed to cool to ~50°C. Methanol (10–20 mL) is added with stirring until the mixture is persistently cloudy, indicating precipitation of the product NQDs. The solid is collected by centrifugation, and redissolved in hexane (10 mL). The NQDs are reprecipitated by addition of methanol (10 mL) and *n*-butanol (1 mL); this reprecipitation removes more of the passivating ligands and allows removal of more unreacted precursors, but also can result in loss of some passivating ligands, which can reduce the solubility and PL intensity of the NQDs. Yield: 60 mg (9%, reckoned as PbSe).

Properties

PbSe NQDs prepared by this procedure are photoluminescent and emit between 2.5 and 2.7 µm. The emission is somewhat broad because a relatively wide range of NQD particle sizes is produced by this procedure. As prepared, the NQDs are hydrophobic and therefore soluble in common solvents such as hexane, toluene, and chloroform. Both the powder form and solutions of the PbSe NQDs appear brown in color. PbSe NQDs tend to oxidize over time, causing the emission bands to broaden and shift to lower wavelengths (indicative of an increase in particle-size dispersion). For best retention of optical properties, the particles should be stored cold in the dark and under an inert atmosphere.

B. LEAD SELENIDE NQDs EMITTING AT 2.8 µm (0.44 eV)

Procedure

Two flasks are constructed and loaded with reactants and solvents exactly as described in Section 37.A. The four-necked flask is loaded with diphenyl ether (10 mL), while the three-necked flask is loaded with lead(II) acetate trihydrate (0.98 g, 2.6 mmol), oleic acid (2.3 mL, 7.2 mmol), diphenyl ether (2 mL), and TOP (10 mL). Both flasks are heated under vacuum for 1 h, exactly as above. After being refilled with inert gas, the three-necked flask is cooled to room temperature, and the four-necked flask is heated to 200°C. After TOPSe stock solution (2.6 mL) is added to the lead-containing flask, an aliquot of this mixture (6 mL) is removed and placed in the addition funnel on the four-necked flask. The remaining reactant solution is quickly injected into the hot diphenyl ether. After 4 min, the contents of the addition funnel are slowly added over ~10 min. Sixteen minutes after the initial injection, the flask is removed from heat, and the reaction is quenched by addition of cold (−90°C) toluene (15 mL). Precipitation and collection of product is carried out by the same method as above. The second precipitation can result in some loss of solubility and photoluminescent properties. Yield: 90 mg (12%).

Properties

NQDs prepared by this route are characterized by an emission wavelength of 2.8 µm. Otherwise, properties are similar to those described for 2.5 µm emitting PbSe NQDs.

C. LEAD SELENIDE NQDs EMITTING AT 3.3 µm (0.38 eV)

Procedure

Two flasks are constructed as described in Section 37.A. The three-necked flask is charged with lead acetate trihydrate (0.98 g, 2.6 mmol), oleic acid (2.5 mL, 7.9 mmol), diphenyl ether (2 mL), and TOP (9 mL). The four-necked flask is loaded with diphenyl ether (10 mL). Both flasks are heated under vacuum, and then refilled with inert gas as above. The three-necked flask is cooled to room temperature, while the four-necked flask is heated to 220°C. TOPSe stock solution (2.6 mL) is added to the lead-containing flask, and an aliquot of this mixture (5.5 mL) is placed in the addition funnel on the four-necked flask. The remaining reactant solution is quickly injected into the hot diphenyl ether. After 3.5 min, the contents of the addition funnel are slowly added over ~8.5 min. Fifteen minutes after the initial injection, the flask is removed from heat, and the reaction is quenched by addition of cold (−90°C) toluene (20 mL). Precipitation and collection of product is carried out by the same method as above, with the same loss of photoluminescent properties and solubility after the second precipitation. Yield: 155 mg (21%).

Properties

NQDs prepared by this route are characterized by an emission wavelength of 3.3 µm. Otherwise, properties are similar to those described for 2.5 µm emitting PbSe NQDs.

D. LEAD SELENIDE NQDs EMITTING AT 3.5 µm (0.35 eV)

Procedure

Two flasks are constructed as described in Section 37.A. The three-necked flask is charged with lead acetate trihydrate (1.56 g, 4.1 mmol), oleic acid (5 mL,

11 mmol), diphenyl ether (3.5 mL), and TOP (11 mL). The four-necked flask is loaded with diphenyl ether (10 mL). Both flasks are heated under vacuum, and then refilled with inert gas as above. The three-necked flask is cooled to room temperature, while the four-necked flask is heated to 205°C. TOPSe stock solution (4.5 mL) is added to the lead-containing flask, and an aliquot of this mixture (13 mL) is placed in the addition funnel on the four-necked flask. The remaining reactant solution is quickly injected into the hot diphenyl ether. After 4 min, the contents of the addition funnel are slowly added over ~15 min. Twenty-three minutes after the initial injection, the flask is removed from heat, and the reaction quenched by addition of cold toluene (20 mL). Precipitation and collection of product is carried out by the same method, but care must be exercised to avoid adding too much methanol, as the product very quickly loses solubility if over-rinsed. Photoluminescent properties and solubility both suffer even more with each subsequent precipitation step. Yield: 270 mg (23%).

Properties

NQDs prepared by this route are characterized by an emission wavelength of 3.5 μm. Otherwise, properties are similar to those described for 2.5 μm emitting PbSe NQDs.

References

1. J. A. Hollingsworth and V. I. Klimov, in *Semiconductor and Metal Nanocrystals: Synthesis, Electronic and Optical Properties*, V. I. Klimov, ed., Marcel Dekker, New York, 2003, Chapter 1.
2. M. Hines and P. Guyot-Sionnest, *J. Phys. Chem. B* **102**, 3655 (1998).
3. (a) C. B. Murray, D. J. Norris, and M. G. Bawendi, *J. Am. Chem. Soc.* **115**, 8706 (1993); (b) L. H. Qu and X. G. Peng, *J. Am. Chem. Soc.* **124**, 2049 (2002).
4. (a) C. B. Murray, S. Sun, W. Gaschler, H. Doyle, T. A. Betley, and C. R. Kagan, *IBM J. Res. Dev.* **45**, 47 (2001); (b) H. Du, C. Chen, R. Krishnan, T. D. Krauss, J. M. Harbold, F. W. Wise, M. G. Thomas, and J. Silcox, *Nano Lett.* **2**, 1321 (2002); (c) B. L. Wehrenberg, C. Wang, and P. Guyot-Sionnest, *J. Phys. Chem. B* **106**, 10634 (2002).
5. J. M. Pietryga, R. D. Schaller, D. Werder, M. H. Stewart, V. I. Klimov, and J. A. Hollingsworth, *J. Am. Chem. Soc.* **126**, 11752 (2004).

Chapter Six

TEACHING LABORATORY EXPERIMENTS

38. TETRA(ACETATO)DICHROMIUM(II) DIHYDRATE

Submitted by ALAN M. STOLZENBERG[*]
Checked by RAM P. NEUPANE[†] and STANTON CHING[†]

$$2CrCl_3 \cdot 6H_2O + Zn \rightarrow 2[Cr(H_2O)_6]Cl_2 + ZnCl_2$$

$$2[Cr(H_2O)_6]Cl_2 + 4NaO_2CCH_3 \rightarrow Cr_2(O_2CCH_3)_4 \cdot 2H_2O + 4NaCl + 10H_2O$$

Tetra(acetato)dichromium(II), also known as chromium(II) acetate, is one of the most accessible compounds of the strongly reducing Cr^{II} ion and is used as the starting material in the preparation of other chromium(II) compounds. Its empirical formula is deceptively simple and does not disclose that the compound is a metal–metal bonded dimer. However, the notable contrast between the red color of hydrated chromium(II) acetate and the blue color of the aqueous Cr^{II} ion and the near diamagnetism of the acetate reveal that the compound is not simply a salt of 1:2 stoichiometry.

The synthesis of chromium(II) acetate can serve many instructional purposes. It can be used to introduce and demonstrate the concepts of metal–metal bonding and of antiferromagnetic coupling. The synthetic procedure requires students to handle materials that are extremely air sensitive (but not pyrophoric!) and demonstrates

[*]Department of Chemistry, West Virginia University, Morgantown, WV 26506.
[†]Department of Chemistry, Connecticut College, New London, CT 06320.

Inorganic Syntheses, Volume 36, First Edition. Edited by Gregory S. Girolami and Alfred P. Sattelberger

features of heterogeneous reductions by active metals. Characterization of the complex provides exposure to magnetic susceptibility measurements and can be used to demonstrate the IR spectra of bridging carboxylates. Of equal importance, students enjoy the synthesis. They are impressed by the dramatic color changes from green to blue to red. The experiment also provides immediate visual feedback on the success of their technique.

Procedures for the preparation of chromium(II) acetate monohydrate have been published in four earlier volumes of *Inorganic Syntheses*[1-4] and in a classic laboratory textbook.[5] The procedures all consist of reduction of an acidic aqueous chromium(III) chloride solution to chromium(II), which is added to a suspension of sodium acetate, but all require complicated glassware or a dry box. A more recent procedure reverses the last step (the acetate to the chromium rather than vice versa) and controls the pH to grow larger crystals of the product over 24 h.[6]

The current recipe has the advantage that it employs equipment commonly found in most teaching laboratories, with two minor exceptions: a commercially available transfer tube and a custom but easily fabricated reaction vessel shown in Fig. 1.

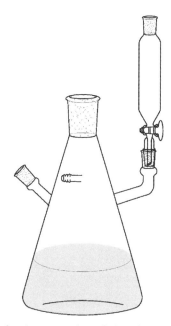

Figure 1. Apparatus for the preparation of chromium(II) acetate monohydrate.

Procedure

The equipment for this experiment includes a modified 500 mL Erlenmeyer filter flask (Fig. 1),[*] a solid rubber stopper to fit the main neck of the flask, a standard taper jointed addition funnel, a tubing adapter for the top joint of the addition funnel, a septum (serum stopper) for the straight sidearm of the flask, two T-tubes, a hose clamp, several pieces of rubber or vinyl tubing, and an oil bubbler. The straight sidearm on the filter flask is sealed with the septum, and the flask is charged with small pieces of mossy zinc (9.0 g, 0.14 mol), a 5 cm long magnetic stir bar, and a solution of $CrCl_3 \cdot 6H_2O$ (10.0 g, 0.037 mol) dissolved in water (40 mL). The flask is clamped securely on top of a magnetic stirrer, and the main neck of the Erlenmeyer flask is sealed with the rubber stopper. The curved sidearm of the flask is fitted with the addition funnel that is topped by a tubing adapter. The tubing adapter is connected with rubber or plastic tubing to a T-tube, and one of the other two arms of the T-tube is connected to a nitrogen source. The nipple on the filter flask is likewise connected with tubing to a second T-tube, and one of the other two arms on that T-tube is connected to an oil bubbler. The remaining arms of the two T-tubes are connected to one other. A hose clamp is placed loosely on the tubing between the bubbler and the T-tube.

The septum is removed from the sidearm on the reaction apparatus, and the stopcock of the addition funnel is opened. The hose clamp is tightened to pinch off the hose to the oil bubbler, and the apparatus is flushed with a strong flow of N_2 gas for several minutes. After the flushing is complete, the hose clamp is loosened and the septum is replaced firmly in the sidearm on the flask. The nitrogen flow should be decreased as necessary to obtain a bubble rate of one or two bubbles per second in the oil bubbler.

At the first opportunity when there is some free time (e.g., while flushing the reaction apparatus with N_2 as described above), water (150 mL) for a later washing step should be placed in a separate Erlenmeyer flask. Nitrogen gas should be bubbled through the water for at least 15 min, while the flask is being cooled in an ice bath. In addition, a slurry of sodium acetate trihydrate (60.0 g, 0.73 mol) in water (55 mL) should be prepared in a 250 mL Erlenmeyer flask.[†] Nitrogen gas should be bubbled through the slurry until it is used below.

The stopcock of the addition funnel is closed, the tubing adapter is removed, and concentrated hydrochloric acid (20 mL) is poured into the addition funnel. Finally, the tubing adapter (still connected with hosing to the T-tube and a flow of N_2) is replaced on top of the addition funnel. The green $CrCl_3$ solution is stirred as rapidly as possible without splashing the solution into the sidearm and its attached

---

[*]The checkers successfully used a 500 mL three-necked round-bottomed flask instead of the modified Erlenmeyer flask.
[†]The checkers heated the mixture to aid dissolution of the sodium acetate. Complete dissolution of the solid, however, does not appear necessary.

tubing. Then about a quarter of the acid is added to the green solution over about 1 min. Hydrogen gas is generated and will escape through the bubbler. (■ **Caution.** *No open flames!*) Continue stirring and adding acid until the solution becomes clear blue. With good stirring, this step typically can be accomplished in about 10–15 min without addition of all of the acid. If necessary, the stirring rate is adjusted or the flask is manually agitated to recover any drops of green liquid that adhere to the walls of the flask.

The next step should be carried out only after the chromium solution has turned clear blue but before the zinc is completely consumed and hydrogen evolution ceases.[*] One end of a CHEM-FLEX transfer line is inserted through the septum of the reaction flask, leaving the point above the solution for the moment.[†] Next, the hose clamp is closed to pinch off the tubing next to the bubbler, causing the N_2 gas to pressurize the flask and to flush the transfer line. The free end of the transfer line is inserted under the surface of the sodium acetate solution; gas bubbles should be visible at the tip. Finally, the stirrer is stopped and end of the transfer line is pushed under the surface of the chromium chloride solution. The N_2 gas flow should be increased as needed to force the solution through the line and transfer the blue chromium(II) chloride solution into the sodium acetate slurry. It may be necessary to hold the solid rubber stopper in place to keep the flask pressurized. A red precipitate of chromium(II) acetate forms when the blue solution contacts the sodium acetate solution. The reaction flask is tilted and the transfer line is manipulated to transfer as much of the blue solution as possible, avoiding any black particulate scum that might be present from impurities in the zinc metal.

Immediately after the transfer is complete, the transfer line is removed from the flask containing the mixed solutions, and the flask that contains the product is capped with a rubber stopper. The flask is then cooled by swirling it in a mixture of ice and water. The transfer line should be cleaned by using a vacuum to pull water and then ethanol through it, and then air is pulled through to dry it.

The following filtration and washing steps must be accomplished in one uninterrupted process to avoid oxidizing the product. Each successive wash portion must be added before the liquid level drops below the top of the solid when it is wet with water or alcohol. Thus, portions of wash solvent should be measured out before starting and the filter flask must be large enough to contain the entire volume of filtrate (500 mL or greater). The solid is collected on a 60 mL medium-porosity sintered glass funnel and washed four times with cold, air-free water (35 mL each time). The speed of the filtration will be greater if the precipitate is stirred up as little as possible with the wash water. The solid is washed with 95%

---

[*]A muddy blue or greenish blue color indicates the presence of chromium(III), which will react with the sodium acetate solution to afford a chromium(III) hydroxide gel that will make the final filtration step proceed very slowly.
[†]The checkers employed a stainless steel cannula. However, cannulas of the large gauge necessary for reasonable flow rates and to prevent clogging (at least 14 gauge) can be unwieldy in student hands.

ethanol (50 mL) and diethyl ether (50 mL), and then the crystals are dried by pulling air through them on the funnel. The brick red product is then spread thinly on a watch glass and is allowed to dry at room temperature until it no longer smells of ether (30 min or less). Occasionally, part of the material will turn the gray-green color characteristic of the oxidized material while drying. The discoloration will spread throughout the sample if the gray-green material is not mechanically removed from the remaining red material. The color change can be accompanied by heat evolution, reportedly sufficient to crack a watch glass. After the solid is dry, it is transferred to a tared test tube, and the tube is flushed by gently blowing N_2 gas over the solid (■ **Caution.** *Do not scatter and inhale the solid.*) The tube is sealed tightly with a tared rubber stopper. The tared tube and stopper are then weighed to determine the yield.

Properties

Samples of chromium(II) acetate properly sealed in a test tube should remain brick red indefinitely. However, if air is admitted, the sample will gradually turn gray-green. If a sample is removed from the test tube for characterization, be sure to gently flush the tube with N_2 gas and cap it tightly afterward.

Characterization

For the following measurements, grind the sample and make the measurement quickly so as to minimize oxidation of the sample. Record the IR spectrum of chromium(II) acetate monohydrate either as a KBr pellet or as a Nujol mull between KBr plates. If apparatus is available, record the magnetic susceptibility.

Questions

1. What is the chemical structure of chromium(II) acetate monohydrate?
2. What magnetic moment per Cr atom is expected for a high-spin coordination complex of chromium(II)? Is this value consistent with the magnetic properties of chromium(II) acetate monohydrate?
3. What are the $C-O$ stretching frequencies for the free (unligated) acetate anion? Are these values different from those seen for chromium(II) acetate monohydrate?
4. Suggest an explanation for any discrepancy that you note in question 2 or 3.

Notes for Instructors

For this experiment, we find it advantageous to use a reaction flask fabricated from a 500 mL Erlenmeyer filter flask and two female standard taper joints (Fig. 1). We used 19/22 joints to match other glassware on hand. The joints are blown onto opposite sides of the flask, 90° from the nipple (hose barb) and at a height above the 350 mL volume line. One joint is mounted on a straight arm inclined roughly 40° from vertical. The other is mounted on a curved sidearm so arranged that the top of the standard taper joint is horizontal. The flask offers the advantages of a flat bottom for stability and efficient stirring and thick walls that are safer when the internal pressure is raised above atmospheric pressure. A three-neck 500 mL flask can be substituted if one does not have access to glassblowing services. The other specialty item we use is a 12-gauge, 75 cm long CHEM-FLEX transfer line (Aldrich # Z23,102-9). It is not necessary to equip the laboratory with one transfer line per student because students typically do not complete the reduction at the same time, and a single line can be used by several students.

Typical advanced undergraduates can easily complete the synthesis in one 3 h laboratory period. The most common delay results from not preparing the slurry of degassed water and sodium acetate while setting up the apparatus and reaction. The time required for the reduction depends primarily on the rate of stirring. Rapid, turbulent stirring can result in complete reduction in about 10 min. In contrast, reduction may require more than 1 h with slow, laminar stirring.

The synthesis can easily be adapted to groups of two students by splitting responsibilities for the tasks. If students work individually, an extra set of hands can be helpful during the transfer.

Yields typically range between 65 and 86% when product is not lost to oxidation. The most significant factors that affect the yield are losses during transfer steps and losses due to oxidation while filtering or drying. Some samples prepared by our undergraduates and stored under nitrogen have remained unoxidized after as long as 10 years!

The compound can be characterized by IR spectroscopy and by magnetic susceptibility measurements. X-ray powder diffraction can also be used if the equipment is available. The IR spectrum is somewhat dependent on the sample preparation technique employed. The literature values are for a sample mulled in Nujol.[7] Notable features include the $O-H$ stretches of the coordinated water at 3486, 3371, and 3271 cm^{-1}, bridging carboxylate CO bands at 1586, 1576, and 1423 cm^{-1}, and a CH_3 deformation at 1453 cm^{-1}.[*] Student samples mulled in paraffin oil give spectra in good agreement with these values. Samples prepared as KBr disks show a broad featureless OH band, a CO stretch at 1577 cm^{-1}, and strong bands at 1419, 1049, and 1031 cm^{-1}. Spectra obtained using ATR sampling

---

[*]These are the correct IR frequencies for chromium(II) acetate hydrate. The data reported in *Chem. Abstr.*, **52**, 9937 (1958) and the *Gmelin Handbook* are incorrect because these two references interchanged the frequencies for chromium(II) and copper(II) acetate hydrates and their dehydrated forms as reported in Ref. 7.

show bands at 3500, 3366, 3272, 1648 (wk), 1570, 1451, 1418, 1355, 1050, 1031, and 685 cm^{-1}.

The corrected magnetic susceptibility of chromium(II) acetate hydrate[*] determined by the Gouy method (using an aqueous slurry covered with a layer of petroleum ether to exclude air) is reported to be $\chi_{mol} = 94 \times 10^{-6}$ cm^3 mol^{-1}.[8] Another study reported $\chi_{mol} = 113 \times 10^{-6}$ cm^3 mol^{-1}.[9] The weak paramagnetism of the samples has been attributed to a chromium(III) impurity. However, variable-temperature NMR studies have established that partial occupation of a low-lying triplet state of the Cr−Cr quadruple bond is responsible for inherent paramagnetism of chromium(II) acetate.[10] Susceptibilities are measured in our teaching laboratories using a Johnson Matthey susceptibility balance, which requires that the samples be ground to a fine powder. The best student samples show corrected χ_{mol} values of 740×10^{-6} cm^3 mol^{-1}, which corresponds to 0.66 unpaired electron per chromium atom. Typical student samples have $\chi_{mol} = 1400$–2200×10^{-6} cm^3 mol^{-1}, which corresponds to 1.1–1.5 unpaired electrons per chromium. The increased susceptibilities likely reflect partial oxidation to chromium(III) during sample grinding. Nonetheless, the measured susceptibility is still significantly less than the value expected for four unpaired electrons and permits the students to conclude that the compounds they prepared cannot contain high-spin chromium(II).

This synthesis of chromium(II) acetate can be combined with other syntheses to create projects with broader enquiries. In our laboratory class, it is paired with the synthesis of copper(II) acetate, another acetate-bridged, dimeric complex in which there is a magnetic interaction between the metal centers. Similarly, it could be paired with the synthesis of molybdenum(II) acetate, $Mo_2(O_2CCH_3)_4$, a quadruply metal–metal bonded compound.[11] The structural diversity of metal acetates could also be explored by combining this synthesis with that of $Cr_3O(O_2CCH_3)_6$[12] or other trimeric "basic metal acetates," which are known for many of the transition metals.[13]

References

1. J. H. Balthis Jr. and J. C. Bailar Jr., *Inorg. Synth.* **1**, 122 (1939).
2. M. R. Hatfield, *Inorg. Synth.* **3**, 148 (1950).
3. M. Kranz and A. Witkowska, *Inorg. Synth.* **6**, 144 (1960).
4. L. R. Ocone and B. P. Block, *Inorg. Synth.* **8**, 125 (1962).
5. W. L. Jolly, *The Synthesis and Characterization of Inorganic Compounds*, Prentice Hall, Englewood Cliffs, NJ, 1970, pp. 442–444.
6. J. C. Reeve, *J. Chem. Educ.* **62**, 444 (1985).
7. G. Costa, E. Pauluzzi, and A. Puxeddu, *Gazz. Chim. Ital.* **87**, 885 (1957).
8. J. W. R. King and C. S. Garner, *J. Chem. Phys.* **18**, 689 (1950).

[*]The susceptibility values are per chromium atom.

9. C. Furlani, *Gazz. Chim. Ital.* **87**, 876 (1957).
10. F. A. Cotton, H. Chen, L. M. Daniels, and X. Feng, *J. Am. Chem. Soc.* **114**, 8980 (1992).
11. R. A. Walton, P. E. Fanwick, and G. S. Girolami, *Inorg. Synth.*, **36**, 78 (2014).
12. M. Eshel and A. Bino, *Inorg. Chim. Acta* **320**, 127 (2001), and references therein.
13. R. C. Mehrotra and R. Bohra, *Metal Carboxylates*, Academic Press, London, 1983.

39. KEGGIN STRUCTURE POLYOXOMETALATES

Submitted by JOSÉ ALVES DIAS, [*] **SÍLVIA CLÁUDIA LOUREIRO DIAS** [*] **, and EDNÉIA CALIMAN** [*]
Checked by JUDIT BARTIS [†] **and LYNN FRANCESCONI** [†]

Polyoxometalates (POMs) are polymeric oxoanions represented by the general formula $[X_xM_mO_y]^{q-}$ ($0 \leq x \leq m$) formed by the condensation of one or more different oxoanions.[1] Of the elements constituting the structure, M is usually molybdenum, tungsten, vanadium, niobium, tantalum, or mixtures of these elements in their highest oxidation states (d^0, d^1), whereas x may be absent (forming the isopolyanions) or may belong to a wide variety of group 1–17 elements (forming the heteropolyanions).[2] A variety of polyanion structures exist with different formulas, but one of the earliest and most studied is the Keggin-type $[XM_{12}O_{40}]^{q-}$, which is named after the British crystallographer James F. Keggin who determined its structure in 1933. The α-Keggin structure, which has T_d symmetry, consists of a central XO_4 tetrahedron surrounded by 12 fused MO_6 octahedra (Fig. 1). The latter are arranged into four triangular M_3O_{13} units, each consisting of three edge-shared MO_6 octahedra; the M_3O_{13} units are linked to one another by sharing corners, and each M_3O_{13} unit also shares an oxygen atom with the central heteroatom.[2] Chemical modifications can tailor the properties of Keggin ions, such as their size, mass, thermal stability, lability, Brönsted acidity, redox potentials, and solubility.[3,4] As a result of this tunability, Keggin ions find uses in many fields, including analytical chemistry, biochemistry, materials science, and catalysis.

The following syntheses involve some of the most common heteropoly compounds with the Keggin structure. These compounds can help students develop a good understanding of how subunits can be connected together to form condensed structures. In addition, the preparations expose the students to a variety of useful laboratory techniques such as vacuum filtration, solvent

[*]Laboratório de Catálise, Instituto de Química, Universidade de Brasília, Brasília, DF 70919-970, Brazil.
[†]Department of Chemistry, Hunter College, 695 Park Avenue, New York City, NY 10065.

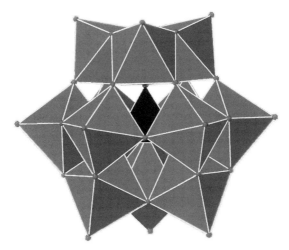

Figure 1. The α-Keggin structure. One face of the central XO_4 tetrahedron is partly visible. (Image from Crystal Impact GbR, Bonn, Germany).

extraction, crystallization, recrystallization, and drying processes. Moreover, because Keggin polyoxometalates can be characterized in solution or the solid state by a variety of methods (e.g., FTIR, XRD, UV–vis, and NMR), student technique can be evaluated by assessing the purity of the samples they obtain.

A. TRI(AMMONIUM) 12-MOLYBDOPHOSPHATE

$$12(NH_4)_6(Mo_7O_{24}) \cdot 4H_2O + 65HNO_3 + 7Na_2HPO_4 \cdot H_2O \rightarrow$$

$$7(NH_4)_3[PMo_{12}O_{40}] \cdot 5H_2O + 14NaNO_3 + 51NH_4NO_3 + 56H_2O$$

Procedure[5]

■   **Caution.**   *Concentrated nitric acid (HNO₃) is highly corrosive to the skin and eyes and should be handled in a fume hood with proper eye and skin protection. Prepare the mixtures of concentrated HNO₃ and water by slowly adding the acid to the water. Let the mixtures cool before using.*

In a 100 mL beaker, Solution A is prepared by dissolving ammonium heptamolybdate tetrahydrate (5.0 g, 4.1 mmol) in H_2O (15 mL). In a 50 mL beaker, Solution B is prepared by dissolving disodium hydrogen phosphate hydrate $Na_2HPO_4 \cdot H_2O$ (1.25 g, 7.8 mmol) in a cooled 1:1 mixture of concentrated HNO_3 and H_2O (15 mL). Then, Solution B is slowly added to Solution A with stirring. A yellow precipitate forms, which is collected by vacuum filtration using a fine Gooch filter. Acidified water is prepared by adding concentrated HNO_3 (6–8 drops, ~0.4 mL) to hot, ~50°C water (10 mL), and the precipitate is washed with the acidified hot water in two portions. The precipitate is allowed to dry for 30 min on the filter, and then is transferred to a Petri dish to dry further in air. The number of water molecules of hydration varies between 5 and 8 per Keggin ion, depending on how long the product is air dried. The typical yield is higher than 90%, based on Mo.[*]

Properties

$(NH_4)_3[PMo_{12}O_{40}] \cdot 5H_2O$ is a yellow crystalline solid that has a low solubility in water, ethanol, acetone, chloroform, and cyclohexane. It decomposes in basic water solution (e.g., 0.5 M NaOH). FTIR (cm^{-1}): 3200 $\nu(N–H)$, 1406 $\delta(N–H)$, 1063 $\nu_{as}(P–O)$, 963 $\nu_{as}(Mo=O)$, 867 $\nu_{as}(Mo–O–Mo)$, 793 $\nu_{as}(Mo–O–Mo)$. Powder XRD (2θ in degrees, Cu Kα radiation, $\lambda = 1.5418$ Å): 10.7, 15.3, 18.5, 21.1, 24.0, 26.2. MAS ^{31}P NMR: δ–4.0 (s). A ^{31}P NMR peak at δ 0.5 indicates that the sample is contaminated with Na_2HPO_4.

Characterization

Characterize your sample by one or more of the following techniques: elemental analysis (N, P, and/or Mo), FTIR (KBr pellet or Nujol mull), powder XRD, or MAS ^{31}P NMR (single-pulse excitation, 8.0 µs duration, without ^{1}H decoupling, recycle delay of 3 s, spin rate of 5 kHz, and 64 acquisitions).

B. 12-TUNGSTOSILICIC ACID

$$12Na_2WO_4 \cdot 2H_2O + Na_2SiO_3 \cdot 5H_2O + 26HCl \rightarrow$$

$$H_4[SiW_{12}O_{40}] \cdot 29H_2O + 26NaCl + 11H_2O$$

Procedure[6,7]

■ **Caution.** *Hydrochloric acid is corrosive to the skin and eyes and should be handled in a fume hood with proper eye and skin protection. Prepare the mixtures of hydrochloric acid and water by slowly adding the acid to the water. Let the mixtures cool before using.*

■ **Caution.** *Diethyl ether is highly flammable. Open flames and sparks should be avoided.*

---

[*]The checkers obtained a yield of 85%.

In a 50 mL beaker, Solution A is prepared by dissolving sodium silicate pentahydrate $Na_2SiO_3 \cdot 5H_2O$ (2.2 g, 10.0 mmol) in water (20 mL). In a 200 mL beaker, Solution B is prepared by dissolving sodium tungstate dihydrate $Na_2WO_4 \cdot 2H_2O$ (36.4 g, 100 mmol) in boiling water (60 mL) and then slowly adding (dropwise!) with stirring 4 M HCl (33 mL, 132 mmol). Then, Solution A is added to solution B with stirring, and the pH of the solution is adjusted to between 5 and 6 by the dropwise addition of 4 M HCl ($\sim$8 mL is usually needed), using a pH meter or pH paper to monitor the pH. The stirred solution is heated to boiling for about 15 min, while continuing to keep the pH between 5 and 6 by the dropwise addition of 4 M HCl as necessary. After the 15 min period, while the solution is still at 100°C, a solution of $Na_2WO_4 \cdot 2H_2O$ (3.64 g, 10 mmol) in water (10 mL) is added, followed by 4 M HCl (16 mL, 64 mmol). The solution is cooled to room temperature with stirring, and then is filtered through fluted filter paper to remove any solids. The filtered solution is transferred to a 250 mL separatory funnel, into which is also placed diethyl ether (16 mL) and a mixture of diethyl ether (12 mL) and concentrated HCl (12 mL) that has been chilled to -12°C. The mixture is shaken, and the lowest layer, which contains the product, is collected. This layer is allowed to crystallize by slow evaporation in a Petri dish in a fume hood. Crystals are obtained after a few days to a week. The product can be recrystallized from water to remove acid and ether residues. The last traces of acid can be removed by heating the product at 150–200°C in a tube furnace under a flow of air. The typical yield is about 70% based on W.[*]

Properties

$H_4[SiW_{12}O_{40}] \cdot xH_2O$ is a white crystalline solid with about 29 water molecules per Keggin unit. In order to reach the hexahydrate state, the product is dried over $CaCl_2$ in a desiccator for 3–5 days. The compound is highly soluble in water, ethanol, acetonitrile, and acetone, but is insoluble in chloroform and hydrocarbons, and decomposes in basic aqueous solutions (e.g., 0.5 M NaOH). FTIR (cm^{-1}): 1019 ν_s(W–O), 980 ν_{as}(W=O), 926 ν_{as}(Si–O), 880 ν_{as}(W–O–W), 781 ν_{as}(W–O–W). Powder XRD (2θ in degrees, Cu Kα radiation, $\lambda = 1.5418$ Å): 10.3, 14.7, 17.9, 20.8, 23.3, 25.5, 29.5, 31.3, 34.8, 37.9. These peaks are characteristic of samples containing about six water molecules per Keggin unit; other peaks are seen in more highly hydrated samples.

Characterization

Characterize your sample by one or more of the following techniques: elemental analysis (Si and/or W), UV–vis spectroscopy (2×10^{-5} M in 0.1 M HCl), FTIR

---

[*]The checkers collected crystals after 1 week and obtained a 53% yield.

(KBr pellet or Nujol mull), powder XRD, or thermogravimetric analysis to determine the number of waters of hydration (heat the sample at $10°C\,min^{-1}$ from room temperature to $700°C$ under air or nitrogen atmosphere).

C. 12-TUNGSTOPHOSPHORIC ACID

$$12Na_2WO_4 \cdot 2H_2O + Na_2HPO_4 \cdot H_2O + 26HCl \rightarrow$$

$$H_3[PW_{12}O_{40}] \cdot 24H_2O + 26NaCl + 13H_2O$$

Procedure[7,8]

■ **Caution.** *Hydrochloric acid is corrosive to the skin and eyes and should be handled in a fume hood with proper eye and skin protection.*

In a 100 mL beaker, sodium tungstate dihydrate $Na_2WO_4 \cdot 2H_2O$ (16.2 g, 49.1 mmol) and disodium hydrogen phosphate hydrate $Na_2HPO_4 \cdot H_2O$ (2.6 g, 16.2 mmol) are dissolved in boiling water (25 mL) with stirring. Then, concentrated hydrochloric acid (13 mL, 156 mmol) is added dropwise with stirring. The product begins to precipitate as a white solid when most of the acid is added. The suspension is cooled in an ice bath, and the solid is collected by filtration and washed quickly with cold water.[*] Yield: 3.0 g (30%).

Properties

$H_3[PW_{12}O_{40}]$ is a white crystalline solid that, as isolated, possesses about 20–29 water molecules per Keggin unit. In order to reach the hexahydrate state, the final product is dried over $CaCl_2$ in a desiccator for several days. It is highly soluble in water, ethanol, ether, acetonitrile, and acetone, but is insoluble in chloroform and hydrocarbons, and decomposes in basic aqueous solution (e.g., 0.5 M NaOH). IR (cm^{-1}): 1080 $\nu_{as}(P-O)$, 990 $\nu(W=O)$, 890 $\nu_{as}(W-O-W)$, 804 $\nu_{as}(W-O-W)$. Powder XRD (2θ in degrees, Cu Kα radiation, $\lambda = 1.5418\,Å$): 10.2, 14.5, 17.8,

---

[*]The yield can be increased to 50–80% with some difficulty as follows. After the acid addition step, the resulting suspension is cooled and transferred to a 250 mL separatory funnel. Diethyl ether (2 mL) is added and the mixture is shaken. Additional diethyl ether is added in 2 mL aliquots, followed by shaking each time, until 10 mL of ether has been added in total. At this point, if solid is present in addition to the liquid layers, water (20 mL) is added and the mixture is vigorously shaken until three liquid layers are present. The lowest layer is an oil and contains the heteropoly acid as an etherate complex, the middle layer is water, and the upper layer is diethyl ether. The lowest layer is separated and the other two layers are discarded. The lowest layer is washed twice with water (20 mL each time). Then, the lower layer is separated, placed in a Petri dish, and evaporated to dryness under a flow of air inside a fume hood. Crystals form in about 2 days. The product may be recrystallized from a minimum amount of water and dried at 150–200°C in air to remove acid and ether residues.

20.6, 23.1, 25.4, 29.5, 31.1, 34.4, 37.5. These peaks are characteristic of samples containing about six water molecules per Keggin unit; other peaks are seen for more highly hydrated samples. MAS ^{31}P NMR: δ −15.2 (s); the exact chemical shift can vary by ~0.3 ppm depending on the degree of hydration.

Characterization[9]

Characterize your sample by one or more of the following methods: elemental analysis (P and/or W), UV–vis spectroscopy (2×10^{-5} M in 0.1 M HCl), FTIR (KBr pellet or Nujol mull), MAS ^{31}P NMR (single-pulse excitation, 8.0 μs pulse, recycle delay of 10 s, spin rate of 5 kHz, and 64 acquisitions), or thermogravimetric analysis to determine the number of waters of hydration (heat the sample at 10°C min^{-1} from room temperature to 700°C under air or nitrogen atmosphere).

D. 12-MOLYBDOPHOSPHORIC ACID

$$12MoO_3 + H_3PO_4 + 12H_2O \rightarrow H_3[PMo_{12}O_{40}] \cdot 12H_2O$$

Procedure[5]

In a 250 mL beaker, molybdenum trioxide MoO_3 (10.0 g, 69.5 mmol) is mixed with H_2O (100 mL) and the resulting slurry is heated to boiling. Then, concentrated H_3PO_4 (12 drops, ~0.6 mL, 9 mmol) is added, and the mixture is heated to reflux for 30 min. The solution is vacuum filtered through a fine frit, and the solid is discarded. If the solution has a greenish color, 10% (v/v) H_2O_2 is added dropwise until the solution turns yellow. The solution is filtered again to remove any solid material. Then the filtered solution is evaporated in a rotary evaporator at 90°C until the volume is reduced to 5 mL, at which point a yellow powder or yellow crystals begin to form.* The solid is collected by filtration and air dried. The typical yield is about 80–85% based on Mo.†

Properties

$H_3[PMo_{12}O_{40}] \cdot 12H_2O$ is a yellow crystalline solid; a yellow-green color indicates the presence of lower-valent Mo^V or Mo^{IV} impurities. The compound is highly soluble in water, ethanol, ether, and acetone, but is insoluble in

---

*The checkers find that a yellow precipitate usually forms upon rotary evaporation, but in one instance no precipitate or crystals appeared. In this case, the concentrated solution was placed in a Petri dish, covered with a Kimwipe®, and left in the fume hood or dried over silica gel in a desiccator under a moderate vacuum. Crystals formed after 2 days.

†The checkers obtained yields of 50–60%.

chloroform and hydrocarbons, and decomposes in basic water solution (e.g., 0.5 M NaOH). FTIR (cm^{-1}): 1063 ν_{as}(P–O), 960 ν_{as}(Mo=O), 864 ν_{as}(Mo–O–Mo), 786 ν_{as}(Mo–O–Mo).

Characterization

Characterize your sample by one or more of the following techniques: elemental analysis (P and/or Mo), UV–vis spectroscopy (2×10^{-5} M in 0.1 M HCl), FTIR (KBr pellet or Nujol mull), or thermogravimetric analysis to determine the number of waters of hydration (heat the sample 10°C min^{-1} from room temperature to 700°C under air or nitrogen atmosphere).

Questions

1. For each experiment you did, what was your percentage yield, based on the limiting reagent? Explain the possible reasons for any disagreement between your yield and the expected yield.

2. Based on the FTIR spectra, discuss the strength of the bonds in the Keggin structure.

3. Compare the obtained powder XRD profile with the standard ICDD (International Centre for Diffraction Data) file for this compound.

4. Comment about the ^{31}P NMR chemical shift of the PO_4^{3-} unit in the Keggin unit compared to that of typical alkaline phosphates ($\sim \delta$ 0).

5. Explain the UV–vis spectrum obtained based on the possible electronic transitions.

6. There are isomers of the α-Keggin structure. Describe the structure of one of these isomers and how it differs from the structure of the α-isomer.

Notes for Instructors

These experiments are suitable for students working individually or in pairs. The preparations typically require the following amounts of laboratory time: ammonium 12-molybdophosphate, 1 h to get to the start of the final air-drying step; 12-tungstosilicic acid, 3.5 h to get to the start of the crystallization step; 12-tungstophosphoric acid, about 2 h to get the start of the crystallization step (about 3 h if the diethyl precipitation procedure is followed); 12-molybdophosphoric acid, about 2.5 h to get to the point of drying the solid.

References

1. N. N. Greenwood and A. Earnshaw, *Chemistry of the Elements*, Pergamon Press, Oxford, 1994, pp. 1175–1185.

2. M. T. Pope, *Heteropoly and Isopoly Oxometalates*, Springer, Berlin, 1983, pp. 1–32.
3. C. L. Hill, ed., *Chem. Rev.* **98**, 1 (1998).
4. T. Okuhara, N. Mizuno, and M. Misono, *Adv. Catal.* **41**, 113 (1996).
5. G. B. Kauffman and P. F. Vartanian, *J. Chem. Educ.* **47**, 212 (1970).
6. E. O. North, *Inorg. Synth.* **1**, 129 (1939).
7. A. Tézé and G. Hervé, *Inorg. Synth.* **27**, 93 (1990).
8. J. C. Bailar Jr., *Inorg. Synth.* **1**, 132 (1939).
9. I. V. Kozhevnikov, *Catalysts for Fine Chemical Synthesis: Catalysis by Polyoxometalates*, Wiley, Chichester, UK, 2002, p. 51.

40. QUADRUPLY METAL–METAL BONDED COMPLEXES OF RHENIUM(III)

Submitted by ANDREW W. MAVERICK,[*] ROBERT P. HAMMER[*], JOHN A. ARNOLD,[†] RICHARD A. WALTON,[‡] and ALFRED P. SATTELBERGER[§]
Checked by LINDSAY T. MCDONALD,[**] SARAH M. LANE[**], and STEVEN C. HAEFNER[**]

One of the landmark discoveries of modern inorganic chemistry was the isolation and structural characterization of transition metal complexes that possess multiple metal–metal bonds. The seminal discoveries of the early 1960s have evolved into a field of considerable breadth and complexity.[1] One particularly important milestone in the field was the synthesis and structural characterization of salts containing the octachlorodirhenate anion, $[Re_2Cl_8]^{2-}$. This anion was first prepared via the high-pressure reduction of the perrhenate ion, $[ReO_4]^-$, by molecular hydrogen in concentrated aqueous hydrochloric acid. This synthesis is interesting from a historical perspective, but is not a particularly convenient method of preparing the binuclear rhenium(III) complex as it requires rather specialized and expensive equipment. It did, however, provide crystals of $K_2Re_2Cl_8 \cdot 2H_2O$, whose X-ray structure revealed an eclipsed orientation of two quasi-square planar $ReCl_4^-$ units as shown below (**I**). The Re–Re separation of 2.24 Å is shorter than the Re–Re separation in rhenium metal.[2] The unusually short distance and eclipsed arrangement of chloride ligands led Cotton et al. to propose the existence of a quadruple metal–metal bond between the rhenium(III) centers.[3]

[*]Department of Chemistry, Louisiana State University, Baton Rouge, LA 70803.
[†]Department of Chemistry, University of California-Berkeley, Berkeley, CA 94920.
[‡]Department of Chemistry, Purdue University, West Lafayette, IN 47907.
[§]Energy Engineering and Systems Analysis Directorate, Argonne National Laboratory, Lemont, IL 60439.
[**]Department of Chemical Sciences, Bridgewater State University, Bridgewater, MA 02325.

I

The synthesis of a tetrabutylammonium salt, $[Bu_4N]_2[Re_2Cl_8]$, was first achieved by the hypophosphorous acid (H_3PO_2) reduction of $[ReO_4]^-$ in aqueous hydrochloric acid.[4] The reaction provided access to a salt of $[Re_2Cl_8]^{2-}$ that has good solubility properties in polar organic solvents and that facilitated numerous studies of the chemical and physical properties of the anion. The major disadvantage of this synthesis is the relatively low yield of the desired salt. A straightforward, high-yield synthesis of $[Bu_4N]_2[Re_2Cl_8]$ was developed some years later,[5] and is the subject of the first synthesis in this section. A solution of $[Bu_4N][ReO_4]$[6] in either benzoyl or *p*-toluoyl chloride (*p*-methylbenzoyl chloride) is heated to reflux for a predetermined period of time and then treated with a concentrated solution of a Bu_4N^+ salt in a mixture of ethanol and hydrochloric acid. The benzoyl chloride or its *para*-methyl-substituted analog reduces and chlorinates the rhenium centers and couples them through the agency of benzoate bridges. Subsequent treatment with hydrochloric acid protonates off the benzoate bridges and yields the desired product.

General Procedures

Reagent-grade benzoyl chloride, *p*-toluoyl chloride, tetrabutylammonium bromide, acetic acid, and acetic anhydride are purchased from Aldrich and used as received. Potassium perrhenate is purchased from Strem Chemicals, Inc. and used as received.

■   **Caution.**   *Proper laboratory attire, including safety glasses and lab coats, should be worn at all times. Rubber gloves should be used when weighing solids and dispensing liquids. Students should familiarize themselves with the hazards associated with the chemicals they use. Material safety data sheets (MSDS) are available for all of the chemicals used in the following syntheses. Direct skin contact with any of the reagents and products should be avoided. If skin contact does occur, wash the affected area with copious amounts of water.*

A. TETRABUTYLAMMONIUM PERRHENATE(VII)

$$KReO_4 + Bu_4NBr \rightarrow [Bu_4N][ReO_4] + KBr$$

Procedure[6]

In a 250 mL beaker equipped with a magnetic stir bar, $KReO_4$ (1.00 g, 3.4 mmol) is suspended in distilled water (50 mL) and the suspension is heated to 80°C, with stirring, until the solid dissolves. If necessary, some additional water may be added to dissolve the solid. Bu_4NBr (1.3 g, 4.0 mmol) is dissolved in the minimum volume of warm water, and this solution is added dropwise to the warm $KReO_4$ solution, with continued stirring. The resulting white suspension is allowed to cool to room temperature (an ice bath will accelerate this step). The solid is vacuum filtered (a water aspirator is adequate for the filtration) through a 60 mL medium-porosity sintered glass frit. The white solid is rinsed with a small amount of ice water and allowed to dry by pulling air through the solid for a few minutes. The solid is then scraped from the frit into a 100 mL one-neck (24/40) Schlenk flask. The flask is fitted with a 24/40 gas inlet tube and evacuated using a mechanical vacuum pump. A dry ice- or liquid nitrogen-cooled trap is used between the flask and the vacuum pump to catch any solvent vapors. Drying the $[Bu_4N][ReO_4]$ is important to the success of the $[Bu_4N]_2[Re_2Cl_8]$ synthesis, and the drying process can be accelerated by placing the Schlenk flask in a warm water bath (50–60°C). Yield: 1.5–1.6 g (90–95%).

Properties

Tetrabutylammonium perrhenate is a white solid that is stable in air. It is soluble in polar organic solvents such as acetonitrile.

B. BIS(TETRABUTYLAMMONIUM) OCTACHLORODIRHENATE(III)

$$2[Bu_4N][ReO_4] + 16PhCOCl \rightarrow [Bu_4N]_2[Re_2Cl_8] + 8(PhCO)_2O + 4Cl_2$$

Procedure[5]

■ **Caution.** *Benzoyl chloride and p-toluoyl chloride are lachrymators (i.e., the vapors cause your eyes to tear) and should be handled in a well-ventilated fume hood.*

A Schlenk flask is flushed with a slow stream of nitrogen and into the flask is put $[Bu_4N][ReO_4]$ (1.5 g, 3.0 mmol). Benzoyl chloride (bp 198°C)[*] (15 mL,

---

[*]*p*-Toluoyl chloride has a higher boiling point (225°C) than benzoyl chloride (198°C) and can be used to shorten the reaction time. It is only slightly more expensive than benzoyl chloride. As described above, the Schlenk flask containing $[Bu_4N][ReO_4]$ (1.5 g, 3.0 mmol) is flushed with a slow stream of nitrogen and *p*-toluoyl chloride (15 mL, 115 mmol) is syringed onto the solid. The flask is fitted with a reflux condenser and the initially pale yellow solution is heated to gentle reflux for 2 h. The workup of the intensely green to green-brown solution is similar to that described above.

130 mmol) is syringed into the flask, which is then fitted with a reflux condenser[*] and all other joints are stoppered. The initially pale yellow solution is heated to gentle reflux overnight (12–14 h).[†] The intensely dark green solution is allowed to cool to room temperature and then treated with a solution of Bu_4NBr (2.5 g, 7.8 mmol) in a mixture of ethanol (35 mL) and concentrated aqueous hydrochloric acid (5 mL). The mixture is then heated to reflux for an additional 1 h and the resulting solution is evaporated to roughly half its volume (passing a stream of N_2 through the apparatus during the evaporation step can help speed the process). A rotary evaporator can also be used for this volume reduction. The resulting feathery blue crystals are filtered in air through a 30 mL medium-porosity sintered glass frit, washed with ethanol (3×5 mL) and diethyl ether (10 mL), and dried under vacuum. Crude yield: 1.5 g (85%).

Into a 150 mL beaker is placed methanol (50 mL) and one drop of concentrated hydrochloric acid, and the solution is heated in air to boiling while stirring with a magnetic stir bar. The crude $[Bu_4N]_2[Re_2Cl_8]$ (about 1 g) is added to the stirred boiling solution. The solution is quickly filtered through a 30 mL medium-porosity sintered glass frit concentrated aqueous hydrochloric acid (50 mL). Some crystalline material precipitates at this stage and may be filtered off. Evaporation of the methanol provides an additional crop of dark blue crystals. A rotary evaporator, if available, is convenient for this step. Alternatively, the methanol can be evaporated using an N_2 stream after the flask is placed in a hot water bath. The total recovery is typically greater than 80%.

Properties

The blue salt is soluble in polar organic solvents such as acetonitrile, dichloromethane, and methanol. The solid is stable in air and can be stored indefinitely without special precautions. It is readily converted into other compounds with Re−Re quadruple bonds. The synthesis of one such derivative, $Re_2(O_2CCH_3)_4Cl_2$, is described below.[7]

Characterization

Record the UV–vis spectrum of $[Bu_4N]_2[Re_2Cl_8]$ in acetonitrile and calculate the extinction coefficient of the lowest-lying absorption band, the $\delta \rightarrow \delta^*$ transition, at about 680 nm. If possible, record the Raman spectrum of the recrystallized material and identify the Re−Re stretching frequency.

---

[*]The authors have not had problems with the hot benzoyl chloride leaching the grease out of the joint between the reaction flask and the condenser. If this does occur, the reaction can be set up again with the joint sealed either with a Teflon® sleeve or with Apiezon H grease.
[†]The checkers refluxed the solution for 20 h with no significant reduction in yield.

C. TETRA(ACETATO)DICHLORODIRHENIUM(III)

$$[Bu_4N]_2[Re_2Cl_8] + 4HO_2CCH_3 \rightarrow Re_2(O_2CCH_3)_4Cl_2 + 2[Bu_4N]Cl + 4HCl$$

Procedure[7]

A 100 mL Schlenk flask is charged with $[Bu_4N]_2[Re_2Cl_8]$ (0.5 g, 0.56 mmol), acetic acid (20 mL), and acetic anhydride (5 mL). The flask is evacuated and backfilled with nitrogen several times to remove dissolved oxygen. The solution is then heated to reflux under a slow stream of for 1.5 h volume (passing a stream of N_2 through the apparatus during this step can help sweep out the evolved HCl). The acetate complex precipitates as the reaction proceeds. The reaction mixture is allowed to cool to room temperature (an ice bath can be used to accelerate the cooling) and the product is filtered in air through a 15 mL medium-porosity sintered glass frit. The orange powder is washed with ethanol (2 × 10 mL) and diethyl ether (10 mL) and dried in vacuum. Yield: 0.34 g (90%).

Properties

Orange $Re_2(O_2CCH_3)_4Cl_2$ is stable in air and virtually insoluble in common organic solvents. It reacts with boiling concentrated aqueous hydrochloric acid to regenerate the octachlorodirhenate anion, which can be recovered as the tetrabutylammonium, potassium, or cesium salt.

Characterization

Record the IR spectrum of $Re_2(O_2CCH_3)_4Cl_2$ and identify the symmetric and antisymmetric O−C−O stretches. If possible, record the Raman spectrum of $Re_2(O_2CCH_3)_4Cl_2$ and identify the Re−Re stretching frequency. Compare this frequency with that obtained for $[Bu_4N]_2[Re_2Cl_8]$.

Questions

1. Which orbitals are involved in forming the metal–metal bonds in $[Re_2Cl_8]^{2-}$? Sketch them out and draw a qualitative molecular orbital diagram for the anion. Label the σ, π, and δ bonding orbitals and their antibonding counterparts. Based on your MO diagram, explain why $[Re_2Cl_8]^{2-}$ possesses a metal–metal quadruple bond.

2. The intensely green intermediate that is initially generated in Section 40.B has the formula $Re_2(O_2CPh)_2Cl_4$. Write balanced equations for the formation of this intermediate from $[Bu_4N][ReO_4]$, and for the conversion of this

intermediate to the final product. Which step (the first, the second, or both) is a redox reaction?

3. $[Re_2Cl_8]^{2-}$ can be reduced to $[Re_2Cl_8]^{3-}$ in solution. Based on your MO diagram in question 1, predict the metal-metal bond order of this species.

Notes for Instructors

The syntheses described above should provide students with an appreciation of the synthetic chemistry and some of the spectroscopic properties of the first quadruply metal–metal bonded complexes that were isolated and that shaped the future of the field. Here, we describe some additional experiments that might be included as part of the laboratory work.

If rhenium metal powder is available, it is readily converted to perrhenic acid, $HReO_4$(aq), and subsequently to $[Bu_4N][ReO_4]$. A procedure to convert rhenium metal powder to perrhenate salts is described in an earlier volume in this series.[8] Typically, rhenium metal powder (3 g, 16 mmol, Strem) is covered with distilled water (10 mL) in a 250 mL Erlenmeyer flask and treated dropwise with 30% hydrogen peroxide (10 mL). Gas evolution is immediate and cooling of the flask may be required if the peroxide is added too quickly. After the addition is complete, the solution is diluted to 50 mL with water and boiled for several minutes to destroy any remaining peroxide. The warm solution is filtered, if necessary, to remove any insoluble material. The solution is allowed to cool and treated with a 40% aqueous solution of $[Bu_4N]OH$ (Aldrich) to neutral pH (a pH meter can be used, but the disposable pH paper works fine). The voluminous white precipitate of $[Bu_4N][ReO_4]$ is filtered off as described above, washed with ice water, and dried in vacuum. Yield: 7.1–7.5 g (90–95%).

The decision of which method to use to convert $[Bu_4N][ReO_4]$ to $[Bu_4N]_2$ $[Re_2Cl_8]$ is left to the discretion of the instructor. If dry $[Bu_4N][ReO_4]$ is available and p-toluoyl chloride is used as a reagent, the synthesis, purification, and at least some of the characterization of $[Bu_4N]_2[Re_2Cl_8]$ can be accomplished in one 4 h laboratory period.

In the early days of the field, there was some confusion regarding the presence and extent of metal–metal bonding in complexes containing bridging ligands. The ready interconversion of $Re_2(O_2CCH_3)_4Cl_2$ and $[Re_2Cl_8]^{2-}$ quickly ended the debate for this pair of rhenium(III) compounds. Orange $Re_2(O_2CCH_3)_4Cl_2$ can be converted back to blue $[Re_2Cl_8]^{2-}$ by refluxing suspensions of acetate in concentrated hydrochloric acid containing $[Bu_4N]Cl$.[9]

References

1. F. A. Cotton, C. A. Murillo, and R. A. Walton, eds., *Multiple Bonds Between Metal Atoms*, 3rd ed., Springer, New York, 2005.

2. F. A. Cotton and C. B. Harris, *Inorg. Chem.* **4**, 330 (1964).
3. (a) F. A. Cotton, N. F. Curtis, C. B. Harris, B. F. G. Johnson, S. J. Lippard, J. T. Mague, W. R. Robinson, and J. S. Wood, *Science* **145**, 1305 (1964); (b) F. A. Cotton, *Inorg. Chem.* **4**, 334 (1964).
4. F. A. Cotton, N. F. Curtis, and W. R. Robinson, *Inorg. Chem.* **4**, 1696 (1964).
5. T. J. Barder and R. A. Walton, *Inorg. Chem.* **21**, 2510 (1982); *Inorg. Synth.* **28**, 332 (1990).
6. J. R. Dilworth, W. Hussain, A. J. Hutson, C. J. Jones, and F. S. McQuillan, *Inorg. Synth.* **31**, 257 (2009).
7. F. A. Cotton, C. Oldham, and W. R. Robinson, *Inorg. Chem.* **5**, 1798 (1966).
8. N. P. Johnson, C. J. L. Lock, and G. Wilkinson, *Inorg. Synth.* **9**, 145 (1970).
9. H. D. Glicksman and R. A. Walton, *Inorg. Chem.* **17**, 3197 (1978).

41. BIS[BIS(TRIPHENYLPHOSPHORANYLIDENE)-AMMONIUM] UNDECACARBONYLTRIFERRATE(2 −)

Submitted by STANTON CHING[*]
Checked by DARIO PRIETO-CENTURION[†], JOHN L. KIAPPES JR.[†], and KENTON H. WHITMIRE[†]

$$Fe_3(CO)_{12} + 4KOH \rightarrow K_2[Fe_3(CO)_{11}] + 2H_2O + K_2CO_3$$

$$K_2[Fe_3(CO)_{11}] + 2[PPN]Cl \rightarrow [PPN]_2[Fe_3(CO)_{11}] + 2KCl$$

A technical challenge often faced in the syntheses of inorganic compounds, and especially organometallic species, is the difficulty of carrying out synthetic procedures in which the starting material, intermediates, and products are all susceptible to reactions with atmospheric oxygen and/or moisture. Although many organic reactions are performed in the absence of air (e.g., those with organo-lithium and Grignard reagents), the main reactants and products are almost always unreactive toward oxygen and water. More sophisticated inert atmosphere manipulations are needed when the reactants and products are air sensitive.

Transition metal carbonyls are a fundamentally important class of organo-metallic complexes that receive significant attention in all standard inorganic chemistry textbooks. Their structure and bonding, chemical reactivity, and physical and spectroscopic properties are used to introduce a wide variety of inorganic topics. The anionic cluster $[Fe_3(CO)_{11}]^{2-}$ (Fig. 1) and its precursor $Fe_3(CO)_{12}$ (Fig. 2) are excellent examples of metal carbonyl compounds that are useful for instruction in both the classroom and laboratory. They illustrate concepts such as metal–metal bonding, metal–ligand back-donation, variable ligand coordination

[*]Department of Chemistry, Connecticut College, New London, CT 06320.
[†]Department of Chemistry, Rice University, 6100 Main Street, Houston, TX 77005-1892.

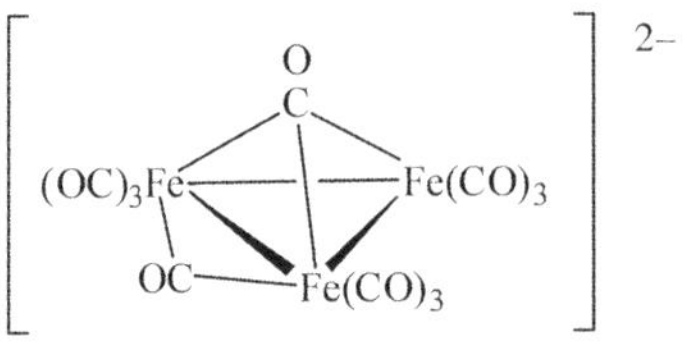

Figure 1. The structure of $[Fe_3(CO)_{11}]^{2-}$.

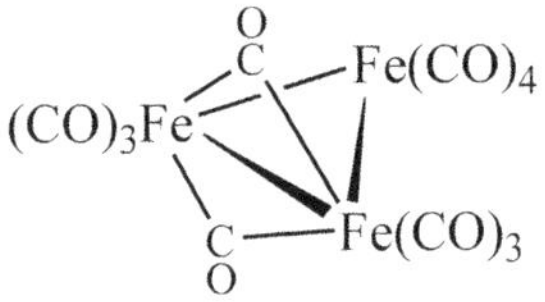

Figure 2. The structure of $Fe_3(CO)_{12}$.

modes, ligand fluxionality, electron counting, and symmetry. The synthesis of $[Fe_3(CO)_{11}]^{2-}$ from $Fe_3(CO)_{12}$ is also an excellent procedure for teaching the manipulation of air-sensitive compounds.

The bis(triphenylphosphoranylidene)ammonium salt $[PPN]_2[Fe_3(CO)_{11}]$ is synthesized in two steps using a simplified variation of the procedure reported by Hodali and Shriver.[1] The first reaction involves attack of KOH on the $Fe_3(CO)_{12}$ starting material, which yields $K_2[Fe_3(CO)_{11}]$. The second reaction involves simple metathesis of K^+ with the large PPN^+ cation.

General Procedures

Air-free glassware with 14/20 standard tapered joints are used throughout. Both $Fe_3(CO)_{12}$ and $[PPN]_2[Fe_3(CO)_{11}]$ can be handled briefly in air as solids if a glove box is not available, but must be stored under nitrogen. The reaction can easily be scaled up or down.

Procedure

■ **Caution.** *Transition metal carbonyl compounds can evolve carbon monoxide, a toxic and flammable gas, when heated or exposed to acid.*

In a glove box, combine $Fe_3(CO)_{12}$ (0.50 g, 0.99 mmol) and KOH (0.56 g, 10 mmol), along with a magnetic stir bar, in a 200 mL Schlenk flask capped with a rubber septum. Attach the flask to a Schlenk line and add deaerated methanol (20 mL) via syringe through the septum while inserting a separate needle to maintain a nitrogen purge. Stir the reaction mixture. A color change from dark green to wine red is observed within a few minutes. Continue stirring for 30 min. While waiting, prepare a solution of bis(triphenylphosphoranylidene)ammonium chloride, [PPN]Cl (1.2 g, 2.1 mmol), dissolved in deaerated methanol (15 mL).

Attach to the reaction flask a nitrogen-purged, medium-porosity frit that has been coupled with a 100 mL Schlenk flask. Apply a slight vacuum to the empty side of the apparatus to provide a good seal on all the joints. Invert the apparatus to filter the reaction mixture.

Add the solution of [PPN]Cl via syringe through the stopcock of the receiving flask (use a straight-tip syringe needle) to precipitate wine red microcrystalline $[PPN]_2[Fe_3(CO)_{11}]$. The crystals shimmer vividly when displayed under a flashlight beam. If crystals are not evident after a few minutes, agitate the flask to induce crystallization.

Isolate the $[PPN]_2[Fe_3(CO)_{11}]$ by filtering back through the same frit. At this stage, the solid should remain in the air-free glassware, but can be washed with solvents that are not deaerated. Wash thoroughly with three 10 mL portions of water, followed by two washings with methanol (10 mL each) and two washings with ether (10 mL each). Remove the 200 mL Schlenk flask containing the washings and replace it with either a cap or a small Schlenk flask. Dry the solid under vacuum for 30 min. Finally, weigh the product in the glove box. Typical yields from undergraduate student syntheses are in the range from 50 to 70%, although higher yields are possible.

Properties

$[PPN]_2[Fe_3(CO)_{11}]$ is a dark red-brown crystalline material that is insoluble in methanol but soluble in dichloromethane. It is air sensitive, but can be handled in air briefly without decomposition. IR (KBr, cm^{-1}): ν(CO) 2013 (w), 1998 (w), 1937 (s), 1902 (s), 1868 (w, sh), 1665 (m), 1636 (m, sh).

Questions

1. Rationalize the differing solubility of $K_2[Fe_3(CO)_{11}]$ and $[PPN]_2[Fe_3(CO)_{11}]$ in methanol.

2. What purpose is served by washing with water, methanol, and ether? Does the order of the washings matter?

3. The initial reaction between $Fe_3(CO)_{12}$ and KOH is an oxidation–reduction reaction. What is being oxidized and reduced? Determine the formal oxidation states of all elements that undergo oxidation and reduction.

4. Identify the IR stretching frequencies associated with the terminal, doubly bridging, and triply bridging carbonyl ligands.

5. Explain why the CO stretching frequencies seen for $[Fe_3(CO)_{11}]^{2-}$ are lower than those for $Fe_3(CO)_{12}$.

Notes for Instructors

The synthesis of $[PPN]_2[Fe_3(CO)_{11}]$, where PPN = bis(triphenylphosphoranyli-dene)ammonium, is straightforward and well suited for instruction on inert atmosphere techniques with glove box, Schlenk line, and syringe techniques.[2] The synthesis of $[PPN]_2[Fe_3(CO)_{11}]$ is chosen because both the $Fe_3(CO)_{12}$ reactant and $[Fe_3(CO)_{11}]^{2-}$ product are stable toward water and just mildly reactive with oxygen. The procedure is therefore very forgiving of mistakes that inevitably arise from inexperience with air-free techniques. The synthesis can also be accommodated within a 4 h laboratory period and uses commercially available glassware. Students find the synthesis to be challenging, but also interesting and enjoyable.

References

1. H. A. Hodali and D. F. Shriver, *Inorg. Synth.* **20**, 222 (1980).
2. D. F. Shriver and M. A. Drezdzon, *The Manipulation of Air-Sensitive Compounds*, 2nd ed., Wiley, New York, 1986.

42. ACETYLIDE COMPLEXES OF RUTHENIUM

Submitted by FREDERICK R. LEMKE[*†]
Checked by TRACEY A. HITT,[‡] JENNIFER L. STEELE,[‡] and GREGORY S. GIROLAMI[‡]

Acetylenes can react with metal complexes in a number of ways. Internal acetylenes usually coordinate to metal centers in a η^2-fashion, in which the C≡C triple bond donates electrons to the metal center, and both carbon atoms form metal-carbon bonds. Terminal acetylenes can also coordinate to metal centers in this fashion, but can react in other ways as well. In particular, because the acetylenic hydrogen atom is relatively acidic for a hydrocarbon, the C−H bond of terminal acetylenes can also add oxidatively to metal centers to form metal acetylide complexes. In some cases, however, a third kind of reaction is seen, in which the terminal acetylene rearranges to a vinylidene complex, M=C=CHR.

[*]Science and Math Division, Mott Community College, 1401 East Court Street, Flint, MI 48503.
[†]Deceased December 20, 2013.
[‡]School of Chemical Sciences, University of Illinois at Urbana-Champaign, Urbana, IL 61801.

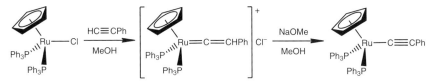

Scheme 1. Reaction sequence to form the ruthenium phenylacetylide complex.

An example of this rearrangement is the reaction of the ruthenium cyclopenta-dienyl complex $(\eta^5\text{-}C_5H_5)RuCl(PPh_3)_2$ with phenylacetylene to give $[(\eta^5\text{-}C_5H_5)Ru(=C=CHPh)(PPh_3)_2]^+$.

Closely related to this vinylidene complex is the ruthenium acetylide compound $(\eta^5\text{-}C_5H_5)Ru(C\equiv CPh)(PPh_3)_2$, which was originally prepared by treatment of the chloro complex $(\eta^5\text{-}C_5H_5)RuCl(PPh_3)_2$ with copper phenylacetylide, $CuC\equiv CPh$.[1] Later, the acetylide complex was prepared by accident during attempts to purify the vinylidene complex $[(\eta^5\text{-}C_5H_5)Ru(=C=CHPh)(PPh_3)_2]^+$ by column chromatog-raphy on alumina.[2] The alumina acted as a base and deprotonated the vinylidene ligand, and subsequently it was found that the vinylidene complexes could easily be deprotonated by addition of many different kinds of bases such as sodium methoxide, methyllithium, or sodium bicarbonate. The deprotonation of the monosubstituted vinylidene complexes to the corresponding acetylide complexes proceeds in high yield, and consequently this reaction sequence is the synthetic method used in the following preparation of the acetylide complexes (Scheme 1). The experimental procedure is a scaled-down and simplified version of a prepara-tion previously published in *Inorganic Syntheses*.[3]

The vinylidene/acetylido interconversions carried out in this experiment are similar to those seen for other metal complexes[4,5] and are relevant to the known ability of $(\eta^5\text{-}C_5H_5)RuCl(PPh_3)_2$ to catalyze the dimerization of phenylacetylene to a mixture of *Z*- and *E*-1,4-diphenyl-1-buten-3-yne.[6] The chemistry of ruthenium cyclopentadienyl complexes has been reviewed.[7]

A. (CYCLOPENTADIENYL)BIS(TRIPHENYLPHOSPHINE)-CHLORORUTHENIUM(II)

$$RuCl_3 \cdot 3H_2O + C_5H_6 + \tfrac{5}{2}PPh_3 \rightarrow (\eta^5\text{-}C_5H_5)RuCl(PPh_3)_2$$

$$+ \tfrac{1}{2}O=PPh_3 + 2HCl + \tfrac{5}{2}H_2O$$

Procedure[3]

The reaction is carried out in a 500 mL two-necked round-bottomed flask equipped with a magnetic stir bar, dropping funnel, and a reflux condenser topped with a nitrogen bypass. The apparatus is purged with nitrogen and charged with triphenylphosphine (8.4 g, 32 mmol) and degassed ethanol (300 mL).[*] The triphenylphosphine is dissolved in the ethanol by heating to reflux. Hydrated ruthenium trichloride (2.1 g, 8 mmol) is dissolved in boiling ethanol (40 mL), and then allowed to cool. Freshly distilled[†] cyclopentadiene (4 mL, 50 mmol) is added to the ruthenium trichloride solution, and the mixture is transferred to the dropping funnel. The dark brown solution is then added to the refluxing triphenylphosphine solution over 10 min. The reaction mixture is dark brown in color, which after 1 h lightens to a dark red-orange with an orange precipitate.[‡] The solution is now safe to expose to air. The solution is cooled to room temperature to complete the precipitation. The solid is collected by filtration, washed with ethanol (4×10 mL) and hexanes (4×10 mL), and dried in vacuum. Yield: 5.14 g (89%).

Anal. Calcd. for $C_{41}H_{35}ClP_2Ru$: C, 67.7; H, 4.8. Found: C, 67.8; H 5.3.

Properties

$(\eta^5\text{-}C_5H_5)Ru(PPh_3)_2Cl$ is an orange solid that is stable to air and moisture. It is sparingly soluble in saturated hydrocarbons and alcohols, but is soluble in aromatic hydrocarbons, chlorocarbons, acetonitrile, and acetone. ^1H NMR ($CDCl_3$): δ 7.36 (br, $J_{HH} = 7.5$ Hz, o-CH), 7.23 (t, $J_{HH} = 7.5$ Hz, p-CH), 7.14 (t, $J_{HH} = 7.5$ Hz, m-CH), 4.08 (s, C_5H_5). $^{31}P\{^1H\}$ NMR ($CDCl_3$): δ 38.6 (s). Melting point: 130–133°C.

Characterization

Record the yield. Characterize the product by IR and ^1H NMR spectroscopies and melting point.

B. (CYCLOPENTADIENYL)BIS(TRIPHENYLPHOSPHINE)- (PHENYLACETYLIDO)RUTHENIUM(II)

$$(\eta^5\text{-}C_5H_5)RuCl(PPh_3)_2 + HC\equiv CPh \rightarrow [(\eta^5\text{-}C_5H_5)Ru(=C=CHPh)(PPh_3)_2]Cl$$

$$[(\eta^5\text{-}C_5H_5)Ru(=C=CHPh)(PPh_3)_2]Cl + NaOCH_3 \rightarrow$$

$$(\eta^5\text{-}C_5H_5)Ru(C\equiv CPh)(PPh_3)_2 + NaCl + HOCH_3$$

---

[*] The checkers note that the ratio of 150 mL of ethanol per gram of hydrated $RuCl_3$ is important to get a good yield in the final precipitation step: for example, the yield drops considerably if 200 mL g^{-1} is used.
[†] The checkers cracked the dicyclopentadiene to cyclopentadiene by heating it in a flask immersed in a silicone oil bath kept at 200°C and distilling through a 10 in. tall Vigreux column at atmospheric pressure.
[‡] The checkers did not see the orange precipitate until after the solution is allowed to cool. They allowed the solution to stand overnight before filtering it.

Procedure[3]

$(\eta^5\text{-}C_5H_5)RuCl(PPh_3)_2$ (1.45 g, 2 mmol) is suspended in methanol (100 mL) in a 200 mL two-necked flask equipped with a reflux condenser and a nitrogen bypass. Phenylacetylene (0.33 mL, 3 mmol) is added dropwise to the suspension, which is kept strictly under nitrogen, and then the mixture is heated to reflux temperature for 30 min. The $(\eta^5\text{-}C_5H_5)RuCl(PPh_3)_2$ gradually dissolves to form a dark red solution. After the solution has cooled, sodium metal (0.2 g, 9 mmol) is added in small pieces whereupon a yellow crystalline precipitate forms as the sodium dissolves. The solid is filtered from the mother liquor, washed with methanol (4 × 3 mL) and hexanes (4 × 3 mL), and dried in vacuum. Yield: 1.37 g (86%).

Anal. Calcd. for $C_{49}H_{40}P_2Ru$: C, 74.3; H, 5.1; P, 7.8. Found: C, 73.7; H, 5.4; P, 7.6.

Properties

$(\eta^5\text{-}C_5H_5)Ru(C{\equiv}CPh)(PPh_3)_2$ is a yellow solid that is only mildly sensitive to air and moisture. It can be recrystallized from $CH_2Cl_2/MeOH$. IR: 2065 cm^{-1} (sharp, C≡C stretch). ^1H NMR (CDCl$_3$): δ 4.30 (s, C$_5$H$_5$), 7.06 (m, *m*-CH and *p*-CH), 7.52 (br, *o*-CH). $^{31}P\{^1H\}$ NMR (CDCl$_3$): δ 51.0 (s). $^{13}C\{^1H\}$ NMR (CDCl$_3$): δ 85.2 (s, C$_5$H$_5$), 123.5 (s, CPh), 127.3–140.1 (Ph).* Mass spectrum: *m/e* 792 (parent ion). Melting point: 202–205°C (dec.).

Many related compounds with other acetylenes can be made, including $(\eta^5\text{-}C_5H_5)Ru(C{\equiv}CR)(PPh_3)_2$, where R = *p*-tolyl, *n*-pentyl, *n*-hexyl, and so on.

Characterization

Record the yield. Characterize the product by IR and ^1H NMR spectroscopies and melting point.

Questions

1. Cyclopentadiene, which is a reagent in the preparation of $(\eta^5\text{-}C_5H_5)RuCl(PPh_3)_2$, is not commercially available. Why not, and how is it obtained for this experiment?

2. Propose a mechanism for the reaction of $(\eta^5\text{-}C_5H_5)RuCl(PPh_3)_2$ with HC≡CPh to give the vinylidene compound $[(\eta^5\text{-}C_5H_5)Ru(=C=CHPh)(PPh_3)_2]^+$.

---

*Although the ^{13}C NMR resonance for the ruthenium-bound carbon of the phenylacetylide ligand has not been reported, in similar compounds this resonance appears near δ 230.

3. In the IR spectrum, the C≡C stretching frequencies of phenylacetylene and $(\eta^5\text{-}C_5H_5)Ru(C\equiv CPh)(PPh_3)_2$ are 2111 and 2068 cm^{-1}, respectively. Explain the difference in terms of the bonding between Ru and the acetylide group. It may be of interest to know that the C≡C stretch in lithium phenylacetylide, LiC≡CPh, appears at 2030 cm^{-1}.

4. What would be the product of adding 1 equiv of HCl to $(\eta^5\text{-}C_5H_5)Ru\text{-}(C\equiv CPh)(PPh_3)_2$?

5. Which compound should be easier to oxidize by one electron, $(\eta^5\text{-}C_5H_5)\text{-}RuCl(PPh_3)_2$ or $(\eta^5\text{-}C_5H_5)Ru(C\equiv CPh)(PPh_3)_2$?

Notes for Instructors

Cyclopentadiene is prepared by the thermal cracking of its Diels–Alder dimer, dicyclopentadiene (3a,4,7,7a-tetrahydro-4,7-methanoindene). It can be prepared by placing dicyclopentadiene (50 mL) into a 500 mL two-necked round-bottomed flask fitted with a 15 cm Vigreux column topped by a water-cooled distillation head. The flask should have at least 10 times the volume of dicyclopentadiene, so as to contain the large amounts of foam generated during heating, which can easily travel up the Vigreux column and contaminate the desired cyclopentadiene product. A thermometer is placed in the thermometer well of the distillation head, and the distillation head is also attached to a 100 mL receiving flask. The distillation is conducted under nitrogen with a heating bath at 130°C. The cyclopentadiene distills over at a head temperature of 40°C. Cyclopentadiene is stable for several hours at room temperature, but it should be used quickly because eventually it redimerizes. Dimerized cyclopentadiene can be saved and reused in the cracking apparatus.

As a special project or extension of this experiment, the students could isolate and characterize the intermediate vinylidene complex, either by working up the reaction of $(\eta^5\text{-}C_5H_5)RuCl(PPh_3)_2$ with phenylacetylene without adding sodium methoxide or by protonation of the isolated phenylacetylide complex with NH_4PF_6.[3]

Another special project would be to employ different *para*-substituted phenylacetylenes ($HC\equiv CC_6H_4X$, where $X = H$, F, Me, OMe) and evaluate how the π-bonding between the ruthenium fragment and the acetylide depends on the electronic properties of the *para*-substituent.[8]

References

1. O. M. Abu Salah and M. I. Bruce, *J. Chem. Soc., Dalton Trans.* 2311 (1975).
2. M. I. Bruce and R. C. Wallis, *Aust. J. Chem.* **32**, 1471 (1979).
3. M. I. Bruce, C. Hameister, A. G. Swincer, and R. C. Wallis, *Inorg. Synth.* **21**, 78 (1982).
4. A. Davison and J. P. Selegue, *J. Am. Chem. Soc.* **100**, 7763 (1978).

5. A. Davison and J. P. Selegue, *J. Am. Chem. Soc.* **102**, 2455 (1980).
6. R. U. Kirss, R. D. Ernst, and A. M. Arif, *J. Organomet. Chem.* **689**, 419 (2004).
7. M. O. Albers, D. J. Robinson, and E. Singleton, *Coord. Chem. Rev.* **79**, 1 (1987).
8. F. Paul, B. G. Ellis, M. I. Bruce, L. Toupet, T. Roisnel, K. Costuas, J. F. Halet, C. Lapinte, *Organometallics* **25**, 649 (2006).

43. *N,N'*-BIS(MERCAPTOETHYL)-1,4-DIAZACYCLOHEPTANE (H₂BME-DACH) AND ITS NICKEL COMPLEX: A MODEL FOR BIOINORGANIC CHEMISTRY

Submitted by MELISSA GOLDEN[*†] and MARCETTA DARENSBOURG[*]
Checked by JENNIFER IRWIN[‡] and BRIAN FROST[†]

The impressive sulfur-based reactivity of square planar nickel complexes containing tetradentate N_2S_2 ligands has been known for many years. Interest has recently resurfaced because of the discovery of similar donor sites in metalloproteins that bind nickel, iron, and cobalt.[1–3] The *N,N'*-bis(mercaptoethyl)-1,5-diazacyclooctane ligand H_2(BME-DACO) and its nickel complex have been particularly useful in establishing the scope of S-based reactivity with electrophiles as displayed in the reaction summary shown in Scheme 1.[4–9] The fundamental features of this reactivity include templated macrocycle production,[4] S-oxygenation as contrasted to oxidation,[5] Lewis acid/base adduct formation,[6] metal-ion capture, and the synthesis of heterodi- and polymetallic complexes.[7]

Whereas the diazacyclooctane (DACO) framework leads to almost perfect 90° angles in the N_2S_2 cavity, its synthesis is tedious and is of low yield.[10] In contrast, the seven-membered ring analog diazacycloheptane (DACH) is commercially available and reacts almost as readily as DACO with ethylene sulfide to produce the bis(mercaptoethyl) derivative, H_2(BME-DACH). Due to the shorter $N \cdots N$ distance in the DACH ring, the $N-Ni-N$ angle is slightly pinched, and the $S-Ni-S$ angle is opened by 5° from the regular ~90° angles of Ni(BME-DACO) (Fig. 1). This structural mismatch results in a few but important differences in the types of reactivity demonstrated for Ni(BME-DACO) versus Ni(BME-DACH).

The highest occupied molecular orbital (HOMO) of Ni(BME-DACH) and Ni(BME-DACO) is sulfur based, and the orbital overlap of the sulfur lone pairs with the filled d-orbitals of nickel results in an antibonding π-interaction.[5] This

[*]Department of Chemistry, Texas A&M University, College Station, TX 77843.
[†]Department of Chemistry, California State University, Fresno, CA 93740.
[‡]Department of Chemistry, University of Nevada, Reno, NV 89557.

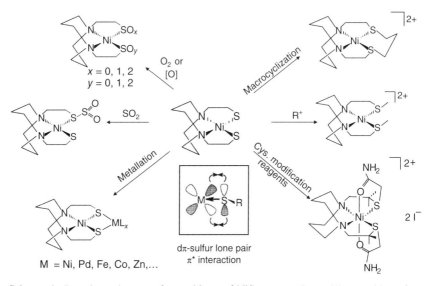

Scheme 1. Reaction scheme to form adducts of NiS_2N_2 complexes, illustrated here for Ni(BME-DACO).

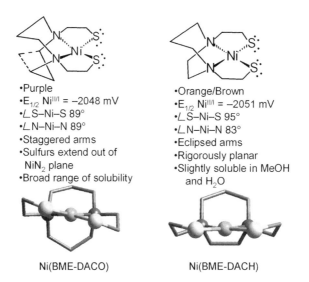

- Purple
- $E_{1/2}$ $Ni^{II/I}$ = −2048 mV
- ∠S–Ni–S 89°
- ∠N–Ni–N 89°
- Staggered arms
- Sulfurs extend out of NiN$_2$ plane
- Broad range of solubility

- Orange/Brown
- $E_{1/2}$ $Ni^{II/I}$ = −2051 mV
- ∠S–Ni–S 95°
- ∠N–Ni–N 83°
- Eclipsed arms
- Rigorously planar
- Slightly soluble in MeOH and H$_2$O

Ni(BME-DACO)   Ni(BME-DACH)

Figure 1. Comparison of Ni(BME-DACO) and Ni(BME-DACH).

antibonding interaction makes both the sulfur atoms and the nickel atom highly reactive. For example, the sulfur atoms are susceptible to attack by Lewis acids.

Here we describe the synthesis of the H_2(BME-DACH) ligand and its nickel complex. The reactions of Ni(BME-DACH) are particularly amenable to independent design of experiments by undergraduates in the inorganic synthesis laboratory. The following options may be used to explore the reactivity of Ni (BME-DACH): (1) uptake of SO_2; (2) metallation reactions with Ni, Cu, or Zn; (3) oxygenation; (4) macrocyclizations; and (5) alkylations and acetylations. Specific instructions for two follow-up studies of reactivity, uptake of SO_2 and acetylation, are given below; examples of the other types of reactivity are available in the primary research literature, although most of it describes the reactivity of the eight-membered ring analog Ni(BME-DACO).

Sulfur dioxide can bind to a metal center as a Lewis base in at least five different coordination modes, and it can also act as a Lewis acid by interacting with an electron-rich metal or with available lone pairs such as those on a thiolate, SR^-, ligand.[11] Examples of complexes capable of forming ligand-based SO_2 adducts are Ni(BME-DACO), Ni($Ph_2PCH_2CH_2S$)$_2$, and Pd(BME-DACO). Sulfur dioxide adducts of these complexes have two ν(SO) stretching frequencies of 1325–1210 and 1145–1060 cm^{-1}.[6] A common characteristic of sulfur-bound SO_2 adducts is that the binding is reversible.[11] Gently heating the solid causes dissociation of the SO_2 ligand and conversion to the precursor thiolate complex, and reformation of the SO_2 adduct occurs upon exposure of the solid to SO_2 gas.[6] The SO_2 ligand may also be removed from the adduct by purging solutions with an inert gas. These adducts are ideal to study by thermogravimetric analysis.

Besides understanding the fundamental chemistry of small-molecule gaseous interactions, a reason to study the reactivity of transition metal thiolates with SO_2 is the possibility of developing a chemical sensor for SO_2. There are several ways in which chemical sensors can respond to the presence of a specific analyte, but the most practical way is a resulting color change that is easily perceptible to the naked eye.[12] In this experiment, students will use color changes to detect SO_2 uptake by Ni(BME-DACH) at levels as low as 100 ppm of SO_2 in air.

Also described below is another example of reactivity at sulfur: the acylation of Ni(BME-DACH).

General Procedures

All materials are commercially available and used as received.

■ **Caution.** *Proper laboratory attire, including eyewear, gloves, and protective clothing, should be worn at all times, and as with any laboratory experiment, responsible practices for the disposal of chemical waste should be followed. Students should familiarize themselves with the hazards associated with* *the chemicals they use; a search of the Internet will find applicable material safety data sheets (MSDS). Nickel complexes are toxic, and direct contact with skin should be avoided.*

A. [*N,N'*-BIS(2-MERCAPTIDOETHYL)-1,4-DIAZACYCLOHEPTANE]-NICKEL(II)

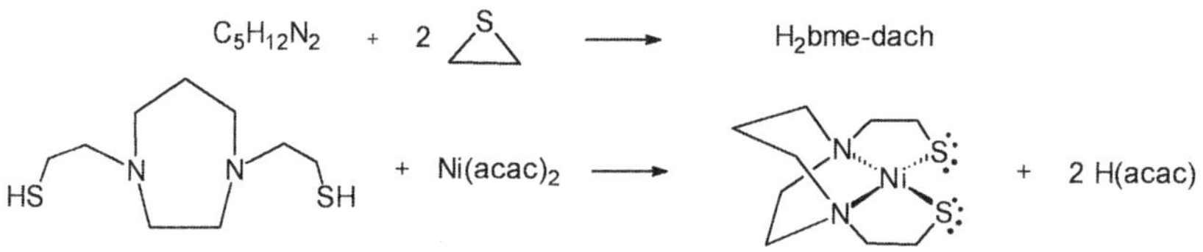

$$C_5H_{12}N_2 \ + \ 2 \ \overset{S}{\triangle} \ \longrightarrow \ H_2bme\text{-}dach$$

Procedure[13]

■ **Caution.** *Ethylene sulfide is toxic and should be handled only in a fume hood, while wearing the proper protective garments and gloves.*

A 250 mL long-necked Schlenk flask containing a magnetic stir bar is connected to a Schlenk line containing an overpressure valve, and the flask is flushed with N_2.[*] The flask is charged with 1,4-diazacycloheptane[†] (5.24 g, 52.4 mmol) and dry benzene (100 mL). The solution is heated to 65°C under N_2. Ethylene sulfide (8.0 mL, 130 mmol) is added by syringe, and the solution is stirred for 48 h at 65°C. If a white precipitate develops, the cooled solution should be anaerobically filtered through Celite® in a glass-fritted funnel. The solvent is removed in vacuum while maintaining the temperature between 60 and 65°C; continued heating under vacuum removes any unreacted DACH. The resulting pale yellow to colorless liquid is dissolved in dry toluene (50 mL) and the solution is transferred to a 500 mL Schlenk flask. To this solution is slowly added (dropwise if possible)[‡] a

---

[*]Because the reaction temperature will be maintained well below the boiling point of the solvent, benzene (80°C), a reflux condenser is optional.

[†]1,4-Diazacycloheptane (also called homopiperazine) is a pale yellow to colorless crystalline compound that should be stored and transferred under nitrogen. Although DACH may be handled briefly in air, this results in significantly decreased yields of H_2(BME-DACH) and the precipitation of an undesired by-product, the white solid that is referred to in the procedure.

[‡]The checkers note that this solution must be slowly added to the ligand (never vice versa): dropwise addition at a rate of 1 or 2 mL/5 min is ideal.

solution of anhydrous[*] nickel(II) acetylacetonate, $Ni(acac)_2$ (13.5 g, 52.4 mmol), completely dissolved in dry toluene (200–300 mL). A dark tan-colored precipitate forms. Upon complete addition of the $Ni(acac)_2$, the mixture is stirred for an additional 1 h.

The crude Ni(BME-DACH) precipitate is collected on a Büchner funnel (anaerobic conditions are not required) and is then purified by recrystallization. The crude product is dissolved in a hot (60°C) mixture of 50:50 $MeOH/H_2O$ (generally about 800–1000 mL of solvent is required) and the resulting solution is filtered in air to remove any solids. The filtrate solution is then stored in a refrigerator overnight to afford dark brown crystals. Yield: 10.1–10.8 g (70–75%).[†]

Anal. Calcd. for $C_9H_{18}N_2NiS_2$: C, 39.02; H, 6.55; N, 10.11. Found: C, 39.08; H, 6.47; N, 10.13.

Properties

UV–vis spectrum in MeOH [λ_{max} (nm) (ε ($M^{-1}cm^{-1}$))]: 457 (290), 278 (12,500), 236 (11,500). Mass spectrum (*m/z*): 277.

Characterization

The Ni(BME-DACH) complex is a useful chemical model of enzymatic active sites because its sulfur-based reaction chemistry resembles that seen in some naturally occurring enzymes with cysteine-rich metal binding sites. Options for studying the reactivity of Ni(BME-DACH) include (1) uptake of SO_2, (2) acylation, or (3) the design of a new experiment. If the last option is chosen, the experimental design should be submitted in advance to the course instructor so that the necessary reagents can be obtained. A good strategy for devising a simple new experiment is to try a reaction that has been reported for the *N,N'*-bis(mercaptoethyl)-1,5-diazacyclooctane analog, Ni(BME-DACO), and to carry out the same reaction with Ni(BME-DACH) and compare the results. To do this, each student should find a literature article on Ni(BME-DACO), adapt it to Ni(BME-DACH), and develop a well thought-out hypothesis and a way to test it.

All products should be isolated and characterized. Among the characterization tools that could be used are IR, UV–vis, and NMR spectroscopy, thermogravimetric analysis (TGA), cyclic voltammetry (CV), and X-ray crystallography.

---

[*]The checkers note that the $Ni(acac)_2$ must be very dry to produce the desired clear emerald green toluene solution.

[†]The checkers report yields of 30–45%, which are sufficient for the subsequent experiments.

B. SULFUR DIOXIDE ADDUCT OF Ni^{II}(BME-DACH)

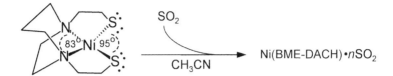

■   **Caution.** *Sulfur dioxide is an inhalation hazard. Thus, the SO$_2$ reactions should be carried out in a fume hood, and all reaction flasks should be thoroughly rinsed before removal from the hood. Any remaining SO$_2$ can be converted to sulfurous acid by bubbling through water, and dilution with copious water renders the solution innocuous.*

Procedure

Using standard Schlenk techniques, SO$_2$ gas is bubbled through a slurry of Ni (BME-DACH) (100 mg, 362 mmol) in methanol (15 mL). The orange-brown solid is transformed almost immediately into a bright red precipitate. The mixture is transferred using a positive pressure of SO$_2$ onto a medium-porosity glass frit and the product is collected by filtration. The product is then dried with a stream of SO$_2$.[*]

Anal. Calcd. for C$_9$H$_{18}$N$_2$NiO$_4$S$_4$: C, 26.7; N, 6.91; H, 4.48. Found: C, 27.0; N, 6.93; H, 4.54.

Properties

The IR spectrum of the sulfur dioxide adduct as a KBr pellet shows characteristic ν(SO) values of 1229 and 1079 cm^{-1}, which are slightly higher than the 1217 and 1075 cm^{-1} frequencies exhibited by the analogous DACO complex, indicating a weaker SO$_2$ adduct.[6] In fact, vigorous grinding to make a KBr pellet for IR studies will result in loss of SO$_2$. The thermogravimetric pyrolysis curve indicates the loss of 2 equiv of SO$_2$. The first is lost between 38 and 60°C, and the second is lost between 80 and 110°C as compared to the higher temperatures (97–133°C) required to drive off the SO$_2$ from the DACO analog. An X-ray crystal structure confirms that two SO$_2$ molecules bind per Ni(BME-DACH) unit. One of the two molecules is more tightly bound and exhibits a S(thiolate) $\cdots$ S(SO$_2$) distance of

---

[*]Ruby red X-ray quality crystals can be grown over 2 days by slow diffusion of ether saturated with SO$_2$ into an acetonitrile solution of Ni(BME-DACH) · 2SO$_2$ at room temperature.

2.557 Å; the other molecule bridges between Ni(BME-DACH) units and exhibits S (thiolate) $\cdots$ S(SO$_2$) distances of 2.660 and 3.450 Å. UV–vis in acetonitrile [λ_{max} (nm) (ε (M^{-1} cm^{-1}))]: 356 (18,000).

Characterization

The sensitivity of Ni(BME-DACH) as a sensor for SO$_2$ may be estimated using test strips made by soaking a piece of filter paper with a slurry of Ni(BME-DACH) (9 mg) in acetonitrile (2 mL). The test strips are suspended in a flask whose volume is precisely determined and which is sealed with a rubber septum. Volumes of SO$_2$ on the order of 25–100 µL are added by using a gas-tight syringe. The flask is swirled, and color changes (tan to yellow-orange) are noted after 2–3 min. The test strips can be regenerated by warming them in a 120°C oven for 1 min and "rewetting" them with acetonitrile. These test strips can be recycled at least 10 times with no noticeable degradation.

C. ACETYLATION OF Ni^{II}(BME-DACH)

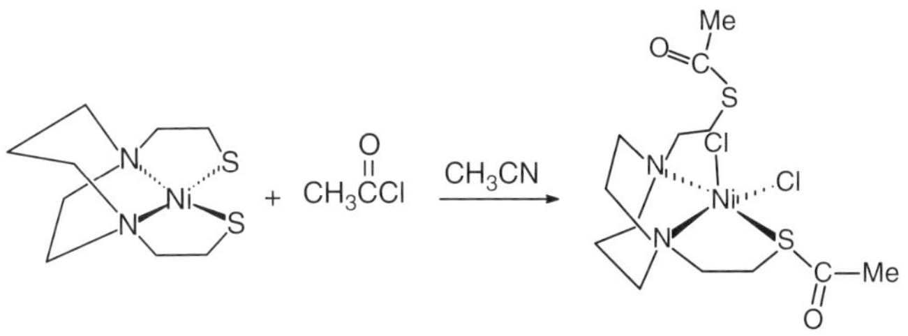

Procedure

■   **Caution.**   *Acetyl chloride is a corrosive liquid that reacts violently with water and amounts used should be kept to a minimum. All alkylating or acetylation reagents are known or suspected carcinogens. No contact with skin should ever occur.*

Under an inert atmosphere, Ni(BME-DACH) (443 mg, 1.60 mmol) is added to a Schlenk flask along with 30 mL of CH$_3$CN. To this deep tan slurry, acetyl chloride (230 µL, 3.23 mmol) is added with stirring resulting in a red solution. Using large test tubes within even larger test tubes for vapor diffusion of ether into the CH$_3$CN solution, red crystals are obtained. The crystalline product is collected on a medium-porosity glass frit open to air and washed with ether. Yield: 473 mg (68%).

Anal. Calcd. for $C_{13}H_{24}N_2Cl_2NiO_2S_2$: C, 36.0; N, 6.45; H, 5.57. Found: C, 37.5; N, 6.28; H, 5.19.

Properties

The X-ray crystal structure of the product demonstrates acetylation of both thiolates, with one thioester remaining bound to Ni and the other displaced by Cl^-, thus illustrating the weak binding ability of the resulting thioester. The structure of the acetylated Ni(BME-DACH) compound is one of the few examples of a five-coordinate Ni^{2+}. Conductivity studies in acetonitrile indicate that the acetyl dichloro complex does not ionize in solution, thereby maintaining its structural integrity. The molar conductance is $6.85\,cm^2\,\Omega^{-1}\,mol^{-1}$ at 25°C. The expected range for a 1:1 electrolyte is $120–160\,cm^2\,\Omega^{-1}\,mol^{-1}$.[14] The magnetic moment μ_{eff} as measured by the Evans method[15] is $3.1\mu_B$, which is within the expected range for two unpaired electrons in a distorted square pyramidal Ni^{2+} complex.[16] The IR spectrum in acetonitrile exhibits an intense $\nu(C{=}O)$ stretching band at $1695\,cm^{-1}$, a medium intensity band at $1737\,cm^{-1}$, and a shoulder at $1710\,cm^{-1}$. Mass spectrum (*m/z*): 457 (acetylated complex + Na)$^+$. UV–vis in acetonitrile [λ_{max} (nm) (ε ($M^{-1}\,cm^{-1}$))]: 194 (10,300), 230 (7820), 260 (4906), 404 (456), 544 (281).

Questions

1. What type of organic reaction does the synthesis of H$_2$(BME-DACH) represent?

2. Why not use NiCl$_2$ instead of Ni(acac)$_2$ in the preparation of Ni(BME-DACH)?

3. In the synthesis of Ni(BME-DACH), does order of reagent addition matter (metal and ligand)? Explain.

4. Describe the orbitals responsible for the Ni$-$SR bond.

5. Does Ni$^{II}-$SR react as a nucleophile or an electrophile? Explain.

6. What type of donor ligands stabilize the Ni^I redox state relative to Ni^{III}?

7. How does attack of electrophiles at sulfur affect the electronic character of the Ni center in Ni(BME-DACH), as indicated by such things as redox potential, geometry, color, and magnetic properties?

8. How do optical absorption bands due to charge transfer and d–d transitions differ?

9. What factors influence the geometry of NiN$_2$S$_2$ complexes and derivatives?

10. For what bioinorganic systems might NiN$_2$S$_2$ complexes be used to model?

11. What is a chemical sensor?

12. Why is SO_2 used rather than CO_2 for adduct formation with Ni(BME-DACH)?

13. What Lewis acids in addition to SO_2 might be used to form adducts with Ni(BME-DACH)?

14. What are the coordination modes for SO_2 to a metal? How does SO_2 binding to thiolate sulfur differ?

Notes for Instructors

This set of experiments can be easily completed within four sessions of a 3 h laboratory period, or it can be an extended laboratory that would include more independent research. We suggest having the students work in pairs, and the work plan along with the division of tasks should be clearly laid out by the team in a pre-laboratory session. We also suggest that the students download the manuscript template from the *Inorganic Chemistry* journal website for the preparation of their reports. The list of questions contains ideas to incorporate into the introduction and the discussion of results.

In preparation for the laboratory, we ask the students (working in pairs) to select a line of reactivity to study, to create a hypothesis of expected reactivity and results, and to propose necessary criteria for assessment. Students will need to draw upon the original literature and their previous experience in the organic laboratory for design of the synthesis, glassware assembly, isolation of the product, and its purification. A wide variety of characterization techniques can be employed depending upon what is available to the laboratory such as IR, NMR, EPR, and UV−vis spectroscopies; magnetic susceptibility; conductivity; thermogravimetric analysis; cyclic voltam-metry; mass spectrometry; and X-ray crystallography.

To facilitate completion of this laboratory in a timely fashion, we suggest that students be provided with a sample of Ni(BME-DACH) so that they can carry out studies of its reactivity while they simultaneously prepare Ni(BME-DACH) for the next group to use. As with any reaction, there are numerous aspects of the synthesis that influence the success of the reaction or lack thereof. For example, students should consider the order of addition of reagents. Specifically, the addition of ligand to metal in the Ni(BME-DACH) synthesis results in the formation of the trimetallic due to the presence of excess metal ions in solution.

This laboratory project is an exercise in developing basic research and laboratory skills, including literature searches. Another goal of this project is to develop the skill of writing a research proposal, in which students must use their knowledge of a system to design a new experiment and to hypothesize what their results might be. An important aspect of the present system is that the exact results of some of the possible experiments are not published, thus helping instructors to evaluate the student's ability to put the scientific method into practice.

References

1. C. Darnault, A. Volbeda, E. J. Kim, P. Legrand, X. Vernède, P. A. Lindahl, and J. C. Fontecilla-Camps, *Nat. Struct. Biol.* **10**, 271 (2003).
2. P. K. Mascharak, *Coord. Chem. Rev.* **225**, 201 (2002).
3. W. Huang, J. Jia, J. Cummings, M. Nelson, G. Schneider, and Y. Lindqvist, *Structure* **5**, 691 (1997).
4. M. Y. Darensbourg, I. Font, D. K. Mills, M. Pala, and J. Reibenspies, *Inorg. Chem.* **31**, 4965 (1992).
5. C. A. Grapperhaus and M. Y. Darensbourg, *Acc. Chem. Res.* **31**, 451 (1998).
6. M. Y. Darensbourg, T. Tuntulani, and J. H. Reibenspies, *Inorg. Chem.* **34**, 6287 (1995).
7. M. L. Golden, S. P. Jeffery, M. L. Miller, J. H. Reibenspies, and M. Y. Darensbourg, *Eur. J. Inorg. Chem.* 231 (2004).
8. P. J. Farmer, J. H. Reibenspies, P. A. Lindahl, and M. Y. Darensbourg, *J. Am. Chem. Soc.* **115**, 4665 (1993).
9. J. J. Smee, D. C. Goodman, J. H. Reibenspies, and M. Y. Darensbourg, *Eur. J. Inorg. Chem.* 539 (1999).
10. D. K. Mills, I. Font, P. J. Farmer, Y. Hsiao, T. Tuntulani, R. M. Buonomo, D. C. Goodman, G. Musie, C. A. Grapperhaus, M. J. Maguire, C. Lai, M. L. Hatley, J. J. Smee, J. A. Bellefeuille, and M. Y. Darensbourg, *Inorg. Synth.* **32**, 89 (1998).
11. G. J. Kubas, *Inorg. Chem.* **18**, 182 (1979).
12. J. Holtz and S. Asher, *Nature* **389**, 829 (1997).
13. J. J. Smee, M. L. Miller, C. A. Grapperhaus, J. H. Reibenspies, and M. Y. Darensbourg, *Inorg. Chem.* **40**, 3601 (2001).
14. W. J. Geary, *Coord. Chem. Rev.* **7**, 81 (1971).
15. D. H. Grant, *J. Chem. Educ.* **72**, 39 (1995).
16. R. S. Drago, *Physical Methods for Chemists*, 2nd ed., Surfside Scientific Publishers, Gainesville, FL, 1992.

44. TIN(II) IODIDE

Submitted by ALAN M. STOLZENBERG[*]
Checked by TIMOTHY P. GRAY[†] and STANTON CHING[†]

$$Sn + I_2 \rightarrow SnI_2$$

Tin(II) iodide is a less commonly encountered compound than other tin halides such as tin(II) fluoride, tin(II) chloride, or tin(IV) iodide. Although SnI_2 has a simple stoichiometry, its chemical structure is rather complex. Pale yellow acidic aqueous solutions of SnI_2 afford brilliant red-orange crystals that contain an extended array in which some tin(II) ions are surrounded by six iodide anions, and others by seven.[1] In the gas phase or in solution, monomeric tin(II) halides are

[*]Department of Chemistry, West Virginia University, Morgantown, WV 26506.
[†]Department of Chemistry, Connecticut College, New London, CT 06320.

electronic analogs of singlet carbene, having both a vacant orbital and a lone pair. Thus, they are simultaneously Lewis acids and Lewis bases.

The synthesis of tin(II) iodide serves multiple educational purposes. Students develop proficiency in handling moderately air-sensitive materials, and learn about the properties of supersaturated solutions and the effect of nucleation. When paired with the synthesis of tin(IV) iodide, it shows that the same reagents can give different products when reaction conditions are varied. Moreover, the chemistry of tin illustrates important aspects of the periodic behavior of the elements. Unlike most main group elements, whose chemistry is dominated by a single oxidation state, tin has two common oxidation states, "stannous" tin(II) and "stannic" tin-(IV), which are of near equal stability. This behavior contrasts with that of other group 14 (IVA) elements. Carbon, silicon, and germanium are nearly always found in the IV oxidation state in inorganic compounds. Lead is most often found in the II oxidation state, although strongly oxidizing compounds of lead(IV) are known. Tin can easily expand its octet and achieve coordination numbers greater than 4, unlike carbon but like the other heavier group 14 elements. Finally, comparisons of the appearance and properties of SnI_2 and SnI_4 test the students' observation abilities and provide an illustration of Fajan's rules about how the iconicity of a bond between a cation and anion depends on the charge of the cation and the relative sizes of the cation and anion.

Several procedures for the preparation of tin(II) iodide have been reported.[2] These include the reaction of aqueous $SnCl_2$ with a HI or an iodide salt, electrolysis of a CuI or Hg_2I_2 solution in a cell that uses a tin anode, reaction of metallic tin with elemental iodine, reaction of metallic tin with HI, and reduction or photo-dissociation of SnI_4. Many of these approaches afford products that are contaminated by tin(IV) iodide, even when performed anaerobically. We have found procedures that involve the metathesis of aqueous solutions of $SnCl_2$ and ZnI_2, freshly prepared by dissolution of the tin or zinc metal in hydrochloric acid or aqueous I_2, respectively, to be unsatisfactory in a teaching laboratory. Most students obtain little or no product at all. In contrast, students are consistently successful in preparing SnI_2 by reaction of metallic tin with elemental iodine in refluxing 2 M HCl. Described below is a preparation directly from the elements that works very well.

Procedure

Into a 100 mL one-necked round-bottomed flask is placed 20 mesh tin powder (1.5 g). The flask is clamped to a ring stand or other support, and placed on a 100 mL heating mantle. The flask is fitted with a reflux condenser topped with a Claisen adapter, and the reflux condenser is attached with rubber or plastic tubing to a cold water tap and a drain. The straight neck of the Claisen adapter is stoppered, and the curved neck is fitted with a gas inlet adapter (either a vacuum adapter with stopcock or a thermometer adapter with a piece of glass tubing inserted through the bushing). The gas inlet is connected with rubber or plastic tubing to one arm of a T- or Y-tube. The other arms of the T- or Y-tube are connected by rubber or plastic tubing to a nitrogen gas source and to an oil bubbler. A hose clamp is positioned loosely on the rubber or plastic tube between the T-tube and the bubbler; this clamp will be used to regulate the flow of nitrogen to the apparatus. The stopper on the Claisen adapter is removed, the nitrogen gas source is turned on, and the hose clamp is closed so that the apparatus is flushed for a few minutes with nitrogen. After the flushing is complete, the nitrogen flow through the apparatus is decreased by opening the hose clamp, the stopper is replaced, and the nitrogen flow is adjusted to give a slow, gentle bubbling in the oil bubbler.

Iodine (2.1 g) and 2 M hydrochloric acid (15 mL) are measured out. The stopper on the Claisen adapter is removed, and the nitrogen flow to the flask is temporarily increased by closing the hose clamp. Into the open neck of the Claisen adapter is inserted a funnel, through which the iodine and dilute hydrochloric acid are added to the flask. Care should be taken so that the iodine crystals do not become stuck in the greased joint or in the condenser. After the addition is complete, the funnel is removed, the hose clamp is opened, and the stopper is replaced. The water to the reflux condenser is turned on, and the solution in the flask is then heated to reflux.[*] As the solid iodine is consumed, the solution becomes very pale yellow.[†] The remaining tin should maintain a bright appearance for 5–10 min.

While the solution is heating, a 100 mL one-necked round-bottomed flask is clamped to a ring stand or other support and the bottom of the flask is positioned in a water bath. The water bath is heated to boiling and the flask flushed briefly with nitrogen through a rubber or plastic hose connected to a nitrogen gas source. Into the open neck of the flask is placed a small glass funnel with a short stem, and into the funnel is placed a loose plug of glass wool. Then the hot tin iodide solution is filtered rapidly through the glass wool plug and funnel. (■ **Caution.** *Tip the flask and pour the solution by holding onto the clamp on the flask neck. Do not touch the hot flask itself.*) The filtrate is collected in the one-necked round-bottomed flask, which is again flushed briefly with nitrogen, and then the flask is stoppered and the heating of the water bath is discontinued. The flask and bath are allowed to cool slowly to room temperature, during which time the SnI_2 crystallizes. The orange-red crystals of SnI_2 are collected on a Büchner funnel and washed with cold water (5 mL) containing a drop of 1 M hydrochloric acid. The crystals are dried briefly with suction, and then are dried by pressing them between two pieces of dry filter paper. The product is weighed to determine the yield. The SnI_2 should be stored in a tightly capped, nitrogen-flushed vial.

---

[*]The reaction mixture should not be stirred. Stirring causes Sn(II) to be formed while the solution is still cold, thus causing premature precipitation of the product.
[†]The checkers noted some loss of I_2 through sublimation during the reaction.

Properties

When stored under nitrogen, the orange-red crystals of SnI_2 slowly change color over several weeks to months. Solid exposed to air changes even more rapidly.

Characterization

Measure the melting point of the product.

Questions

1. SnI_4 has a lower melting point and is more soluble in organic solvents than SnI_2. What conclusions can you draw about the nature of the bonding in the tin iodides from these observations?
2. Calculate the mole ratio of Sn to I_2 used in the procedure. Which is the limiting reagent? Why is the reaction carried out with this ratio of reactants and under a nitrogen atmosphere?

Notes to Instructors

Typical advanced undergraduate students can complete the synthesis in less than one 3 h laboratory period. Student yields are usually 30–50%. The moderate yield is probably due in part to the partial solubility of SnI_2 in water at room temperature. Another factor that can reduce the yield is premature crystallization of the product during the filtration through glass wool; this can happen if the solution partly cools before or during the filtration step.

In addition to its melting point, SnI_2 can also be characterized by X-ray powder diffraction, ^{119}Sn NMR, and Mössbauer spectroscopy if the appropriate instrumentation is available. Major diffraction peaks are observed at d-spacings of 3.274, 3.170, 5.83, 3.540, 3.090, 2.292, and 2.164 Å, in order of decreasing intensity.[1] The ^{119}Sn chemical shift of SnI_2 is $\delta -779$ relative to tetramethyltin in DMSO-d_6.[3] The lines observed in the Mössbauer spectrum of SnI_2 have isomer shifts of 1.65 and 1.27 mm s^{-1} relative to β-Sn.[1]

It is instructive to have the students also prepare SnI_4.[*] Observant students will notice that, although the needle shape of SnI_2 crystals is substantially different from the crystal form adopted by SnI_4, the colors of the two solids are sufficiently similar that a direct comparison is necessary to distinguish the somewhat redder SnI_2 from SnI_4. The two compounds have substantially different melting points. SnI_4 melts at 144°C. In contrast, SnI_2 melts at 320°C,[2] which is above the range of a

---

[*]SnI_4 can be prepared by the method in Ref. 4, with the exception of substituting dichloromethane for carbon tetrachloride.

melting point apparatus with a conventional Hg thermometer. Comparisons of the solubilities of the compounds are also informative. Unlike SnI_4, SnI_2 is only slightly soluble in $CHCl_3$, benzene, and acetone.

References

1. R. A. Howie, W. Moser, and I. C. Trevena, *Acta. Crystallogr. B* **28**, 2965 (1972).
2. A. Aman, *Gmelin Handbook of Inorganic Chemistry*, System Number 46, Part C-1, Tin: Compounds with Hydrogen, Oxygen, Nitrogen, and the Halogens, Springer, Heidelberg, 1972, pp. 446–456.
3. R. W. Schaeffer, B. Chan, M. Molinaro, S. Morissey, C. H. Yoder, C. S. Yoder, and S. Shenk, *J. Chem. Educ.* **74**, 576 (1997).
4. T. Moeller and D. C. Edwards, *Inorg. Synth.* **4**, 119 (1953).

45. *N-tert*-BUTYL-3,5-DIMETHYLANILINE

Submitted by MIRCEA D. GHEORGHIU[*]
Checked by MEE-KYUNG CHUNG[†] **and JEFFREY M. STRYKER**[‡]

The industrial chemical reaction that arguably has had the greatest effect on society and human health is the Haber–Bosch process for converting atmospheric nitrogen to ammonia.[1] This reaction, which was discovered in the early twentieth century, is the basis of the modern artificial fertilizer industry and has made it possible for global food production to keep pace with the increase in the world's population over the last century. To proceed at reasonable rates, the Haber–Bosch reaction requires very high temperatures (over 300°C) and pressures (over 150 bar), and so is a very energy intensive process despite the fact that the reaction of N_2 and H_2 to give ammonia is downhill thermodynamically at room temperature and pressure.

Some bacteria are able to convert atmospheric dinitrogen into nitrogen-containing products, and these enzymatic processes necessarily operate at room temperature and pressure. It had long been an aim of inorganic chemists to discover nonenzymatic catalysts that convert N_2 to ammonia under similarly mild conditions,[2] and in 2003 it was found that certain molybdenum compounds bearing bulky amido ligands could catalyze this reaction.[3] The use of sterically demanding ligands to reduce the coordination number of metal complexes and increase their reactivity toward small molecules is an important theme in inorganic chemistry. The first transition metal complex with bulky amido ligands was

[*]Department of Chemistry, Massachusetts Institute of Technology, Cambridge, MA 02139.
[†]Department of Chemistry, University of Alberta, Edmonton, Alberta, Canada T6G 2G2.

prepared in 1934,[4] and since the 1950s such compounds have been the subject of intensive research.[5]

The current experiment involves the preparation of the sterically hindered amine *N-tert*-butyl-3,5-dimethylaniline.[6,7] Other preparations of this amine involve addition of methyllithium to *N*-3,5-dimethylphenylacetone imine[8] and the reaction of 1-bromo-2,4-dimethylbenzene with *tert*-butylamine either via aryne formation[9] or by palladium-catalyzed alkylation.[10] The current method, the reaction of *tert*-butylamine with the 2,4,6-trimethylpyrylium cation, involves inexpensive starting materials and proceeds in high yield. The molybdenum(III) complex of the deprotonated form of this amine, $Mo[N(t\text{-}Bu)(3,5\text{-}C_6H_3Me_2)]_3$, splits the $N\equiv N$ triple bond in N_2 to afford molybdenum(VI) nitrido products.[11,12] This latter reaction is the key step in the recently discovered catalytic process to convert N_2 to ammonia under ambient conditions.

A. 2,4,6-TRIMETHYLPYRYLIUM TETRAFLUOROBORATE

Procedure

■   **Caution.** *Diacetone alcohol is an irritant, acetic anhydride is corrosive and a lachrymator (causes the eyes to tear), and fluoroboric acid is corrosive and a lachrymator. This reaction should be carried out in a well-ventilated fume hood.*

The central neck of a 250 mL three-necked (14/20) round-bottomed flask containing a 2.5 cm long stir bar is fitted with a Vigreux column topped with a water-cooled reflux condenser. All joints should be greased and held together with clips, and the reaction flask should be clamped in place before it is charged with reagents. A thermometer and thermometer bushing adapter are placed in one of the side necks, and a pressure-equalizing dropping funnel is placed in the other. The flask is charged with acetic anhydride (75 mL) and the position of the thermometer is adjusted so that its bulb is immersed in the liquid. The flask is cooled in an ice water bath and stirring is started. When the liquid has reached a temperature of ~5°C, diacetone alcohol, 4-hydroxy-4-methyl-2-pentanone (10 mL, 80 mmol), is added in three portions through the top of the condenser. The mixture is stirred for 5–10 min to ensure that the temperature has equilibrated. The ice water bath is removed, and the bottom of the flask is carefully dried with paper towels. The flask is then placed in a 250 mL heating mantle connected to a Variac (which should be *off* for the time being). By using the dropping funnel, 48% fluoroboric acid (10.4 mL, 80 mmol) is added at a rate of one drop every 10–20 s. The temperature should not exceed 95°C; if it does, the addition should be stopped until the temperature has dropped below this value. When all of the HBF$_4$ has been added, the heating mantle should be turned on and the Variac adjusted to keep the temperature at 95–100°C for an additional 30 min. The heating mantle is removed (■ **Caution.** *Both the mantle and the glassware are hot!*) and the round-bottomed flask is placed in an ice water bath. When the temperature of the liquid has reached 5°C, diethyl ether (100 mL) is added through the top of the condenser. Stirring is continued for an additional 10 min. The product is collected by filtration on a fritted (25–50 μm) Büchner funnel without suction. (■ **Caution.** *The filtrate is extremely acidic and should be disposed of properly!*)[*] The brownish precipitate is washed with diethyl ether (3 × 10 mL). At this point, the precipitate should be pale yellow; additional washings may be necessary to obtain a product of this color. The remaining liquid is drained by applying suction briefly to the collected precipitate. The precipitate is dried in a Petri dish overnight in a hood. Yield: 8–10 g (48–60%).[†]

Properties

2,4,6-Trimethylpyrylium tetrafluoroborate is a pale yellow solid. Melting point: 218–220°C (dec.). ^{1}H NMR (CD$_3$CN): δ 2.65 (s, 3H), 2.80 (s, 6H), 7.72 (s, 2H).

Characterization

Record the yield, melting point, and ^{1}H NMR spectrum of the product.

B. *N-tert*-BUTYL-3,5-DIMETHYLANILINE

BF$_4$ + 2 NH$_2$(*t*-Bu) $\longrightarrow$ + H$_2$O + [NH$_3$(*t*-Bu)]BF$_4$

NH(*t*-Bu)

[*]To dispose the highly acidic filtrate safely, a 2.5 cm stir bar should first be placed in an Erlenmeyer flask having a volume two to three times larger than that of filtrate. Then the filtrate is poured into the flask. To the stirred filtrate is added solid Na$_2$CO$_3$ (10 g) in small portions so as to avoid extensive foaming. When the last of the solid has been added and CO$_2$ evolution stops, the pH of the solution is taken with pH paper. If the solution is still acidic, more Na$_2$CO$_3$ should be added. After the solution is neutralized, it should be transferred into a waste container.

[†]The checkers obtained a yield of 11.3 g (67%).

Procedure

■   **Caution.**   *tert-Butylamine is toxic, flammable, corrosive, and has a strong unpleasant smell. Acetonitrile is flammable and toxic. This procedure should be carried out in a well-ventilated fume hood.*

The central neck of a 500 mL three-necked (14/20) round-bottomed flask containing a 2.5 cm long stir bar is fitted with a Vigreux column topped with a water-cooled reflux condenser. All joints should be greased and held together with clips, and the reaction flask should be clamped in place before it is charged with reagents. A thermometer and thermometer bushing adapter are placed in one of the side necks, and a pressure-equalizing dropping funnel is placed in the other. The flask is charged with anhydrous acetonitrile (100 mL) and *tert*-butylamine (21 mL, 200 mmol). The dropping funnel is charged with a solution of 2,4,6-trimethylpyrylium tetrafluoroborate (8.4 g, 40 mmol) in anhydrous acetonitrile (150 mL). (All of this solution may not fit in the dropping funnel. It is acceptable to add portions to the dropping funnel as it empties.) The flask is flushed with nitrogen, and the pyrylium solution is added at a rate of one drop every 2 s with stirring. The addition should take no more than 4 h. The contents of the flask are transferred into a 500 mL single-necked round-bottomed flask. The acetonitrile is removed in a rotary evaporator. The resulting oil is dissolved in petroleum ether (~100 mL), and this solution is washed with water (~100 mL). The organic layer is dried over anhydrous $MgSO_4$ and then filtered through a fritted Büchner filter. The solvent is removed on a rotary evaporator. Yield: 4.7 g (66%).[*]

Properties

Crude *N-tert*-butyl-3,5-dimethylaniline is a yellow oil; it is possible to obtain the pure colorless liquid by distillation under vacuum: bp 45°C at 1.5×10^{-2} mmHg. ^{1}H NMR (400 MHz, C_6D_6): δ 1.18 (s, 9H), 2.20 (s, 6H), 3.06 (br s, 1H, NH), 6.37 (s, 2H), 6.46 (s, 1H). The amine can be deprotonated with *n*-butyllithium in diethyl ether to afford the salt $LiN(t\text{-}Bu)(3,5\text{-}C_6H_3Me_2) \cdot Et_2O$, which is an excellent starting material for the preparation of reactive transition metal complexes.[10–12]

Characterization

Record the yield and analyze the crude sample by GC/MS and ^{1}H NMR spectroscopy, the latter as a solution of the product in C_6D_6.

---

[*]This was the yield obtained by the checkers after distilling their product in vacuum at 85°C and 0.2 mmHg.

Questions

1. The ^{19}F NMR spectrum of 2,4,6-trimethylpyrylium tetrafluoroborate contains two closely spaced singlets in a ratio very close to 1:4. Explain why there are two fluorine signals and why the ratio is 1:4.

2. The nitrogen atom of NH$_2$(*t*-Bu) is a nucleophile that preferentially attacks atoms in the 2,4,6-trimethylpyrylium ring that are depleted in electron density. A density functional theory calculation* on the parent unsubstituted pyrylium cation provides the Merz–Kollman–Singh electrostatic potential-derived charges (MKS EPDC)[13,14] illustrated in Fig. 1. Which atom(s) of the pyrylium ring are preferentially attacked by the amine?

3. Write a plausible mechanism for the formation of *N-tert*-butyl-3,5-dimethylaniline from the reaction of the 2,4,6-trimethylpyrylium salt with NH$_2$(*t*-Bu). *Hint*: It takes several steps.

4. Why does the reaction of N$_2$ with H$_2$ to form ammonia require such high temperatures and pressures, despite the fact that this reaction is downhill thermodynamically at room temperature?

5. What is known about the structure of the enzyme used by certain bacteria to convert atmospheric nitrogen into other nitrogen-containing products?

6. Compare and contrast the bonding of N$_2$ and CO to transition metals. Which ligand is usually bound more strongly and why?

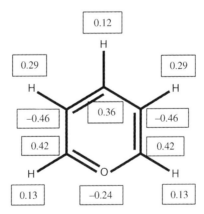

Figure 1. MKS ESPD charges of the pyrylium cation (B3LYP/6-311+G(d,p) level, C$_{2v}$ symmetry).

*The calculations were performed at the B3LYP/6-311+g(d,p) level.

7. The molybdenum(III) compound $Mo[N(t\text{-}Bu)(3,5\text{-}C_6H_3Me_2)]_3$ reacts with N_2 to give the nitrido compound $N{\equiv}Mo[N(t\text{-}Bu)(3,5\text{-}C_6H_3Me_2)]_3$ in which the $N{\equiv}N$ triple bond has been cleaved. What is the mechanism of this reaction?

Notes for Instructors

Procedure B takes more than 4 h; the first good stopping point can be reached in less than 4 h if the students use their time efficiently. Remind students that they must come prepared and arrive promptly at the beginning of laboratory in order to finish this experiment during the available laboratory time. The extraction step can be completed the following day (or the following laboratory period) if the reaction mixture is placed into the refrigerator in a sealed, labeled container.

If you decide to have the students purify their material by vacuum distillation, it is best to do so in bantam glassware with at least 10 mL of crude material, in order to minimize losses on the walls of the apparatus. The crude product obtained by two or more students can be combined to obtain the recommended amount of material. Some of the extraction solvent is often present in the crude oil. The distillation should be carried out with stirring and slow heating, so that this solvent is pumped away before the product starts to distill. If necessary, remind the students that the numbers on the Variac are not temperatures!

If the students have access to the electronic structure program Gaussian, have them reproduce the DFT calculation described in question 2.

With additional laboratory time, it is possible to extend the experiment by having the students deprotonate *N-tert*-butyl-3,5-dimethylaniline with *n*-butyllithium in pentane, followed by addition of diethyl ether to induce crystallization as the monoetherate of $LiN(t\text{-}Bu)(3,5\text{-}C_6H_3Me_2) \cdot Et_2O$.[10,11] Treatment of $MoCl_3(THF)_3$ with 2 equiv (not 3!) of this lithium salt in diethyl ether followed by filtration and crystallization affords the molybdenum(III) compound $Mo[N(t\text{-}Bu)(3,5\text{-}C_6H_3Me_2)]_3$,[10,12] which reacts slowly over 3 days with N_2 at $-35°C$ to give the nitrido complex $N{\equiv}Mo[N(t\text{-}Bu)(3,5\text{-}C_6H_3Me_2)]_3$.[12]

Acknowledgments

The author thanks Prof. C. C. Cummins for his support.[*]

---

[*]The checkers acknowledge the Natural Sciences and Engineering Research Council of Canada (Collaborative Research Opportunities), the University of Alberta, the Killam Trusts, and the Province of Alberta (Alberta Ingenuity) for research funding and fellowships (to M.-K.C.).

References

1. V. Smil, *Nature* **400**, 415 (1999).
2. R. R. Schrock, *Acc. Chem. Res.* **38**, 955 (2005).
3. D. V. Yandulov and R. R. Schrock, *Science* **301**, 76 (2003).
4. O. C. Dermer, and W. C. Fernelius, *Z. Anorg. Allg. Chem.* **221**, 83 (1934).
5. M. F. Lappert, P. P. Power, A. R. Sanger, and R. C. Srivastava, *Metal and Metalloid Amides*, Ellis Horwood, Chichester, UK, 1980.
6. P. Vernaudon, H. G. Rajoharison, and C. Roussel, *Bull. Soc. Chim. Fr.* 205 (1987).
7. Y-C. Tsai, F. H. Stephens, K. Meyer, A. Mendiratta, M. D. Gheorghiu, and C. C. Cummins, *Organometallics* **22**, 2902 (2003).
8. A. R. Johnson, C. C. Cummins, and S. Gambarotta, *Inorg. Synth.* **32**, 123 (1998).
9. C. C. Cummins, *Chem. Commun.* 1777 (1998).
10. W. Zhang, Y. Lu, and J. S. Moore, *Org. Synth.* **84**, 163 (2009).
11. C. E. Laplaza, W. M. Davis, and C. C. Cummins, *Organometallics* **14**, 577 (1995).
12. C. E. Laplaza, M. J. A. Johnson, J. C. Peters, A. L. Odom, E. Kim, C. C. Cummins, G. N. George, and I. J. Pickering, *J. Am. Chem. Soc.* **118**, 8623 (1996).
13. U. C. Singh, and P. A. J. Kollman, *Comput. Chem.* **5**, 129 (1984).
14. B. H. Besler, K. M. Merz Jr., and P. A. J. Kollman, *Comput. Chem.* **11**, 431 (1990).

CUMULATIVE CONTRIBUTOR INDEX
VOLUMES 31–36

Indices for previous volumes appear in *Inorganic Syntheses, Collective Index for Volumes 1–30*, T. E. Sloan, ed., Wiley-Interscience (1997).

Inorganic Syntheses, Volume 36, First Edition. Edited by Gregory S. Girolami and Alfred P. Sattelberger
© 2014 John Wiley & Sons, Inc. Published 2014 by John Wiley & Sons, Inc.

CUMULATIVE SUBJECT INDEX
VOLUMES 31–36

The compound names in this cumulative Subject Index are for the most part simplified versions based on the rules given in *Nomenclature of Inorganic Chemistry, IUPAC Recommendations*, N. G. Connelly and T. Damhus, eds., Royal Society of Chemistry (2005). Inverted forms are used for most entries, with the anionic ligands generally listed first and neutral ligands afterwards, and usually in alphabetical order within each set. Note that the names in this Subject Index may be different from those given in the preparations themselves.

Indices for previous volumes appear in *Inorganic Syntheses, Collective Index for Volumes 1–30*, T. E. Sloan, ed., Wiley-Interscience (1997).

Inorganic Syntheses, Volume 36, First Edition. Edited by Gregory S. Girolami and Alfred P. Sattelberger.
© 2014 John Wiley & Sons, Inc. Published 2014 by John Wiley & Sons, Inc.

CUMULATIVE FORMULA INDEX
VOLUMES 31–36

The chief aim of this formula index is to help the reader locate specific compounds and groups of compounds. The significant or central atom has been placed first in the formula in order to group together as many related compounds as possible. This procedure often involves placing the cation last if it is of relatively minor interest (*e.g.,* an alkali metal cation or tetraalkylammonium ion): $[Cr(CN)_6]K_3$; $[Re_2Cl_8](NBu_4)_2$; $Mo_2(O^tBu)_6$. Wherever it seemed best, formulas have been entered in their usual form (*i.e.,* as used in the text) for easy recognition: $AsCl_3$, N_4S_4, Mo_6Cl_{12}, RuO_4, WCl_4. More complicated compounds are indexed under their compositional formulae, followed by a structural formula and/or a name: $C_{12}H_{16}Si$, $(\eta^5\text{-}C_9H_7)SiMe_3$, 1-(trimethylsilyl)indene. Where there may be almost equal interest in two or more parts of a formula, two or more entries have been made, *e.g.,* $[C_5Me_5]K$ and $K[C_5Me_5]$. A list of abbreviations used in this index follows.

Indices for previous volumes appear in *Inorganic Syntheses, Collective Index for Volumes 1–30*, T. E. Sloan, ed., Wiley-Interscience (1997).

acac = 2,4-pentanedionato or acetylacetonato
bpm = bis(1-pyrazolyl)methane
bpy = 2,2′-bipyridine
C_{60} = fullerene
cat = catecholato
DAniF = *N,N′*-di-*p*-anisylformamidinate
dcbpy = 4,4′-dicarboxy-2,2′-bipyridine
dcmpz = dimethyl-1*H*-pyrazolato-3,5-dicarboxylate
dcpp = 1,3-bis(dicyclohexylphosphino)propane
dctpz = di-*tert*-butyl-1*H*-pyrazolato-3,5-dicarboxylate
dien = diethylenetriamine
diglyme = diethylene glycol dimethyl ether (2,5,8-trioxanonane)
dmpe = 1,2-bis(dimethylphosphino)ethane
dmphen = 2,9-dimethyl-1,10-phenanthroline
dmptacn = 1,4-bis(2-pyridylmethyl)-1,4,7-triazacyclononane
dpb = 2,3-bis(2-pyridyl)benzoquinoxaline

Inorganic Syntheses, Volume 36, First Edition. Edited by Gregory S. Girolami and Alfred P. Sattelberger
© 2014 John Wiley & Sons, Inc. Published 2014 by John Wiley & Sons, Inc.

dpp = 2,3-bis(2-pyridyl)pyrazine

dppe = 1,2-bis(diphenylphosphino)ethane

dppf = bis(diphenylphosphino)ferrocene

dppm = bis(diphenylphosphino)methane

dppz = diphenyl-1H-pyrazolato

dpq = 2,3-bis(2-pyridyl)quinoxaline

DXyl^{2,6}F = N,N'-2,6-xylylformamidinate

(t-HMPA-B)H$_8$ = 1,2,4,5-tetrakis(2-hydroxy-2-methylpropanamido)benzene

hfac = 1,1,1,5,5,5,-hexafluoro-2,4-pentanedionato

(hoz)H = 2-(2'-hydroxyphenyl)-2-oxazoline

L$^{R,R'/R''}$ = an N-aryl β-diketiminate, which have the NC^{R}CC^{R}N backbone with substituted phenyl groups on the nitrogen. In this notation, R is the substituent on the backbone carbon, R' indicates the *ortho* substituents, and R'' the *para* substituents on the phenyl ring. Thus,

    L^{Me,Me2} indicates R=R'=Me, R''=H; [N(2,6-C$_6$H$_3$Me$_2$)CMe]$_2$CH

    L^{Me,Me3} indicates R=R'=R''=Me; [N(2,4,6-C$_6$H$_2$Me$_3$)CMe]$_2$CH

    L^{Me,iPr2} indicates R=Me, R'=^{i}Pr, R''=H; [N(2,6-C$_6$H$_3$^{i}Pr$_2$)CMe]$_2$CH

    L^{tBu,iPr2} indicates R=^{t}Bu, R'=^{i}Pr, R''=H; [N(2,6-C$_6$H$_3$^{i}Pr$_2$)C^{t}Bu]$_2$CH

15-MC$_{Cu(II)N(picHA)}$-5 = a metallo-crown (MC) with 15 total atoms in the crown, the subscript after MC is the ring metal (Cu^{II}), the other heteroatom (N), and the ligand used to make the MC (H$_2$PicHA = picoline hydroxamic acid), with 5 oxygen atoms in the MC

mes = mesityl, 2,4,6-trimethylphenyl

Me$_2$pz = 3,5-dimethyl-1H-pyrazolato

Me$_3$tach = 1,3,5-trimethyl-1,3,5-triazacyclohexane

Me$_3$tacn = 1,4,7-trimethyl-1,4,7-triazacyclononane

N-MeIm = N-methylimidazole

nbac = 1-(4-nitrobenzyl)-4,7,10-tris(carbamoylmethyl)-1,4,7,10-tetraazacyclododecane

(nha)H$_3$ = 3-hydroxy-2-naphthohydroxamic acid

2,3-(OH)$_2$C$_6$H$_3$-TTP(H$_2$) = 5-(2,3-dihydroxyphenyl)-10,15,20-tri-p-tolylporphyrin

3,4-(OH)$_2$C$_6$H$_3$-TTP(H$_2$) = 5-(3,4-dihydroxyphenyl)-10,15,20-tri-p-tolylporphyrin

2,3-(OMe)$_2$C$_6$H$_3$-TTP(H$_2$) = 5-(2,3-dimethoxyphenyl)-10,15,20-tri-p-tolylporphyrin

3,4-(OMe)$_2$C$_6$H$_3$-TTP(H$_2$) = 5-(3,4-dimethoxyphenyl)-10,15,20-tri-p-tolylporphyrin

OTf = SO$_3$CF$_3$

OTs = SO$_3$C$_6$H$_4$-p-Me

pdt = propanedithiolate

phen = 4,7-diphenyl-1,10-phenanthroline

PPN = bis(triphenylphosphoranylidene)ammonium, (Ph$_3$P)$_2$N

py = pyridine

pz = pyrazolyl

QBC = tetrabenzocyclyne, 5,6,11,12,17,18,23,24-octadedydrotetrabenzo[a,e,i,m]
    cyclohexadecene

quin = quinuclidine (1-azabicyclo[2.2.2]octane)

(salen)H$_2$ = N,N'-bis(2-hydroxybenzylidene)ethylenediamine

(salen^{t}Bu)H$_2$ = N,N'-bis(2-hydroxy-3,5-di(*tert*-butyl)benzylidene)ethylenediamine (salmeten)H$_2$ = bis(3-salicylidene aminopropyl)methylamine

[H$_4$(salphen)(OH)$_2$]H$_2$= *N,N'*-bis(salicylidene)-4,5-dihydroxyphenylenediamine

[^{t}Bu$_4$(salphen)(OH)$_2$]H$_2$= *N,N'*-bis[3,3',5,5'-tetra(*tert*-butyl)salicylidene]-4,5-dihydroxyphenylenediamine

[(EtO)$_2$H$_2$(salphen)(OH)$_2$]H$_2$= *N,N'*-bis[3,3'-di(ethoxy)salicylidene]-4,5-dihydroxyphenylenediamine

TBC = tribenzocyclyne, 5,6,11,12,17,18-hexadedydrotribenzo[a,e,i]cyclododecene

tdhp = 1,2,6,7-tetracyano-3,5-diimino-3,5-dihydropyrrolizinido

teeda = *N,N,N',N'*-tetraethylethylenediamine

terpy = 2,2':6',2''-terpyridine

tetraglyme = tetraethylene glycol dimethyl ether (2,5,8,11,14-pentaoxapentadecane)

tetren = tetraethylenepentamine

thf = tetrahydrofuran

tht = tetrahydrothiophene

tmeda = *N,N,N',N'*-tetramethylethylenediamine

tmhd = 2,2,6,6,-tetramethyl-3,5-heptanedionato

Tp = tris(pyrazolyl)hydroborate

Tp* = tris(3,5-dimethylpyrazolyl)hydroborate

tpa = tris(2-pyridylmethyl)amine

tren = triethylenetetramine

triglyme = triethyleneglycol dimethyl ether (2,5,8,11-tetraoxadodecane)